SUSTAINABLE TRANSPORTATION ENERGY PATHWAYS

SUSTAINABLE TRANSPORTATION ENERGY PATHWAYS
A Research Summary for Decision Makers

Edited by Joan Ogden and Lorraine Anderson

Institute of Transportation Studies
University of California, Davis
One Shields Avenue, Davis, California 95616

© 2011 by The Regents of the University of California, Davis campus
All rights reserved. Published 2011

Available under a Creative Commons BY-NC-ND 3.0 license
<http://creativecommons.org/licenses/by-nc-nd/3.0/>.
For information on commercial licensing, contact copyright@ucdavis.edu.

Table of Contents

Introduction: Imagining the Future of Transportation — 1

Part 1: Individual Fuel/Vehicle Pathways — 13

1: The Biofuels Pathway — 15
   Nathan Parker, Bryan Jenkins, Peter Dempster, Brendan Higgins, and Joan Ogden

2: The Plug-in Electric Vehicle Pathway — 38
   Jonn Axsen, Christopher Yang, Ryan McCarthy, Andrew Burke, Kenneth S. Kurani, and Tom Turrentine

3: The Hydrogen Fuel Pathway — 64
   Joan Ogden, Christopher Yang, Joshua Cunningham, Nils Johnson, Xuping Li, Michael Nicholas, Nathan Parker, and Yongling Sun

Part 2: Pathway Comparisons — 95

4: Comparing Fuel Economies and Costs of Advanced vs. Conventional Vehicles — 97
   Andrew Burke, Hengbing Zhao, and Marshall Miller

5: Comparing Infrastructure Requirements — 121
   Joan Ogden, Christopher Yang, Yueyue Fan, and Nathan Parker

6: Comparing Greenhouse Gas Emissions — 133
   Timothy Lipman and Mark A. Delucchi

7: Comparing Land, Water, and Materials Impacts — 171
   Sonia Yeh, Gouri Shankar Mishra, Mark A. Delucchi, and Jacob Teter

Part 3: Scenarios for a Low-Carbon Transportation Future — 187

8: Scenarios for Deep Reductions in Greenhouse Gas Emissions — 189
   Christopher Yang, David McCollum, and Wayne Leighty

9: Transition Scenarios for the U.S. Light-Duty Sector — 209
   Joan Ogden, Christopher Yang, and Nathan Parker

10: Optimizing the Transportation Climate Mitigation Wedge — 234
    Sonia Yeh and David McCollum

Part 4: Policy and Sustainable Transportation — 249

11: Toward a Universal Low-Carbon Fuel Standard — 251
    Daniel Sperling and Sonia Yeh

12: Key Measurement Uncertainties for Biofuel Policy — 263
    Sonia Yeh, Mark A. Delucchi, Alissa Kendall, Julie Witcover, Peter W. Tittmann, and Eric Winford

13: Beyond Life-Cycle Analysis: Developing a Better Tool for Simulating Policy Impacts — 278
    Mark A. Delucchi

Conclusion: Key Findings and Paths Forward — 297

Acknowledgments: STEPS Program Sponsors — 311

Acknowledgments — 312

Authors and Researchers — 315

Introduction: Imagining the Future of Transportation

We stand at the beginning of a revolution in transportation and energy. Over the next several decades, a convergence of growing demand, resource constraints, and environmental imperatives will reshape our energy system. These forces will change the way we travel and the kinds of vehicles we drive, and will challenge the century-long primacy of petroleum and the internal combustion engine. This transformation will unfold over many decades. But it poses urgent questions today because of the long time horizon inherent in developing new technologies and changing the energy system.

Transportation Energy Challenges

Energy supply is a critical concern for the transport sector. Global demand for mobility is growing rapidly, with the number of vehicles projected by the International Energy Agency (IEA) to triple by 2050. This is especially true in the developing world, where the number of vehicles is growing by 5 to 6 percent per year. About 97 percent of transport fuels currently come from petroleum, a large fraction of which is imported by the countries where it is used. Costs for conventional crude oil are rising, and direct substitutes for petroleum (such as unconventional oil from oil shale and tar sands) face economic, technical, and environmental challenges.

Direct combustion of fossil fuels for transportation accounts for a significant fraction of global primary energy use (19 percent), air pollutant emissions (5 to 70 percent, depending on the pollutant and region), and greenhouse gas (GHG) emissions (23 percent for 2005 on a well-to-wheels basis), according to the International Energy Agency.[1] Although improved energy efficiency in buildings or low-carbon electricity generation might offer lower-cost ways of reducing carbon emissions in the near term, decarbonizing the transport sector will be critically important to achieving the long-term, deep cuts in carbon emissions required for climate stabilization.

A host of complex resource issues complicates the path toward a sustainable transportation system. These include availability of low-carbon primary energy resources to make new transportation fuels, availability of land and water to produce these fuels, constraints on critical materials such as platinum for fuel cells or lithium for batteries, and impacts on the broader economy.

THE LONG TRANSITION

How fast can we make a transition to alternative fuels and vehicles? Transitions in the transportation sector take a long time, for several reasons.

First, passenger vehicles have a relatively long lifetime (15 years average in the United States). Even if a new technology were to rapidly capture 100 percent of new vehicle sales, it would take a minimum of 15 years for the vehicle stock to turn over. In practice, adoption of new vehicle technologies occurs much more slowly; it can take 25 to 60 years for an innovation to be used in 35 percent of the on-road fleet.[2] For example, research into gasoline hybrid electric vehicles (HEVs) in the 1970s and 1980s led to a decision to commercialize in 1993, with the first vehicle becoming available for sale in 1997. HEVs still represent only about 3 percent of new car sales nationally in the U.S., 5% in California and fewer than 0.5 percent of the worldwide fleet. This slow turnover rate is also true for relatively modest technology changes such as the adoption of automatic transmissions or fuel injection. The time frame for new technologies relying on electric batteries, fuel cells, or advanced biofuels could be even longer since they all need further RD&D investment before they can be commercialized.

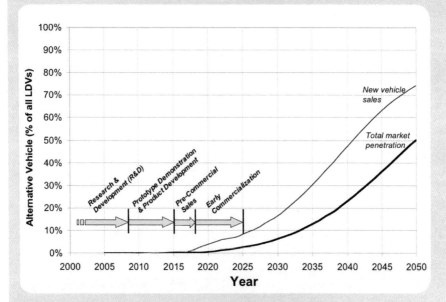

The steps needed to commercialize a new vehicle technology add up to a long time horizon. An alternative vehicle technology for which research and development began in 2000 might not reach 50 percent market penetration or hope to capture 75 percent of new vehicle sales until 2050. Source: Joshua M. Cunningham, Sig Gronich, and Michael A. Nicholas, Why Hydrogen and Fuel Cells Are Needed to Support California Climate Policy, UCD-ITS-RR-08-06 *(Institute of Transportation Studies, University of California, Davis, 2008).*

Second, changing the fuel supply infrastructure, especially if this means switching on a massive scale from liquid fuels to gaseous fuels or electrons, will require both time and a significant amount of capital. Historically, major changes in transport systems such as building canals and railroads, paving highways, and adopting gasoline cars have taken many decades to complete. Transitions will require developing new supply chains using renewable or other low-carbon sources and replacing existing fossil fuel and electricity plants. Such paradigm shifts will require close coordination among fuel suppliers, vehicle manufacturers, and policymakers.

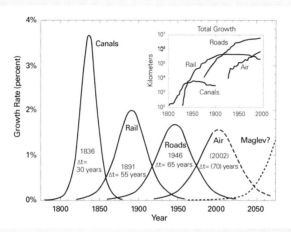

It takes 30 to 70 years to fully implement new infrastructures, judging by historical data on the time it has taken for major U.S. transportation infrastructures to reach their peak market penetration. Source: Jesse H. Ausubel, Cesare Marchetti, Perrin Meyer, Toward Green Mobility: The Evolution of Transport, European Review, Vol. 6, No. 2, 137-156 (1998). Posted with permission on http://phe.rockefeller.edu/green_mobility/.

Each fuel/vehicle pathway faces its own transition challenges, which can vary with region and can slow market penetration. These include infrastructure compatibility, consumer acceptance (based on, for example, limited range or long recharging times for batteries or and limited initial infrastructure for hydrogen fuel cell vehicles), cost, availability of primary resources for fuel production, greenhouse gas emissions, and other environmental and sustainability issues (such as air pollutant emissions, and water, land, and materials use).

USES OF TRANSPORT ENERGY, 2005

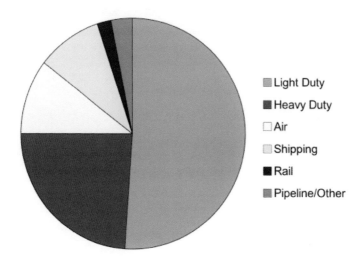

Fraction of global transport energy use in 2005, Source: International Energy Agency, Transport, Energy and CO_2: Moving Towards Sustainability *(Paris, France: IEA, 2009).*

Approaches to Sustainable Transportation

Government and industry are seeking sustainable solutions for the future transportation system. Three approaches are often proposed to reduce transport-related energy use and emissions:

- **Improve efficiency.** This means shifting to more efficient modes of transport, such as from cars to mass transit (bus or rail), or from trucks to rail or ships. Further efficiency improvements could be achieved by reducing vehicle weight, streamlining, and improving designs of engines, transmissions, and drive trains, including hybridization. In the heavy-duty freight movement subsector and in aviation, there is also promise of significant efficiency improvements.
- **Replace petroleum-based fuels with low- or zero-carbon alternative fuels.** These include renewably produced biofuels, and electricity or hydrogen produced from low-carbon sources such as renewables, fossil energy with carbon capture and storage (CCS), or nuclear power. Alternative fuels have had limited success thus far in most countries, with alternative-fuel vehicles currently making up less than 1 percent of the global fleet;[3] however, the context for alternative fuels is rapidly changing and a host of policy initiatives in Europe, North America, and Asia are driving toward lower-carbon fuels and zero-emission vehicles.
- **Reduce vehicle miles traveled.** This might be achieved by encouraging greater use of carpooling, cycling, and walking, combining trips, and telecommuting. In addition, city and regional smart growth practices (planning so that people do not have to travel as far to work, shop, and socialize) could reduce GHG emissions by as much as 25 percent.[4]

The emerging consensus among transportation energy analysts is that all three approaches will be needed if we are to meet stringent societal goals for carbon reduction and energy supply security. In this book we concentrate on the prospects and challenges for large-scale development of alternative vehicles and fuels.

POLICIES DRIVING CHANGE

State and federal policy initiatives in the United States are driving toward lower-carbon fuels and zero-emission vehicles.

In California:
- The Zero-Emission Vehicle (ZEV) Regulation requires automakers to offer 7,500 pure ZEVs for sale by 2014 and 25,000 pure ZEVs by 2017. An expanded program will be proposed in 2011 for requirements through 2025.
- AB 1493 (the Pavley Act) regulates vehicle CO_2 emissions and requires a 30-percent reduction in GHG emissions by 2016. An expanded fleet program will be proposed in 2011 for requirements through 2025, as part of a broad "Clean Cars" program.
- AB 118 provides funding of $200 million per year for the establishment of alternative fuel infrastructure and vehicle rebates through 2015.
- SB 1505 requires that source-to-wheel emissions of GHG from vehicular hydrogen be reduced by 30 percent on a per-mile basis when compared to the average gasoline vehicle.
- AB 32 (the Global Warming Solutions Act) requires California's Air Resources Board to enforce a statewide GHG emissions cap reaching 1990 levels by 2020. One component of this is a GHG cap and trade program that includes transportation fuels "in the cap."
- Executive Order S-3-05 sets GHG emission reduction targets for the state, including the mandate to reduce emissions to 80 percent below 1990 levels by 2050.
- California Air Resources Board Resolution 10-49 (November 18, 2010) establishes a Low Carbon Fuel Standard to reduce the carbon intensity of California's transportation fuels by at least 10 percent by 2020.
- The proposed Clean Fuels Outlet Regulation requires that an alternative fuel supply be provided once 20,000 alternative-fuel vehicles are on the road. This regulation is under consideration by the California's Air Resources Board.
- SB 2, which was signed by Governor Jerry Brown in April 2011, solidifies the requirement for 33% of electricity production to come from renewables by 2020 (California's Renewable Electricity Standard (RPS)).

On the federal level:
- The Renewable Fuel Standard requires 36 billion gallons of biofuel by 2022.
- The Corporate Average Fuel Economy (CAFE) standard requires that new light-duty sales average 35.5 mpg by 2016. An expanded program will be proposed in 2011 for requirements through 2025. This includes both a CAFE and gCO_2/mi fleet requirement.
- Plug-in electric vehicle (PEV) and biofuel tax breaks amount to 45 cents per gallon for purchase of ethanol and up to $7,500 for purchase of a PEV.
- The U.S. Department of Energy budget includes funds for research, development, and demonstration of battery electric vehicles and smart grid technologies, and research on fuel cell vehicles and hydrogen production, delivery, and storage.
- In conjunction with the United Nations Climate Change Conference in Copenhagen in 2009, President Obama put forth a goal of reducing GHG emissions in the United States 83 percent by 2050.

Alternative Fuel and Vehicle Pathways

Although our current transportation system is based almost exclusively on petroleum and the internal combustion engine, there are many other possibilities. A variety of more efficient vehicles (including those with hybrid drive trains and fuel cells along with battery electric vehicles) and alternative fuels (including compressed natural gas, ethanol, methanol, DME, F-T diesel, electricity, and hydrogen) have been proposed to address climate change and energy security concerns.

POSSIBLE VEHICLE TYPE – FUEL SOURCE COMBINATIONS

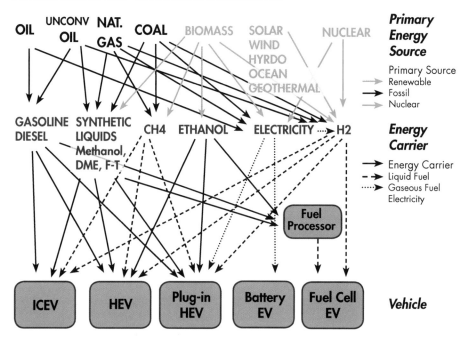

A variety of combinations of vehicle types and fuel sources are possible to meet our transportation needs. Possible fuel/vehicle pathways are shown here, with primary energy sources at the top, energy carriers (fuels) in the middle, and vehicle options at the bottom. F-T= Fischer-Tropsch process, ICEV = internal combustion engine vehicle, HEV=hybrid electric vehicle, EV = electric vehicle.

While many of these pathways offer potential societal benefits in terms of emissions or energy security, the path forward is unclear. Much of the public discourse has been framed as winner-take-all debates among advocates for particular "silver bullet" technologies. Policy proposals and media coverage suffer from a "fuel du jour" syndrome, waves of short-lived enthusiasm for one technology after another. Given the rapidly changing technology and policy landscape, consensus is lacking about which option or options to pursue, and when and where to pursue them.

Scope of This Book: Sustainable Transportation Energy Pathways

The purpose of this book is to help inform decision makers in industry and government about the potential costs and benefits of different fuel/vehicle pathways, and to illuminate viable transition strategies toward a sustainable transportation future. We draw heavily on insights gained from the Sustainable Transportation Energy Pathways (STEPS) research program at the University of California, Davis. STEPS began in 2007, with a goal of performing robust, impartial comparative analyses of different fuel/vehicle pathways drawing on engineering, economics, environmental science, and consumer behavior. An interdisciplinary team of 15 Ph.D.-level researchers and 25 graduate students was formed, with support coming from 22 diverse sponsoring organizations, each of which contributes to the STEPS consortium.

WHAT IS SUSTAINABLE TRANSPORTATION?

Energy sustainability has been defined as "providing for the ability of future generations to supply a set (or basket) of energy services to meet their demands without diminishing the potential for future environmental, economic and social well-being."[5] Sustainability is not necessarily a static concept or an end state: for transportation, "future well-being" could mean providing mobility to growing numbers of people.

How do we define and measure sustainability for transportation? Life-cycle analysis (LCA) is a powerful method for evaluating and comparing fuel/vehicle pathways with respect to a set of sustainability metrics. These could include primary energy use, greenhouse gas emissions, air pollutant emissions, water use, land use, materials requirements, and other factors that might be harder to quantify such as reliability and resiliency. The life cycle of a product encompasses all of the physical and economic processes involved directly or indirectly in its life, from extracting the raw materials used to make it to recycling the product at the end of its life. Life-cycle analysis for transportation analyzes all the steps in producing and using fuels: resource extraction and transport, production of the fuel, fuel delivery to refueling stations, and use in vehicles. (Sometimes the energy and materials used to make vehicles are also included in the life cycle, but these tend to be significantly lower than fuel cycle energy use and emissions.) Emissions, energy use, and other factors can be estimated at each step and added up to give a "well-to-wheels" total.

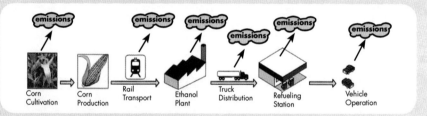

A fuel life-cycle analysis traces a fuel pathway from well to wheels. The corn ethanol pathway is shown here as an example. Source: M. A. Delucchi, Lifecycle Analyses of Biofuels *(Institute of Transportation Studies, University of California, Davis, 2006), UCD-ITS-RR-06-08.*

LCA can also be used as a basis for estimating the societal costs of different fuel/vehicle pathways including externalities, such as health damage from air pollution, climate impacts of greenhouse gas emissions, and economic costs of oil insecurity. When these costs are added to the direct cost of owning and operating the vehicle, low-emission options become more competitive with conventional fuels.[6]

But while LCA is a very useful tool, LCA alone can't define sustainability. Deciding on the acceptable limits for different LCA metrics (for example, allowable well-to-wheels GHG emissions) is a difficult task. Besides, complex social and economic factors come into play when trying to formulate a practical (and enforceable) definition of sustainability. While carbon storage, biodiversity, soil conservation, water use, water quality, and air pollution are amenable to LCA, the socioeconomic principles of welfare of local communities, land-rights issues, and labor welfare are not. Furthermore, sustainability impacts from market-mediated or macroeconomic effects—including indirect land-use change (iLUC), food price, and food availability—can be very important but are hard to measure and predict. Such effects become important when energy and environmental policies affect prices, which in turn affect consumption and hence output, which then changes emissions. (For example, U.S. biofuel policy led to use of the corn crop for fuel ethanol, which caused a spike in Mexico's corn-based food prices.)

Several authors[7] have suggested developing a "sustainability index" incorporating multiple criteria. This work is still nascent, in part because it is difficult to value different attributes on the same scale, and valuation depends on cultural and political norms. In this book, we discuss sustainability based on LCA and cost-benefit concepts, while recognizing this is a major simplification. Our underlying assumption is that pathways that score well on many sustainability metrics are likely to be attractive.

STEPS research is organized around four fuel pathways: hydrogen, biofuels, electricity, and fossil fuels. We have explored technical aspects, cost, market issues, environmental implications, and transition issues for each individual pathway. STEPS research is also organized by thread or project area, allowing us to compare fuels with each other along multiple dimensions—with respect to consumer behavior, infrastructure requirements, well-to-wheels energy use and emissions, policy, and vehicle technology. This allows us to develop integrative scenarios to address goals like reducing greenhouse gas emissions or oil dependency.

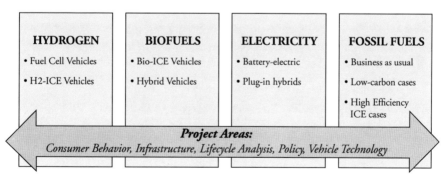

Sustainable Transportation Energy Pathways (STEPS) research is organized by energy pathway, with comparative analysis in project areas.

Analyzing single-fuel pathways gave us a strong basis for comparing different fuels and developing scenarios about how the various fuel/vehicle pathways might be integrated to meet societal goals. The STEPS research has flowed naturally from single pathway analyses to robust comparison of fuel pathways to integrative scenarios and transition analyses for future vehicles and fuels, and increasingly to case studies that inform carbon and alternative fuel policies in California, the United States, and beyond. It addresses these four "big picture" questions that the parts of this book are organized around:

1. **What do individual fuel/vehicle pathways look like?** We characterize individual fuel/vehicle pathways, with chapters on biofuels, electricity, and hydrogen. We explore technical aspects, costs, market issues, environmental implications, and transition issues for each pathway. Our interdisciplinary approach enables us to describe each pathway with depth and sophistication from multiple perspectives.

2. **How do these pathways compare?** Building on single-pathway analyses, we compare different fuel/vehicle pathways with respect to vehicle technology and costs, infrastructure issues, and well-to-wheels environmental impacts. We have done this on an impartial, self-consistent basis, across many dimensions. This allows us to understand when different fuel/vehicle options might be available and how the costs and benefits compare.

3. **How could we combine pathways and approaches to meet societal goals for carbon reduction, energy security, and such?** Drawing on the insights in Parts 1 and 2, we have developed integrative scenarios for reaching societal and policy goals (for example, 80 percent reduction in GHG emissions by 2050). We have studied transition issues as well as interactions between, for example, electricity and transportation. We have found that "silver bullet" solutions won't reach long-term goals and that a portfolio approach is needed, incorporating both near-term and long-term technologies and changes in behavior.

4. **What policy measures and tools are needed to encourage progress toward sustainable transportation?** We discuss policies and strategies for developing a sustainable transportation system, as well as measurement challenges that must be addressed in order for analysts to be able to predict the full impact of potential policies.

Perhaps the single most important insight from the STEPS research is that a portfolio approach will give us the best chance of meeting stringent goals for a sustainable transportation future. Given the uncertainties and the long timelines, it is critical to nurture a portfolio of key technologies toward commercialization. All our work in characterizing pathways and comparing them flows toward this conclusion.

We invite you to explore intriguing pathways, to compare options, and to synthesize these insights into your own vision for the future of transportation.

Notes

1. International Energy Agency (IEA), *Energy Technology Perspectives*, 2008, p. 650.
2. M. A. Kromer and J. B. Heywood, *Electric Powertrains: Opportunities and Challenges in the U.S. Light-Duty Vehicle Fleet*, LEFF 2007-02 RP (Sloan Automotive Laboratory, MIT Laboratory for Energy and the Environment, May 2007), http://web.mit.edu/sloan-auto-lab/research/beforeh2/files/kromer_electric_powertrains.pdf.
3. Exceptions: in Brazil, around 50 percent of transport fuel (by energy content) is ethanol derived from sugar cane; in Sweden, imported ethanol is being encouraged; in India, Pakistan, and Argentina, compressed natural gas (CNG) is widely used; and in the United States, ethanol derived from corn is currently blended with gasoline up to 10 percent by volume in some regions and accounts for 3 percent of U.S. transport energy use, according to the U.S. Energy Information Administration's *Annual Energy Outlook 2009*.
4. A. M. Eaken and D. B. Goldstein, "Quantifying the Third Leg: The Potential for Smart Growth to Reduce Greenhouse Gas Emissions," Proceedings from the 2008 ACEEE Summer Study on Energy Efficiency in Buildings (American Council for an Energy-Efficient Economy, 2008).
5. A. Löschel, J. Johnston, M. A. Delucchi, T. N. Demayo, D. L. Gautier, D. L. Greene, J. Ogden, S. Rayner, and E. Worrell, "Stocks, Flows, and Prospects of Energy," chapter 22 in *Linkages of Sustainability*, ed. T. E. Graedel and E. van der Voet (MIT Press, 2009), http://pubs.its.ucdavis.edu/publication_detail.php?id=13584.
6. See M. A. Delucchi, *A Conceptual Framework for Estimating Bioenergy-Related Land-Use Change and Its Impacts over Time*, UCD-ITS-RR-09-45 (Institute of Transportation Studies, University of California, Davis, 2009); D. Greene, P. Leiby, and D. Bowman, *Integrated Analysis of Market Transformation Scenarios with HyTrans,* ORNL/TM-2007/094 (Oak Ridge National Laboratory, 2007); J. Ogden, R. H. Williams, and E. D. Larson, "A Societal Lifecycle Cost Comparison of Cars with Alternative Fuels/Engines," *Energy Policy* 32 (January 2004): 7-27, http://pubs.its.ucdavis.edu/publication_detail.php?id=; A. Rabl and J. V. Spadaro, "Public Health Impact of Air Pollution and Implications for the Energy System," *Annual Review of Energy and the Environment* 25 (2000): 601–27; J. V. Spadaro, A. Rabl, E. Jourdaint, and P. Coussy, "External Costs of Air Pollution: Case Study and Results for Transport between Paris and Lyon," *International Journal of Vehicle Design* 20 (1998): 274–82; Y. Sun, J. Ogden, and M. A. Delucchi, "Societal Lifetime Cost of Hydrogen Fuel Cell Vehicles," accepted for publication in the *International Journal of Hydrogen Energy* 2010; C. E. Thomas, "Fuel Cell and Battery Electric Vehicles Compared," *International Journal of Hydrogen Energy* 34 (2009): 6005–20.
7. These authors include D. L. McCollum, G. Gould, and D. L. Greene, *Greenhouse Gas Emissions from Aviation and Marine Transportation: Mitigation Potential and Policies*, UCD-ITS-RP-10-01 (Institute of Transportation Studies, University of California, Davis, 2010), http://pubs.its.ucdavis.edu/publication_detail.php?id=1363; and R. Zah, M. Faist, J. Reinhard, and D. Birchmeier, "Standardized and Simplified Life-Cycle Assessment (LCA) as a Driver for More Sustainable Biofuels," *Journal of Cleaner Production* 17, Supplement 1 (2009): S102–S105; S. Yeh, D. A. Sumner, S. R. Kaffka, J. M. Ogden, and B. M. Jenkins, *Implementing Performance-Based Sustainability Requirements for the Low Carbon Fuel Standard—Key Design Elements and Policy Considerations*, UCD-ITS-RR-09-42 (Institute of Transportation Studies, University of California, Davis, 2009).

Part 1: Individual Fuel/Vehicle Pathways

We start by characterizing individual fuel pathways and accompanying vehicle technologies. Biofuels, which are here today and also under development for the future, can be used in internal combustion engine vehicles (ICEVs) as well as in hybrid vehicles. Electricity can be used in battery electric and plug-in hybrid vehicles, which are slowly making their way into showrooms. Hydrogen can be used in ICEVs, but our primary interest here is its use in fuel cell vehicles, a rapidly developing technology that could be available within the next 5 to 10 years. The three chapters in this part examine multiple aspects of the biofuels, electricity, and hydrogen pathways, including technical status and outlook, environmental impacts, infrastructure requirements, transition scenarios, and policies and business strategies needed to support the pathway.

- **Chapter 1** examines biofuels, which have the advantage that they can be made to resemble conventional fuels and in some cases can easily be incorporated into the existing fuel distribution system, easing transition issues. At the same time, biofuels face challenges with respect to resource availability, cost, and environmental and economic impacts. This chapter draws on detailed modeling of future biofuel infrastructure that has been done in response to policy goals for renewable fuels. We describe current and future biofuels production technology and develop biofuel supply curves for the United States.

- **Chapter 2** focuses on electricity and its use in plug-in vehicles of both the hybrid and pure electric variety. Electric-drive technology promises clean skies, quiet cars, and plentiful fuel produced from nonpolluting domestic sources, but it faces a fundamental challenge: how to store energy and supply power. This chapter draws from several streams of research—including testing of battery technology, modeling of the electricity grid, and eliciting consumer data regarding PEV design interests and potential use patterns—to sort through the hype and improve understanding of this pathway and the advances it must make to become competitive with ICEVs.

- **Chapter 3** explores hydrogen, a fuel pathway with the long-term potential to greatly reduce oil dependence as well as transportation emissions of greenhouse gases and air pollutants. Complex technical and logistical challenges must be overcome before a hydrogen-based transportation system can become a reality. This chapter discusses some of the major questions regarding future use of hydrogen in the transportation sector and highlights STEPS research on these issues.

Chapter 1: The Biofuels Pathway

Nathan Parker, Bryan Jenkins, Peter Dempster, Brendan Higgins, and Joan Ogden

Biofuels have been seen as the nearest-term answer to the need for alternatives to petroleum fuels in the transportation sector. Despite recent debate over life-cycle environmental impacts and potential food-sector impacts, much interest remains in the expansion of biofuel production capacity to displace petroleum and provide low-carbon fuels—especially for heavy transport and aviation, where few other sustainable alternatives to liquid fuels exist. Besides, biofuels (and bioenergy production more generally, including heat and power applications) offer opportunities for economic development, diversification of the farm sector, integration of forest management, and diversion of urban wastes from landfills.

But these opportunities come with substantial challenges. With the current state of technology, the lowest-cost biofuels do not provide major environmental benefits, while the biofuels that are expected to provide significant benefits are not yet commercially viable. Resources for the United States have been estimated to be sufficient to produce enough biofuel to meet roughly a third of the nation's transportation fuel demand, but large uncertainties are associated with these estimates and sustainability of production and manufacturing processes is not yet fully understood. Additionally, best uses for biomass—whether to produce the liquid biofuels discussed in this chapter, to generate electricity for electric vehicles, to produce hydrogen for fuel cell vehicles, or to be used in other sectors—are still to be sorted out through technology innovation and market action.

This chapter discusses some of the major questions regarding future use of biofuels in the transportation sector and highlights STEPS research on these issues.
What is the technical outlook for advanced biofuel production technologies?
- To what extent can biofuels contribute to future transportation fuel supply? What are the constraints on feedstock for those biofuels? Where is advanced biofuels production likely to take place in the United States?
- How compatible are biofuels with existing vehicles and infrastructure?
- What are the environmental impacts of biofuels compared to alternatives? How do we measure sustainability for biofuels?
- What policies and business strategies are needed to support biofuels in both the near and long term?

CHALLENGES ON THE BIOFUELS PATHWAY

These complex technical and environmental challenges must be overcome before biofuels can make a major contribution to transportation energy needs:

- **Technical challenges.** The portfolio of advanced technologies that can convert biomass to fuels on a large scale is mostly in the demonstration phase, and the challenges associated with scaling up those technologies lie ahead.
- **Logistical challenges.** Biofuels can be produced from a variety of feedstocks and transported by rail, ship, pipeline, or truck, similar to gasoline. Depending on the biofuel, it may be possible to utilize the existing petroleum infrastructure, either by blending biofuels with conventional fuels or by producing designer biofuels that can drop in to existing supply systems. Other biofuels, such as ethanol (or E85—85 percent ethanol and 15 percent gasoline), would require dedicated storage and transport systems.
- **Resource availability.** The amount of biomass feedstock available, regionally and nationally, for conversion to fuels is uncertain and limited compared to transportation fuel demand. The supply depends in part on yields of energy crops and on market participation of waste and residue biomass suppliers.
- **Environmental and sustainability issues.** Large-scale biofuels production places significant demands on arable land, water, and agricultural inputs. The environmental impact of using these resources must be weighed against the benefits from producing biofuels. The range of impacts is large as a result of the variety of biomass feedstock, regional differences in native ecosystems and crop yields, and the efficiency of biofuel production.
- **Macroeconomic impacts.** Use of biomass for energy can impact markets for other biomass products, especially food and feed. Indirect land-use effects resulting from this market force lead to potentially larger greenhouse gas emissions than the reductions realized by fossil fuel replacement.
- **Transition issues / coordination of stakeholders.** Transitional barriers for biofuels are lower than for other alternative fuels. In the case of E85, greater deployment of flexible-fuel vehicles is needed to stimulate demand. Coordination is needed between suppliers of biomass feedstocks and investors in biorefineries.
- **Policy challenges.** Biofuel policies must be crafted to maximize benefits. This is a dynamic challenge as the impacts are uncertain and highly variable, and they depend on a number of outside forces. This challenge is highlighted by the discussion of indirect land-use change presented in Chapter 12.

Technology Status and Outlook

Biofuels are a diverse set of fuels derived from biomass (material of recent organic origin—for example, plant material, animal products, and organic wastes). These fuels can be alcohols (ethanol, butanol, or methanol), hydrocarbons (similar to gasoline, diesel, and jet fuel), hydrogen, or synthetic natural gas. This chapter focuses only on the liquid fuels.

Biomass can be converted to liquid fuels using many different routes. There is great diversity among these routes in both technological readiness and long-term outlook for meeting transportation energy needs. In large part, the commercial readiness rests with the conversion technology, and the long-term potential depends on the feedstock. First-generation processes are commercially available today, and advanced processes aiming to convert cellulosic materials (such as agricultural, forest, or municipal solid wastes and energy crops) and algae are under development.

The generic pathway for production of a biofuel has five components. First, the biomass is grown and harvested or separated from a waste stream. Then the biomass is stored, either at the site of production, the biorefinery, and/or an intermediate depot. If it's stored at the biorefinery or an intermediate depot, transportation to the conversion facility (biorefinery), which is the third component, precedes this. Fourth, at the biorefinery the biomass is converted to biofuels and coproducts. Fifth, the biofuels are distributed to refueling stations, with possible blending with petroleum fuels at an intermediate fuel terminal. The cost of biorefineries is the largest single capital investment in the supply chain (about 85 percent of the investment), with feedstock production equipment and fuel delivery equipment playing a much smaller role.

COMPONENTS OF A BIOFUEL PRODUCTION PATHWAY

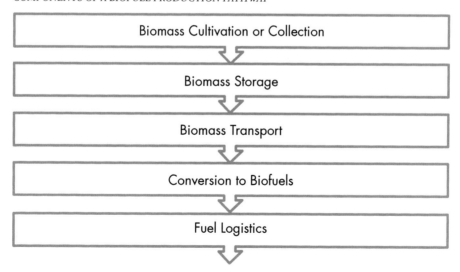

Production of a biofuel has five components. The biomass is (1) grown and harvested, (2) stored, (3) transported to the conversion facility (biorefinery), and (4) converted to biofuels. Then, (5) the fuel is distributed to refueling stations.

Conversion technology is the current roadblock to realizing significant production of advanced biofuels. The development of practical and cost-effective conversion technologies for cellulosic biomass feedstocks is key. Even with significant research investment, no large-scale commercial capacity yet exists for biofuels other than direct sugar and starch fermentation biorefineries to make ethanol based on beverage alcohol production practices, and the transesterification of lipids to make biodiesel. These two types of biofuel—ethanol from yeast fermentation of sugar and starch, and biodiesel from fats, oils, and greases—comprise the class of so-called first-generation biofuels and are made using moderately mature technologies.

Research into exploiting the chemical diversity and energy content of other biomass feedstocks has continued to promise new technologies to expand the scale of biofuel production. So-called second- and later-generation conversion technologies are in development to utilize cellulosic resources either through thermochemical processes that utilize heat, pressure, and catlysts to produce fuels or through biological processes utilitizing organsims and enzymes to reduce the plant material to sugars and then ferment these to the desired products.

Technologies to gather biomass resources and transport them to biorefineries exist in well-established industries of agriculture, forestry, and waste management. However, large-scale development of biorefineries with consistent, dependable feedstock supplies will depend on improvements in storage and transportation technologies. Stability of sugars in storage is a major concern for maintaining year-round feedstock quality for biological conversion processes. More efficient long-distance transport will improve the flexibility of the biorefineries in their feedstock sourcing.

The biofuels that are commercially available use food crops as feedstocks. These feedstocks—sugar, starch, and oils—are relatively easy to convert to ethanol or biodiesel but represent only a small fraction of the biomass of the plant, limiting the yield of fuels per unit of land area. The advantage to using these feedstocks is that established commodity markets provide a reliable supply of the feedstock, and extensive research has been done on improving the yields of these crops. On the other hand, the utilization of feedstocks that are not currently used on any large scale for food or feed such as wood, herbaceous energy crops, and algae avoids direct market impacts on food commodities, although indirect effects may remain.

EXAMPLES OF LIQUID BIOFUEL PRODUCTION PATHWAYS

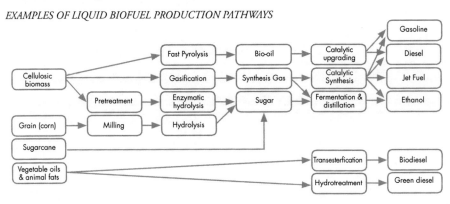

Liquid biofuels are already being produced from sugar/starch crops and oil plants / animal fats with first-generation technologies. Second-generation technologies will produce ethanol and diesel from cellulosic biomass.

Ethanol from sugar/starch crops

Ethanol from sugar is the simplest route for producing biofuels. The sugar source is predominantly sugarcane, but sugar beets or high sugar content food-processing wastes can also be used. Sugarcane ethanol is a mature technology developed in large part in Brazil. The cane is milled to extract the sugars, which are then fermented to ethanol using yeast. Ethanol is distilled from the beer, leaving a liquid by-product (vinasse, which is used as a fertilizer) and the cane fiber (bagasse, which is used to produce heat and power). In 2009, 6.6 billion gallons of ethanol were produced from sugarcane in Brazil.

Ethanol produced from corn is the dominant biofuel pathway in the United States. The corn ethanol industry is well established, with 204 corn ethanol facilities producing 13.2 billion gallons of ethanol in 2010.[1] There are two types of technologies: wet mill and dry mill. Wet mill technologies separate the germ, fiber, gluten, and starch components of the corn kernel through steeping, screens, cyclones, and presses. The starch fraction can then be converted to ethanol. It is the more capital- and energy-intensive process with lower ethanol yields but higher-value coproducts. Dry mill processes first grind the corn, sending the full kernel through the saccharification and fermentation process before separating ethanol from the coproduct, distillers grains (typically dried, in which case it becomes dried distillers grains or DDG). Dry mill ethanol facilities were responsible for more than 86 percent of ethanol production in 2010.[2]

Biodiesel from oils and fats

Biodiesel (which here refers only to fatty acid methyl ester or FAME) is a mature technology for creating diesel-like fuels from oils and animal fats. In 2009, 4.7 billion gallons of biodiesel were produced worldwide, with 540 million gallons produced in the United States (predominantly from soybean oil) and 2.65 billion gallons produced in Europe (predominantly from rapeseed oil). Biodiesel is also produced from waste greases and animal fats at small volume. It is made by transesterfication, a catalyzed chemical conversion of oils or fats and an alcohol (typically methanol) to biodiesel and significant quantities of glycerol coproduct. FAME can be produced from virgin seed oils, waste greases, or animal fats, though the process design is optimized differently for the different resources. The dominant production process in the United States, accounting for approximately 78 percent of biodiesel production in 2008,[3] uses alkali catalyst with virgin soy oil feedstock;[4] an acid catalyzed process is most economic for waste cooking oil.[5] The dominant cost in producing biodiesel is that of the feedstock, especially true for virgin seed oils.

An alternative technology to produce diesel fuels from oils is the hydrotreatment process.[6] In this process, the lipids and hydrogen pass through a hydroprocessing unit where the oxygen is stripped from the lipids through decarboxylation and hydrodeoxygenation reactions. The resulting products are a combination of "green diesel" and lighter hydrocarbons (naphtha and/or propane) with by-products of water and carbon oxides (CO and CO_2). The green diesel fuel is reported to have a number of desirable properties: high cetane number (70–90), energy density equivalent to ultra-low sulfur diesel, sulfur content of less than 1 ppm (USLD < 10 ppm sulfur), and good stability. Green diesel could potentially be used as a premium blendstock allowing for the use of lower-valued light-cycle oil as part of a diesel blend. This technology has recently been commercialized.

Biofuels from cellulosic biomass

Advanced routes of biofuel production take advantage of more of the plant and of cellulosic biomass materials such as wood, grasses, straws and stovers from agriculture, and the organic fraction of municipal wastes (paper, cardboard, wood, textiles, and such). In simple terms, cellulosic biomass is made up of three major components: cellulose, hemicellulose, and lignin. Cellulose and hemicellulose are carbohydrate polymers that can be broken down into component sugars for fermentation. Lignin is inert for biological conversion processes but can be utilized in thermal conversion processes.

There are a host of conceptual designs for creating liquid fuels from cellulosic biomass, but the commercially viable conversion technologies have yet to be determined. Most have not been proven beyond the laboratory scale. A number of pilot, demonstration, and early commercial-scale biorefineries are under development using a broad suite of technologies; these will provide a greater understanding of the commercial viability of the schemes in the near future. The technologies can be classified as utilizing biological conversion processes, thermochemical processes, or a combination of the two.

Estimates of the cost of production rely on a number of engineering studies with process-level modeling of the biorefinery. The majority of studies of cellulosic ethanol consider the biochemical pathway where the cellulose and hemicellulose are converted to sugars through enzymatic hydrolysis and saccharification, then fermented to make ethanol. The thermochemical pathway via gasification and synthesis has been found to be similar in cost and performance to the biochemical pathway at the scale of 45 million gallons of ethanol per year.[7] The biochemical route is taken to be the model cellulosic ethanol technology due to the larger base of supporting literature. The thermochemical pathway may prove to be the technology better suited in certain cases, but the performance is likely to fall in the range studied.

Biological conversion of cellulosic biomass to ethanol (cellulosic ethanol) has been the focus of significant research and is well described in literature. It is the basic technology of 7 of the 14 demonstration and commercial biorefineries receiving funding from the U.S. Department of Energy. Four demonstration biorefineries are currently operational worldwide and are expected to produce more than 3 million gallons in 2011.

The biochemical pathway begins with feedstock pretreatment to make the cellulose available to the enzymes. There are a number of techniques under research and development for this pretreatment, including dilute acid hydrolysis, ammonia fiber explosion, liquid hot water, and steam explosion. In the process of exposing the cellulose, the hemicellulose is broken into its component sugars (xylose, arabinose, and so on). The exposed cellulose is then converted to glucose with cellulase enzymes. Glucose is fermented to ethanol and the five-carbon (C5) sugars are fermented to ethanol either in a combined reactor using recombinant Zymomonas mobilis or in separate reactors using yeast for the C6 sugars and Z. mobilis for the C5 sugars. In some advanced designs, a consolidated bioprocessing (CBP) approach is taken where all biological conversions (enzyme production, enzymatic hydrolysis, and fermentation) occur in the same reactor.[8] This design is attractive, but the enzyme to make it possible has yet to be identified. In most designs, the lignin is separated from the beer, dried, and combusted to produce steam and electricity for the biorefinery with some net export of electricity.

CHAPTER 1: THE BIOFUELS PATHWAY

The projected costs for cellulosic ethanol production using current technology cover a large range, with three main sources of variation. First is the expected yield of ethanol from cellulosic material. Estimates range from 52.4 gallons to 76.4 gallons per dry ton of switchgrass or corn stover. This variation is due to difference in the performance of the pretreatment, cellulase enzymes, and fermentation organisms each study assumes.[9] Second is the capital investment required, where a variety of configurations have been studied and different yields assumed. Within the same study, capital costs varied by 42 percent due to different configurations of pretreatment, hydrolysis, fermentation, and distillation.[10] The third factor is the variable operating cost—mainly the cost of cellulase enzymes. For example, one study[11] projects cellulase enzymes available at $0.32/gal of ethanol where another puts the cost at $1.05/gal. Also of interest is that the estimate for year 2000 technology in an earlier study falls below the more recent estimates of current costs, demonstrating that as more is learned about these technologies, limitations are identified that lead to additional costs.

COMPARISON OF PRODUCTION COST ESTIMATES FOR CELLULOSIC ETHANOL

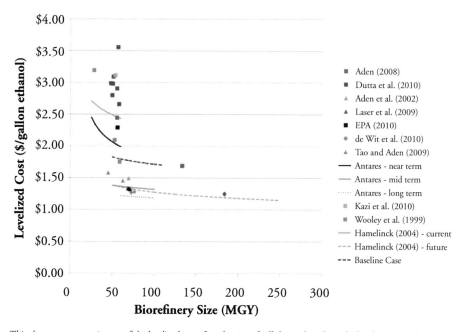

This chart compares estimates of the levelized cost of production of cellulosic ethanol via the biochemical pathway arrived at by a number of different studies.[14] Near-term technology assessments are represented by squares, midterm technology (7–15 years ahead) by triangles, and long-term projections by diamonds.

Thermochemical conversion of cellulosic biomass to fuels can take many routes, borrowing from fossil energy technologies in many cases. The Fischer-Tropsch (F-T) synthesis process is among the most studied and furthest developed. Commercial facilities exist or have existed in the past for production of F-T fuels from both coal and natural gas. In order to commercialize biomass F-T fuels, a number of modifications are needed compared with coal or natural gas-based F-T fuels. An economically viable biomass gasification technology must be developed along with optimization of gas cleanup and the F-T synthesis processes for the resulting biomass-based synthesis gas. A number of biomass gasifier configurations have been studied, each with benefits depending on the context of the project.[15]

Projected costs for current technology F-T diesel production cover a large range, representing some disagreement on which technologies are current and which are unproven, as well as differences in design. One study[16] states that hot gas cleanup (tar cracking) is not yet commercial while all other studies use it. One study[17] uses an indirectly fired atmospheric gasifier while most others use pressurized, oxygen-blown, directly fired gasifiers. In projecting future technology versus current technology, one study[18] foresees no changes in the design but projects reductions in capital and operating costs due to incremental improvements and increases in scale. Another study[19] presents a case with mature technology where a once-through configuration is designed for greater electricity production than the other studies. The EPA projection[20] is significantly lower compared to other studies at similar scale and timeframe, but the study provides little information to support this estimate.

COMPARISON OF PRODUCTION COST ESTIMATES FOR F-T DIESEL

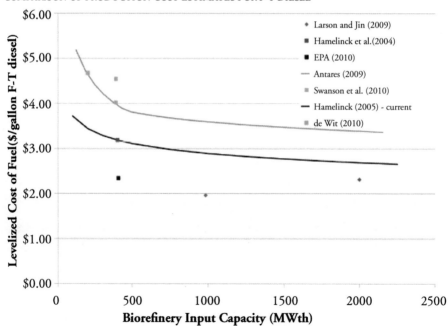

This chart compares estimates of the levelized cost of production of F-T diesel arrived at by a number of different studies.[21] Near-term technology assessments are represented by squares, midterm technology (7–15 years ahead) by triangles, and long-term projections by diamonds.

F-T diesel is just one example of the gasification/synthesis thermochemical production route to produce hydrocarbon biofuels. Other conversion technologies are based on the route with a difference in the catalytic synthesis that takes place after the synthesis gas is produced. Another example is methanol-to-gasoline (MTG) technology, which combines methanol synthesis of the synthesis gas and a catalytic conversion of methanol to gasoline.

An alternative thermochemical route to producing hydrocarbon fuels from cellulosic biomass is fast pyrolysis with upgrading. Fast pyrolysis uses high temperatures in the absence of oxygen to degrade the solid biomass into a bio-oil similar to petroleum but with high oxygen and water contents. The bio-oil can be upgraded to fuels (gasoline, diesel, and/or jet fuel) through a combination of hydrotreating, hydrocracking, and dehydration.

Biofuels from algae

Biofuels could be produced from algae with many advantages over both first-generation biofuels and cellulosic biofuels. Algae produce significantly higher biomass in the same area compared to other energy crops. Some strains also produce high fractions of oils. Also, algae do not require soils to grow, which reduces the pressure biomass production would put on lands with agricultural production. And algae can use degraded water resources.

Algae cultivation can take place in an open pond, a closed photobioreactor (PBR), or a combination of the two. Open ponds are less capital-intensive but lack the environmental control of PBRs. Species control is one of the principal challenges for open-pond technology since native algae strains tend to outcompete the desired strain selected for optimal fuel production.[22] Hence, only three algae species have been successfully cultivated in open ponds for an extended time period.[23] All three species are "extremophiles" in that they can survive under extreme conditions such as high pH or salinity that prohibit the growth of other organisms. PBRs can also suffer from contamination issues, but the controlled environment and closed system makes it easier to prevent contamination.

The harvesting and dewatering step can be capital- and energy-intensive. Open pond biomass concentrations are typically 0.5 g/L[24] while PBR concentrations are on the order of 1 to 12 g/L.[25] Bulk harvesting techniques such as flocculation and settling can be used to increase the biomass concentration to approximately 1 percent,[26] which may be acceptable for some fuel production processes such as anaerobic digestion or alcoholic fermentation, but the water content must be decreased further in order to use other biofuel production pathways. Filtration and centrifugation—both energy-intensive processes—can reduce the water content down to approximately 80 percent, resulting in algae paste. This paste can be dried or used in its wet form, depending on the production process. Filtration can be cost-effective for filamentous or large algae cells, but centrifugation is more cost-effective for small, spherical strains such as those of the genus Chlorella.

Possible thermochemical methods for converting algae biomass into fuel include thermochemical liquefaction, gasification with Fischer-Tropsch processing, and pyrolysis.[27] Alternatively, lipid extraction and transesterification can be used to produce biodiesel. Hydrotreating the lipids could be used to produce renewable diesel.[28] All of these techniques, with the exception of thermochemical liquefaction, require dry or nearly dry biomass, which requires substantial energy inputs. Techniques for lipid extraction from microalgae are not well established.

Biomass Resources for Biofuel Production

Biomass resources are often characterized by their mass, but this can be misleading as the energy content of biomass materials varies significantly, and energy is what is interesting. The solar energy annually captured as biomass by all terrestrial plants is approximately 2,500 exajoules (EJ), or more than fifteen times global petroleum use. Obviously, the vast majority of this will not and should not be made available for energy production. Most of global biomass production either provides greater value to society in its natural state or cannot be economically accessed for bioenergy production. Hence, the fraction of biomass that can advantageously be employed to produce transportation fuels is very small. Global estimates suggest that 10 to 25 percent of transportation fuel needs could be met with biofuels,[29] with biofuels playing a larger role if vehicles are made more efficient and biomass productivity is increased. In 2009, only 2 percent of transportation fuel demands were met using biofuels, predominantly in the form of ethanol from sugarcane and corn, and biodiesel from rapeseed, palm, and soy oils. There is room for growth in biofuels, but they cannot be expected to provide more than 25 percent of transportation energy globally.

In the United States, estimates of biomass that could be developed in the near term range from 208 to 801 million dry tons,[30] of which we estimate that between 156 and 443 million tons could be from waste and residue sources. Growth in waste and residue sources will be limited. Producing more biomass will require the growth of energy crops. By 2030, more than a billion tons of biomass could be sustainably produced from agriculture, forestry, and municipal waste,[31] sufficient to meet roughly a third of transportation fuel demand, and possibly more with advancements in the efficiency of future vehicles.[32]

Underlying the projections for biofuel potential are assumptions regarding technology development (making currently unattractive cellulosic feedstocks economic), land use, food demand, and overall agricultural productivity (including energy crop yields). One main concern is the ability of the land base to support energy production as well as food production. In these arguments food production is always given the priority, but the market does not prioritize food, as a consequence of disparities in purchasing power across the global population. Some small fraction of the biomass resource consists of organic wastes that do not interfere with land markets. These resources are limited relative to transportation fuel demand and will grow minimally in response to demand for biomass, but they avoid some sustainability concerns of crops grown for energy production. Residues from conventional agricultural crops and forestry operations provide significant potential although there is some debate over the sustainable use of these resources.[33] Residue resources may increase over time and in response to market demands for biomass as farmers maximize the total value of their crops. This resource has the potential to lead to adverse impacts on food production. The rest of the resource depends on lands that are currently idle or lands being freed through increases in agricultural productivity. Researchers who developed a global estimate of biofuel potential using abandoned agricultural land that is not currently forested or urbanized found an upper limit on biofuels grown on these marginal lands to be 12 times current production or approximately 17 percent of current global petroleum consumption.[34]

Biofuel production will need to be improved in several ways to meet this potential. First, the development of biofuel technologies that can utilize cellulosic biomass would enable access to the significant waste or residue streams from agriculture and forestry sectors as well as urban wastes.

Additional biomass could come from purpose-grown crops such as switchgrass, Miscanthus, oil seeds, algae, and many others, but the extent to which these can contribute to overall supply is not fully understood and can potentially expand beyond the limits suggested here. Industrial algae production could also significantly expand biomass resources due to high growth rates and yields and could potentially use marginal water, but future production levels and costs also remain highly speculative.[35]

Additionally, best uses for biomass are still to be sorted out through technology innovation and market action. The liquid biofuels discussed in this chapter would mostly be used in internal combustion engine vehicles. Electrification of the light-duty vehicle fleet might provide higher-efficiency use of biomass in this sector compared with liquid fuel production, although heavy transport and aviation will still likely depend on or prefer liquid fuels for economic reasons. Biomass can also be used to produce hydrogen, and if large-scale reliance on hydrogen for transport and other power sectors emerges, liquid biofuel production may serve mostly as an interim market solution. Multiple markets for biomass in the energy and bio-based product sectors are likely to continue to develop, extending the portfolio of conversion options and driving innovation toward more integrated production chains.

CASE STUDY: HOW THE U.S. BIOFUEL SUPPLY MIGHT MEET RFS2

The United States adopted a volumetric mandate for biofuels (the Renewable Fuel Standard or RFS) in 2005 and strengthened it in December 2007 as part of the Energy Independence and Security Act (EISA). The RFS2 mandates annual consumption of biofuels increasing to a quantity equivalent to 36 billion gallons of ethanol on an energy basis (2.9 EJ or 23.7 billion gallons of gasoline equivalent or gge) in 2022. While the law was written as a volumetric mandate, the EPA has interpreted the law as a mandate of energy quantities in order to provide a level playing field for all biofuels.[36] Specific mandates are defined each year for several subcategories of renewable fuels differentiated by feedstock and life-cycle carbon intensity (CI). In 2022, for example, of 21 billion ethanol equivalent gallons of advanced biofuels (not corn ethanol; 50-percent reduction in CI from gasoline required), 16 billion gallons must be cellulosic biofuels (from cellulosic feedstocks; 60-percent reduction in CI from gasoline required); the remaining 15 billion gallons to reach the 36 billion gallon total can be any renewable fuel with a 20-percent reduction in CI, including corn ethanol (existing corn ethanol facilities were given a grandfathered exemption to the CI requirement).

The UC Davis Geospatial Bioenergy Systems Modeling (GBSM) project has developed a spatially explicit model of how future biofuel supply chains in the United States might be constructed, and has applied the model to analyzing the domestic potential to meet RFS2. The GBSM aims to assess the potential U.S. biofuel supply by simulating it and its environmental impacts under scenarios of resource constraints, technology limitations, and policy limitations. The modeling framework incorporates a spatially explicit assessment of biomass resources, engineering-economic models of biorefineries, a GIS-based transportation cost model, and a supply chain optimization model. At the heart of the research is the integrated supply chain model, which maximizes the profit of the biofuel industry over the full supply chain using real-world data on potential biomass supply and including distribution of biofuel to the consumer. The model describes the optimal behavior of an industry to supply biofuels given a fuel demand, a biofuel selling price, and constraints on feedstock supply. Simply put, if biofuel can be delivered to the refueling stations for less than the given selling price, it is profitable for the industry to supply that biofuel, and the infrastructure would be built to reap that profit. If biofuels cannot be delivered for less than the selling price, the fuel demand is met with conventional fuels at the given selling price. In addition, when demand for fuel exceeds the supply of feedstock, the difference is made up with conventional fuels. Results of the model show that the potential for biofuel production in the United States is significant relative to current production.

Biofuel supply is dependent on a number of highly uncertain parameters. First is the resource base—what will be made available to the biofuel industry and at what cost. Second is the conversion technologies—what will be the conversion efficiency and cost of the conversion for unproven technologies. Finally, the demand for biofuels is unknown in two important aspects despite the mandated volume: (1) the amount of each fuel type (ethanol, biodiesel, biomass-based F-T diesel) that will be acceptable to use in the future vehicle fleet, and (2) the price the market will be willing to pay for each. Because of the uncertainty of these three sets of parameters, the study considers a range of outcomes through sensitivity analysis.

The biomass resources considered for this study include agricultural residues, switchgrass, forest residues (including unused mill residues), pulpwood, municipal waste, yellow grease, and animal fats. The model also uses the 2009 USDA long-term projections to describe conventional agriculture including corn, seed oils, and crop acreage for estimating residues,[37] which limits the analysis to 2018. Still, given the uncertainty in all parameters and the stability of the projection in 2015–2018, the analysis can be used to comment on the 2022 supply.

The supply of agricultural residues is constrained in order to maintain soil organic matter and for erosion control. These sustainability parameters will be different for each field and are not fully understood. Switchgrass is only considered to be grown on marginal land to avoid competition with food crops. This assumption is convenient for modeling energy crops with minimal impact on the food sector but may not be the way the market allocates land use. The yields and costs of production for both agricultural residues and switchgrass are not well known. To capture this, we developed three scenarios of each. We also developed three scenarios of availability of municipal wastes to capture the uncertainty in the participation rates of municipalities in making their wastes available for energy conversion. The biofuel conversion technologies considered range from biodiesel from seed oils, yellow grease, and animal fats to cellulosic ethanol or Fischer-Tropsch diesel and naphtha from wastes, residues, and energy crops. Scenarios were run at three levels of optimism for the cellulosic ethanol technology and two levels of optimism for Fischer-Tropsch technology. On the demand side, three scenarios were also developed, with ethanol consumption limited to blends with gasoline of (1) 10 percent or 20 percent for all vehicles, (2) 10 percent for conventional vehicles, and (3) 85 percent for all flexible-fuel vehicles. Aggregate scenarios were developed from all logical combinations of the resource, technology, and demand scenarios.

For each scenario, the research found the optimal design of the biofuel system over a range of prices to produce supply curves. These supply curves show the quantity of biofuels that would be made available at a given market price. The curves indicate biofuel potentials in 2018 ranging from 21 to 46 billion gallons of gasoline equivalent (gge) at prices below $4/gge depending on the resource and technology scenario. These volumes would meet 9 to 21 percent of the projected total transportation gasoline and diesel demand and would represent an increase of 300 percent over 2009 production levels. Below $3/gge, between 12 and 32 billion gge are projected to be feasible. The baseline scenario resulted in 32.5 billion gge at $4/gge and 22 billion gge at $3/gge. Constraints on the sustainable supply of biomass restrict growth of biofuels to not much more than the quantities available at $4/gge. The maximum supply identified under the assumptions used was 50 billion gge at $6/gge in the high feedstock scenario. This maximum supply would increase if cost-competitive production of algae-based biofuels is developed or if yields for energy crops on marginal land are higher than projected.

Biomass from waste and residue resources (municipal solid waste, agricultural residues, and forest residues) can provide quantities of biofuels that assist with policy goals. This resource is especially important as it avoids many sustainability concerns of biofuels produced from crops. Nationally, waste and residue resources are projected by the model to provide 7 to 16 billion gge of biofuel per year, accounting for between 35 and 64 percent of the RFS2 mandate in both 2018 and 2022. The remaining biofuels are predominantly corn ethanol (up to 10 billion gge) and soy biodiesel (up to 1 billion gge) in the 2018 case and expanding to include switchgrass and pulpwood-based biofuels at the higher volumes of the 2022 mandate.

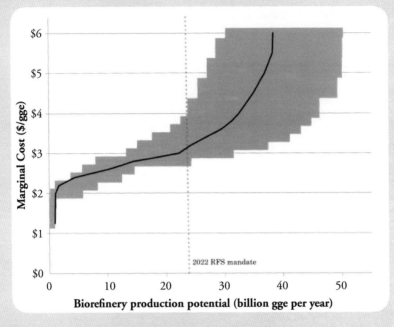

The GBSM arrived at this baseline supply curve for biofuels in 2018. The curve indicates the quantity of biofuels that would be made available at a given market price (gge = gallons of gasoline equivalent). RFS2 requires that 36 billion gallons of ethanol equivalent be sold annually by 2022, which converts to 24 billion gge. The shaded area shows the range of outcomes for the different scenarios evaluated by the model.

Investments in biorefineries required to meet mandated volumes of biofuels are large and depend on the specific pathways chosen. Greater reliance on cellulosic technologies requires higher capital investment than systems that rely on conventional biofuel technologies such as corn ethanol or fatty acid methy ester (FAME) biodiesel; however, these technologies have significant sustainability and energy-balance benefits over corn-based technologies. In addition, systems where the Fischer-Tropsch technology is chosen to convert cellulosic biomass have higher capital costs than systems where cellulosic ethanol is the technology of choice for cellulosic biomass. The total investment in biorefineries required to meet the 2022 RFS2 mandate is between $100 and $360 billion, with a baseline estimate of $160 billion. This would entail an annual investment of $9 to $30 billion ($13 billion baseline) during the years from 2010 to 2022.

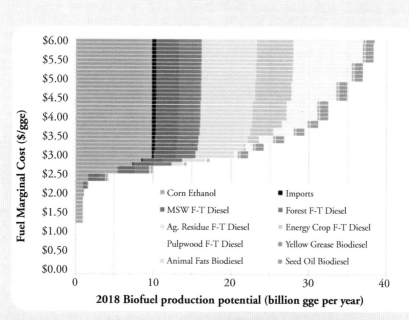

The GBSM arrived at this baseline supply curve for biofuels in 2018 by production pathway (gge = gallons of gasoline equivalent). The biofuel conversion technologies considered range from biodiesel from seed oils, yellow grease, and animal fats to cellulosic ethanol from wastes, residues, and energy crops, and Fischer-Tropsch diesel and naphtha from wastes, residues, and energy crops.

GBSM modeling gives a picture of how the biofuel industry is likely to organize spatially in the United States. Feedstock availability is the dominant force in determining the spatial distribution of biorefineries. Biomass is more expensive to transport than finished fuel products, which leads to the conversion industry locating largely to minimize feedstock transport cost. Cellulosic biofuel production will be predominantly located in the Midwest where the industry can utilize agricultural residues and the Southeast to access forestry residues. Additional biorefineries are sited near population centers to take advantage of both municipal waste resources and local fuel markets. The GBSM predicts large cellulosic biorefineries that draw feedstock from as far as 100 miles away.

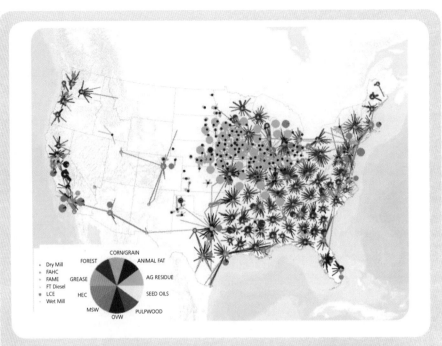

This map shows the spatial distribution of biorefineries in 2018 projected by the GBSM to meet the RFS2-mandated volumes of biofuels. The pie chart shows the relative size of individual refineries and their feedstock portfolio. Lines indicate the source of these feedstocks. Corn ethanol facilities are shown as small blue dots and are not to scale with the other refineries.

The ability of the U.S. transportation fuel sector to utilize ethanol is a significant factor shaping biofuel development. Either an increased ethanol blend maximum (currently at 10 percent but approved by the EPA to increase to 15 percent) or significantly increased flexible-fuel vehicle penetration will be required to utilize future ethanol production, which is likely to exceed 10 percent of the U.S. transportation fuel supply. Conversion of cellulosic biomass to drop-in hydrocarbons through the Fischer-Tropsch or other processes may provide an alternative fuel pathway. A review of the literature on cellulosic conversion technologies did not show a distinct cost advantage for either production method given the current state of technology development. The majority of scenarios result in Fischer-Tropsch diesel as the product biofuel.

In summary, domestically produced biofuels have the potential to achieve the goals set out by RFS2 at costs that are within the range of historical gasoline prices. A significant fraction of these fuels will come from waste and residue resources. Whether this potential will turn into real fuel depends on advancements in conversion technologies and the development of reliable feedstock supply chains in a short time period.

Vehicle and Infrastructure Compatibility

Biofuels are generally compatible with internal combustion engine vehicle (ICEV) technologies and can also be used in hybrid electric drive trains. Many ICEVs already use liquid biofuels, whereas only a small fraction have been adapted to run on gaseous fuels or hydrogen. However, most of the existing fleet of gasoline and diesel ICEVs can only operate on a relatively low biofuel blend—up to 10 percent by volume of ethanol or 5 percent of biodiesel—to avoid adverse effects on vehicle operation and durability. The percentage of ethanol that can be blended into gasoline for conventional vehicles is currently under debate. All vehicles in Brazil must be capable of accepting blends of up to 25 percent ethanol. In the United States, the EPA has recently approved the use of E15 (15 percent ethanol and 85 percent gasoline) for vehicles made after 2001. This decision may not lead to E15 being offered, though, as safeguards must be in place to prevent older vehicles and small off-road engines from mistakenly using E15.

An increasing number of flexible-fuel vehicles (FFVs) in the United States, Brazil, and Sweden can use higher blends of ethanol (up to 85 percent) or 100 percent gasoline. FFVs vary the engine operation depending on the ethanol content of the fuel, measured by the oxygen sensor in the exhaust. In addition, they use larger fuel injectors and different materials in the fuel system to guard against the corrosive nature of ethanol. In Brazil 17 percent of vehicles are FFVs. In the United States, 3.3 percent of vehicles are currently FFVs, but that is expected to grow to 15 percent by 2020. Estimates of the cost of making vehicles flexible-fuel capable range from $50 to $100 per vehicle,[38] which is the cheapest modification for alternative fuels. The issue is getting enough on the road to make E85 a viable fuel option for refueling stations.

Biodiesel can legally be blended at any percentage with petroleum diesel. However, some engine manufacturers do not honor warranties if biodiesel blends are used. The most common blend is B20 (20 percent biodiesel by volume) to avoid issues with cold weather.

"Drop-in" biofuels are hydrocarbon fuels produced from biomass that can be blended freely with petroleum gasoline or diesel and used in conventional vehicles without modification. These fuels provide a seamless transition to alternative fuels as the vehicles and infrastructure require no modification. One drop-in biomass-based diesel fuel produced by the hydrotreatment process is in early commercialization. Other drop-in biofuels are still precommerical, though a few demonstration facilities exist. The cost of these fuels has yet to be determined, and it is unclear whether it will be more costly to develop drop-in fuels or to overcome the infrastructure and vehicle compatibility issues of ethanol. Additionally, these fuels will not be an exact match for petroleum fuels and will require refining to get the fuel properties in line with specifications for gasoline and diesel.

Since liquid biofuels blended in limited amounts are similar to neat gasoline or diesel in terms of vehicle performance and refueling time, and do not require new vehicle types, they can be relatively transparent to the consumer. Fuel costs may therefore be the main factor determining consumer acceptance. In Brazil, for example, FFV users select their fuel based on price. Reduced range and reduced fuel economy with ethanol and, to a lesser extent, biodiesel, can also be a factor in consumer acceptance.

An extensive infrastructure is required to supply liquid biofuels to a refueling station, as explored in greater detail in Chapter 5. Some forms of biofuel might be transported in the existing gasoline and diesel distribution infrastructure, but some forms cannot. If drop-in biofuels were produced, they could be co-transported with existing fuels. Ethanol cannot be transported in gasoline pipelines because of its tendency to absorb water and its corrosiveness. It requires its own distribution and storage systems through the fuel distribution terminal. Gasoline-ethanol blends of 15 percent ethanol or less can be blended at the distribution terminal and used in existing refueling station infrastructure. New and separate storage tanks and dispensing pumps at the refueling station will be needed for blends beyond E15.

Sustainability Aspects of Biofuels

Vigorous debate is going on within the academic community and among government, environmental, and industry groups regarding the sustainability of biofuel production, considering both its environmental impacts and its competition with food production. The sustainability of any biofuel is dependent on the specific pathway used to produce it. However, information that relates sustainability to the supply potential is scarce. The definition of "sustainable biofuels" is neither clear nor agreed upon. Generally, the definition of a sustainable practice is one that meets current needs without compromising the ability of future generations to meet their needs.[39] But the generality of this definition leaves lots of room for interpreting how it applies to the questions surrounding biofuel production.[40]

Biofuel production can be environmentally unsustainable in a number of ways: by causing habitat loss/deforestation, soil degradation, greenhouse gas emissions, pollution of water and air, aquifer depletion, and so on. (See Chapter 7 for more on the direct land-use and water impacts of biofuels production; see Chapter 12 for more on the GHG emissions and indirect land-use impacts.) A significant amount of water is required to grow energy crops and to convert any feedstock into biofuels, and many biofuel pathways can lead to reduction in water quality through intensification of agriculture.[41] Whether a particular biofuel reduces life-cycle air pollutant emissions compared to a baseline petroleum fuel depends on the production pathway, with some pathways yielding a net benefit and others a net detriment.[42] Removing agricultural residue for use in biofuel production also raises concerns about soil quality impacts and carbon emissions.[43] And production of biofuels can pose a threat to biodiversity when it results in habitat loss as well as impacts on water and soil quality.[44]

Competition for land between food and energy crops is also cause for caution. The boom in production of corn-based ethanol in response to both federal mandates and rising gasoline prices played a significant role in the doubling of the price of corn from 2006 to 2008.[45] Most options to produce biofuels on a significant scale will require the use of large quantities of agricultural land. But productive agricultural land is a limited and valuable resource that provides basic nourishment to a growing global population. The question of whether it is a good idea to incentivize the development of another major use for this scarce resource is becoming important, especially since many agricultural practices have negative environmental impacts.

Furthermore, introducing biofuel production that is competitive with petroleum fuels links the global agricultural and land markets to energy markets. It is not likely to be possible to limit production of biofuels to marginal land; biomass, like traditional crops, will grow better and be more profitable on good agricultural land. A potential danger in linking these markets is that it amplifies the impact of petroleum prices on food prices.

Although expanding the quantity of lands in agricultural production can ease the problem of direct food-fuel competition, this expansion often leads to major environmental impacts, including deforestation, habitat loss, and resulting loss in biodiversity,[46] as well as greenhouse gas emissions caused by releasing the carbon stocks of the converted land.[47] These impacts can more than cancel the gains achieved by the production of biofuels.

Despite these serious issues, it is important to note that there is a great deal of variability in the potential impact of biofuel production pathways—on both food production and the environment. Within this variability, the opportunity exists for a limited sustainable biofuels industry. But the viability and extent of such a sustainable biofuels industry depends on the costs of production, primary and coproduct market values, and any subsidies for such production influencing overall profits. The policy basis for subsidies, as well as for mandates and other expressions of government influence, therefore requires extensive information relating to net economic, environmental, and social benefits, if any. The present debate over biofuels in part reflects high levels of uncertainty about these outcomes and the need for more comprehensive information.

Policies and Business Strategies Needed to Support Biofuels

The ease of manufacture for first-generation biofuels has led to various national incentives for large-scale development. Brazil, for example, has built a large biofuel industry under government policy and financial support around ethanol from sugarcane, a historically important crop for the country. The United States has similarly encouraged ethanol production from corn (maize), making it the largest source of biofuel in the nation for both petroleum displacement and motor fuel oxygenates. But both sugarcane and corn ethanol production have been criticized as being less sustainable and environmentally beneficial than government policy might suggest. This is particularly the case for corn ethanol in light of more recent analyses suggesting that increasing crop production in response to rising fuel prices and ethanol market value or to fulfill biofuel mandates can lead to indirect land-use changes that in turn cause excess emissions of greenhouse gases relative to the fossil fuels the ethanol is intended to replace. The subject remains open to debate as neither global modeling nor direct monitoring capabilities are sufficiently well developed to provide definitive understanding around the issue.

Policies intended to promote the development of a sustainable biofuels industry must account for the multitude of factors highlighted in the previous section or accept that unintended consequences will occur. Such policies must allow for a high degree of uncertainty in the impacts and be flexible to respond to new information as it is generated. In addition, some degree of certainty within the policy must be imposed in order to promote a business environment that is friendly to investment.

Past and current biofuel policies have not taken this holistic approach but have instead focused on four policy goals: energy security, rural economic development, criteria air pollutant reduction, and greenhouse gas emission reduction. Policy is in the early phases of incorporating some sustainability aspects in addition to greenhouse gas reduction. At the national level, policies have focused on developing a domestic alternative to petroleum fuels using mandates and subsidies for biofuels as the main policy instruments.

As mentioned earlier, the federal RFS2 program establishes specific annual volume standards for cellulosic biofuels, biomass-based diesel, advanced biofuels, and total renewable fuels that must be used in transportation. To meet RFS2 by 2022, 16 billion gallons of biofuel must come from cellulosic feedstocks, such as agricultural and forest biomass, in addition to the 15 billion gallons of conventional biofuels produced largely from grain. The requirements include definitions and criteria for both fuels and the biomass feedstock used to produce them, including a ceiling for direct emissions and emissions from land-use change during all stages of fuel and feedstock production, distribution, and use by the consumer.

California's Low-Carbon Fuel Standard (LCFS), which mandates a 10-percent decrease in the carbon intensity of transportation fuels sold by 2020 relative to a 2010 baseline, accounts for the indirect effects of land-use change coupled with biomass production.[48] You can read more about this in Chapter 11.

The policy challenge is to promote sustainable biofuels while not promoting biofuels that could cause more harm than good. Government policies aimed at increasing biofuel production and use must accurately assess the associated social, environmental, and economic impacts. Several micro- and macro-level considerations need to be assessed. On a micro scale, the local impacts of the individual biorefinery and its supply chain need to be considered. On the macro scale, the impacts of the biofuels industry as a whole on agricultural markets and scarce global resources of arable land and high-quality water must be considered. Assessing the micro-scale impacts requires meticulous accounting and auditing, leading to additional cost for producing certified sustainable fuels. The macro-scale impacts are more difficult to determine and cannot be directly controlled by the individual producers of biofuels.

Summary and Conclusions

- There are a large number of pathways for biofuels production. The costs and benefits of biofuels vary greatly, depending on the specific pathway taken.

- With the biofuels production technology that is mature now, so-called first-generation technology, the lowest-cost biofuels do not provide major environmental benefits. Some represent marginal improvements over petroleum while others are actually worse than petroleum fuels in terms of environmental impacts.

- The biofuels that are expected to provide significant environmental benefits (advanced biofuels) are not yet commercially viable. Significant quantities of advanced biofuels are expected to be produced before 2015 by the first commercial-scale biorefineries. If the technologies prove to be viable, rapid expansion will take place in response to the existing strong government mandates. These biofuels are expected to have small greenhouse gas footprints but face some of the same indirect land-use change challenges as conventional biofuels if cultivating their feedstocks displaces food crops.

- Biofuels can make limited but significant contributions to a sustainable transportation energy supply. Liquid biofuels have an advantage over other petroleum alternatives (hydrogen and electricity) in serving sectors such as aviation and freight that require easily transportable, energy-dense fuels.

- STEPS research on the supply potential of biofuels shows that advanced biofuels from waste, residues, and energy crops grown on marginal land could provide between 2 percent and 16 percent of transportation energy in the United States in the next decade, with an additional 5 percent from conventional corn and soy-based biofuels. This result depends significantly on advancements in conversion technologies, the development of reliable feedstock supply chains, and the participation of potential biomass suppliers. This includes the participation of farmers in providing residues, waste management companies in providing the organic fraction of municipal solid waste, and forestry operations in collecting more of the timber that's not suitable for sale.

- Balancing sustainability with increasing production is delicate and will require policy intervention. Sustainable exploitation of biomass resources requires the consideration of many factors, some of which are not directly controlled by the biofuels industry. Capturing all factors within a regulatory framework will be difficult. Additionally, such complex regulations will be difficult to translate into a well-defined space in which industry can confidently operate. Chapter 12 explores this topic in more depth.

Notes

1. Renewable Fuels Association, *2011 Ethanol Industry Outlook*, http://www.ethanolrfa.org/pages/annual-industry-outlook.
2. S. Mueller, *Detailed Report: 2008 National Dry Mill Corn Ethanol Survey* (Energy Resources Center, University of Illinois at Chicago, 2010).
3. U.S. Census Bureau, *Fats and Oils: Production, Consumption, and Stocks — 2008* (Washington, DC: U.S. Government Printing Office, 2009).
4. This process is described in M. J. Haas, A. J. McAloon, et al., "A Process Model to Estimate Biodiesel Production Costs," *Bioresource Technology* 97 (2006): 671–78 and Y. Zhang, M. A. Dube, et al. (2003), "Biodiesel Production from Waste Cooking Oil: 1. Process Design and Technological Assessment," *Bioresource Technology* 89 (2003): 1–16.
5. Y. Zhang, M. A. Dube, et al., "Biodiesel Production from Waste Cooking Oil: 2. Economic Assessment and Sensitivity Analysis," *Bioresource Technology* 90 (2003): 229–40.
6. J. Holmgren, C. Gosling, et al., "New Developments in Renewable Fuels Offer More Choices," *Hydrocarbon Processing* (September 2007): 67–71.
7. L. Tao and A. Aden, "The Economics of Current and Future Biofuels," *In Vitro Cellular and Developmental Biology—Plant* 45 (2009): 199–217.
8. C. N. Hamelinck, G. v. Hooijdonk, et al., "Ethanol from Lignocellulosic Biomass: Techno-economic Performance in Short-, Middle-, and Long-Term," *Biomass and Bioenergy* 28 (2005): 384–410; M. Laser, H. Jin, et al., "Projected Mature Technology Scenarios for Conversion of Cellulosic Biomass to Ethanol with Coproduction Thermochemical Fuels, Power, and/or Animal Feed Protein," *Biofuels, Bioproducts and Biorefining* 3 (2009): 231–46.
9. A. Dutta, N. Dowe, et al., "An Economic Comparison of Different Fermentation Configurations to Convert Corn Stover to Ethanol Using *Z. mobilis* and *Saccharomyces*," *Biotechnology Progress* 26 (2010): 64–72, and F. K. Kazi, J. A. Fortman, et al., "Techno-economic Comparison of Process Technologies for Biochemical Ethanol Production from Corn Stover," *Fuel* 89 (Supplement 1, 2010): S20–S28 both use experimentally verified performance measures and result in the highest production costs.
10. Kazi et al., "Techno-economic Comparison of Process Technologies."
11. A. Aden, "Biochemical Production of Ethanol from Corn Stover: 2007 State of Technology Model" (National Renewable Energy Laboratory, 2008).
12. Kazi et al., "Techno-economic Comparison of Process Technologies."
13. R. Wooley, M. Ruth, et al., "Lignocellulosic Biomass to Ethanol Process Design and Economics Utilizing Co-current Dilute Acid Prehydrolysis and Enzymatic Hydrolysis Current and Futuristic Scenarios" (National Renewable Energy Laboratory, 1999).

14. Aden, "Biochemical Production of Ethanol from Corn Stover"; Dutta et al., "An Economic Comparison of Different Fermentation Configurations"; A. Aden, M. Ruth, et al., "Lignocellulosic Biomass to Ethanol Process Design and Economics Utilizing Co-Current Dilute Acid Prehydrolysis and Enzymatic Hydrolysis for Corn Stover," NREL/TP-510-32438 (National Renewable Energy Laboratory, 2002); Laser et al., "Projected Mature Technology Scenarios for Conversion of Cellulosic Biomass to Ethanol"; U.S. Environmental Protection Agency (EPA), *Renewable Fuel Standard Program (RFS2) Regulatory Impact Analysis* (2010): 1120; M. de Wit, M. Junginger, et al., "Competition Between Biofuels: Modeling Technological Learning and Cost Reductions over Time," *Biomass and Bioenergy* 34 (2010): 203–17; Tao and Aden, "The Economics of Current and Future Biofuels"; Antares Group, Inc., *National Biorefinery Siting Model, Draft Final Report—Task 2: Technologies for Biofuels Production* (2009); Kazi et al., "Techno-economic Comparison of Process Technologies"; Wooley et al., "Lignocellulosic Biomass to Ethanol Process Design and Economics"; C. N. Hamelinck, A.P.C. Faaij, et al., "Production of FT Transportation Fuels from Biomass; Technical Options, Process Analysis and Optimisation, and Development Potential," *Energy* 29 (2004): 1743–71.
15. For details see Hamelinck et al., "Production of FT Transportation Fuels from Biomass"; E. D. Larson, H. Jin, et al., "Large-scale Gasification-based Coproduction of Fuels and Electricity from Switchgrass," *Biofuels, Bioproducts and Biorefining* 3 (2009): 174–94; R. M. Swanson, A. Platon, et al., "Techno-economic Analysis of Biomass-to-Liquids Production Based on Gasification," *Fuel* 89 (Supplement 1, 2010): S11–S19.
16. Swanson et al., "Techno-economic Analysis of Biomass-to-Liquids Production Based on Gasification."
17. Antares Group, *National Biorefinery Siting Model.*
18. Hamelinck et al., "Production of FT Transportation Fuels from Biomass."
19. Larson et al., "Large-scale Gasification-based Coproduction of Fuels and Electricity from Switchgrass."
20. EPA, *Renewable Fuel Standard Program (RFS2) Regulatory Impact Analysis.*
21. Larson et al., "Large-scale Gasification-based Coproduction of Fuels and Electricity from Switchgrass"; Hamelinck et al., "Ethanol from Lignocellulosic Biomass"; U.S. Environmental Protection Agency (EPA), *Renewable Fuel Standard Program (RFS2) Regulatory Impact Analysis* (2010): 1120; Antares Group, *National Biorefinery Siting Model*; Swanson et al., "Techno-economic Analysis of Biomass-to-Liquids Production Based on Gasification"; de Wit, Junginger, et al., "Competition Between Biofuels."
22. J. Sheehan, T. Dunahay, et al., "A Look Back at the U.S. Department of Energy's Aquatic Species Program—Biodiesel from Algae," NREL/TP-580-24190 (National Renewable Energy Laboratory, 1998).
23. Y.-K. Lee, "Microalgal Mass Culture Systems and Methods: Their Limitation and Potential," *Journal of Applied Phycology* 13 (2001): 307–15.
24. C. Jimenez, B. R. Cossio, et al., "The Feasibility of Industrial Production of Spirulina (*Arthrospira*) in Southern Spain," *Aquaculture* 217 (2003): 179–90.
25. H. Qiang, Y. Zarmi, et al., "Combined Effects of Light Intensity, Light-path and Culture Density on Output Rate of *Spirulina platensis* (Cyanobacteria)," *European Journal of Phycology* 33 (1998): 165–71.
26. U.S. Department of Energy, National Algal Biofuels Technology Roadmap (draft, 2009).
27. L. Brennan, and P. Owende, "Biofuels from Microalgae—A Review of Technologies for Production, Processing, and Extractions of Biofuels and Co-products," *Renewable and Sustainable Energy Reviews* 14 (2009): 557–77.
28. Antares Group, *National Biorefinery Siting Model.*
29. D. Tilman, J. Hill, et al., "Carbon-Negative Biofuels from Low-Input High-Diversity Grassland Biomass," *Science* 314 (2006): 1598–1600; M. Hoogwijk, A. Faaij, et al. (2003), "Exploration of the Ranges of the Global Potential of Biomass for Energy," *Biomass and Bioenergy* 25 (2003): 119–33; M. Parikka, "Global Biomass Fuel Resources," *Biomass and Bioenergy* 27 (2004): 613–20; CONCAWE, *Well-to-Wheels Analysis of Future Automotive Fuels and Powertrains in the European Context, WELL-to-WHEELS Report Version 2c* (March 2007); International Energy Agency, *Energy Technology Perspectives 2008* (OECD/IEA 2008).
30. N. C. Parker, *Modeling Future Biofuel Supply Chains Using Spatially Explicit Infrastructure Optimization*, PhD dissertation, Transportation Technology and Policy program, University of California, Davis, 2011; EPA, *Renewable Fuel Standard Program (RFS2) Regulatory Impact Analysis:* 1120; National Academy of Sciences, *Liquid Transportation Fuels from Coal and Biomass: Technological Status, Costs, and Environmental Impacts* (Washington, DC: National Academies Press, 2009).
31. R. Perlack, L. Wright, et al., "Biomass as Feedstock for a Bioenergy and Bioproducts Industry: The Technical Feasibility of a Billion-Ton Annual Supply," joint report prepared by the U.S. Department of Energy and the U.S. Department of Agriculture, Environmental Sciences Division (Oak Ridge National Laboratory, 2005), p. 75.
32. B. Jenkins, R. Williams, et al., "Sustainable Use of California Biomass Resources Can Help Meet State and National Bioenergy Targets," *California Agriculture* 63 (2009): 168–77.

33. R. Lal, "World Crop Residues Production and Implications of Its Use as a Biofuel," *Environment International* 31 (2005): 575–84; W. W. Wilhelm, J.M.F. Johnson, D. L. Karlen, and D. T. Lightle, "Corn Stover to Sustain Soil Organic Carbon Further Constrains Biomass Supply," *Agronomy Journal* 99 (2007): 1665–67.
34. C. B. Field, J. E. Campbell, et al., "Biomass Energy: The Scale of the Potential Resource," *Trends in Ecology and Evolution* 23 (2008): 65–72.
35. Sheehan et al., "A Look Back at the U.S. Department of Energy's Aquatic Species Program."
36. Even though the RFS2 mandate is written in units of ethanol gallon equivalent energy, gasoline gallon equivalents are used in this text for ease of comparison with other fuels discussed in the book. 1 ethanol gallon = 0.66 gge.
37. U.S. Department of Agriculture (USDA), *USDA Agricultural Projections to 2018*, Long-Term Projections Report OCE-2009-1 (Washington, DC: USDA, Office of the Chief Economist, 2009).
38. EPA, *Renewable Fuel Standard Program (RFS2) Regulatory Impact Analysis*.
39. World Commission on Environment and Development (WCED), *Our Common Future* (Oxford: Oxford University Press, 1987).
40. S. Yeh, D. A. Sumner, et al., "Implementing Performance-Based Sustainability Requirements for the Low Carbon Fuel Standard—Key Design Elements and Policy Considerations" (Institute of Transportation Studies, University of California, Davis, 2009); Roundtable on Sustainable Biofuels, "RSB Principles and Criteria for Sustainable Biofuel Production" (version 2.0, 2010).
41. National Research Council and The National Academies, *Water Implications of Biofuels Production in the United States* (Washington, DC: National Academies Press, 2008).
42. M. Wu, Y. Wu, et al., Mobility Chains Analysis of Technologies for Passenger Cars and Light-Duty Vehicles with Biofuels: Application of the GREET Model to the Role of Biomas in America's Energy Future (RBAEF) Project (Center for Transportation Research, Argonne National Laboratory, 2005).
43. R. Lal, "World Crop Residues Production and Implications of Its Use as a Biofuel," *Environment International* 31 (2005): 575–84.
44. J. H. Cook, J. Beyea, et al., "Potential Impacts of Biomass Reduction in the U.S. on Biological Diversity," *Annual Reviews of Energy and Environment* 16 (1991): 401–31.
45. B. A. Babcock, "Statement Before the U.S. Senate Committee on Homeland Security and Government Affairs," Hearing on Fuel Subsidies and Impact on Food Prices, Washington, DC, 2008.
46. Cook et al., "Potential Impacts of Biomass Reduction."
47. J. Fargione, J. Hill, et al., "Land Clearing and the Biofuel Carbon Debt," *Science* 319 (2008): 1235–38; T. Searchinger, R. Heimlich, et al., "Use of U.S. Croplands for Biofuels Increases Greenhouse Gases Through Emissions from Land-Use Change," *Science* 319 (2008): 1238–40.
48. A. Farrell and D. Sperling, *A Low-Carbon Fuel Standard for California, Part 1: Technical Analysis* (Institute of Transportation Studies, University of California, Davis, 2007).

Chapter 2: The Plug-in Electric Vehicle Pathway

Jonn Axsen, Christopher Yang, Ryan McCarthy, Andrew Burke, Kenneth S. Kurani, and Tom Turrentine

While biofuels seem to represent the nearest-term answer to the demand for alternative fuels, electricity is closing in as a viable choice. Electric-drive technology continues to pique the imagination of motorists with its promise of clean skies, quiet cars, and plentiful fuel produced from nonpolluting domestic sources. In the designs they have dangled before us, automakers have shown us variations in plug-in electric vehicle (PEV) size, performance, and definition in efforts to overcome the fundamental challenge of electric drive: how to store energy and supply power. PEVs (a category that includes plug-in hybrid electric vehicles or PHEVs as well as battery electric vehicles or BEVs) are powered at least in part by electricity from the grid—a fuel that under certain conditions is less costly and more environmentally friendly than gasoline. Because vehicle electrification can improve the total energy efficiency (MJ/mile) of the vehicle and may allow lowering of the carbon intensity (gCO_2/MJ) of the fuel used in vehicles over time, PEVs offer a form of transportation with the potential for very low greenhouse gas (GHG) emissions.

PEVs have now entered the marketplace with models from several manufacturers. However, PEV technology has yet to achieve widespread market success. In this chapter we sort through the hype and improve understanding of the PEV pathway and what it will take to become competitive with internal combustion engine vehicles (ICEVs). We draw from several streams of research—including testing of battery technology, modeling of the electricity grid, and eliciting consumer data regarding PEV design interests and potential use patterns—to address these questions:

- What is the technical outlook for PEV technology and batteries?
- How will widespread charging of PEVs influence the operation and evolution of the electricity grid, and how does infrastructure need to develop for our transportation system to transition to the PEV pathway?
- What are the expected environmental impacts of electricity use for charging vehicles, and how can we minimize them?
- How do PEVs fit into long-term deep GHG-reduction scenarios?
- What policies and business strategies are needed to support PEVs in both the near and long terms?

CHALLENGES ON THE PEV PATHWAY

These complex technical and logistical challenges must be overcome if an electricity-based transportation system is to become widespread:

- **Technical challenges.** PEVs face high costs and limited all-electric range due to inherent energy storage limitations of batteries. There are also trade-offs among different battery chemistries regarding power, energy, cost, safety, and longevity. The present state of battery technology is sufficient for early market formation, but costs and range may need to improve for markets to expand.

- **Infrastructure challenges.** Current 110-volt recharge potential at home may be suitable for PHEVs and low-range BEVs. However, widespread commercialization of longer-range BEVs will require at-home 220-volt charging, and potentially workplace and public charging at 220 volts or higher. A significant fraction of people, mainly in urban areas, do not have access to off-street parking, which may limit adoption of PEVs.

- **Transition issues / coordination of stakeholders.** The electrification of transportation could start with giving consumers what they want: less technologically ambitious PHEVs. Near-term commercialization of such less-electrified PEVs could pave the way for future sales of longer-range BEVs by increasing manufacturing experience and whetting consumer appetites for PEVs. Utilities will need to provide the appropriate incentives to consumers to charge during less-expensive off-peak hours.

- **Policy challenges.** Policy makers could better support a gradual transition to electric-drive technology—for example, starting with greater hybridization and low-range PHEVs to stimulate further vehicle electrification in the future. However, policy should be sure to address well-to-wheel PEV emissions—that is, account for regional variations and future expectations of electricity grid carbon intensity. The role of PEVs and electricity needs to be examined in the context of a broader suite of vehicle and fuels-related policies (such as CAFE and the LCFS). California policymakers are already developing this portfolio of policies for transportation, as are other regions around the world.

Technology Status and Outlook

PEVs have followed a tortuous pathway of development. Spurred by disruptions in petroleum supply and price, and by policies on air pollution and climate change, much effort and many resources have been devoted to PEVs over the past three decades. In the United States, the Hybrid and Electric Vehicle Act of 1976 laid the groundwork for battery, motor, and power-and-control electronics technologies that emerged during the 1990s.[1] Battery electric vehicles garnered renewed attention in the 1990s, stimulated by General Motors' development of the EV-1 (a.k.a. Impact) and California's zero-emission vehicle (ZEV) mandate. Automakers eventually produced a limited number of BEVs in California to meet the modified ZEV Mandate. Then after years of further technology development and policy debate, policy makers were convinced by automobile manufacturers in the late 1990s that battery technology was not ready to meet manufacturers' EV performance goals. Some battery technologies, namely NiMH, later proved successful in less-demanding hybrid electric vehicles.

Today attention is increasingly turning toward PHEVs, which use both grid electricity and gasoline as fuels. Policymakers are increasingly giving attention to PEV pathways.[2] For instance, President Obama set a national target of 1 million PEVs on the road by 2015, and a federal tax credit is available, beginning in 2009, and will be in place for a number of years.[3] Several states offer additional advanced vehicle rebates and charging infrastructure subsidies.

Battery technology remains the largest technological challenge on the PEV pathway. Although breakthroughs in advanced battery chemistries since 2000 allow for more ambitious PEV designs than those available in the 1990s, important limitations remain. In this section we address those technological limitations and prospects for batteries and then consider them in light of the PEV design preferences expressed by potential PEV buyers in a survey. We also summarize key issues for other PEV components and recharge devices.

Battery technology goals, capabilities, and prospects

The commercial success of the PEV depends on the development of appropriate battery technologies. Much uncertainty exists about the battery parameters to best power a PEV and where different battery technologies stand in meeting such requirements. While electric-drive advocates claim that battery technology is sufficiently advanced to achieve commercial success, critics counter that substantial technological breakthroughs are required to realize mass market adoption. Further, there is disagreement on what a PEV is or should be.

DIFFERENTIATING PEVS BASED ON THEIR BATTERY DISCHARGE PATTERNS

There are many different designs for PEVs based on their battery discharge pattern. Here's a breakdown of the differences:

- While a BEV is designed to operate only in charge-depleting (CD) mode, a PHEV can operate in CD *or* charge-sustaining (CS) mode. Driving the PEV in CD mode depletes the battery's state of charge (SOC), and CD range is the distance a fully charged PEV can be driven before depleting its battery. While a BEV would need to be recharged, a PHEV switches to CS mode, which then relies on gasoline energy as with a conventional HEV; the gasoline energy maintains the battery's SOC—but the vehicle does not use grid electricity until recharged.

- PHEVs can be further differentiated based on whether their CD mode is designed for all-electric (AE) operation (using only electricity from the battery) or for blended (B) operation (using both electricity and gasoline in almost any proportions). In this chapter, we denote CD range and operation for PHEVs as AE-X or B-X, where X is the CD range in miles. We use BEV-X to denote the range of electric vehicles.

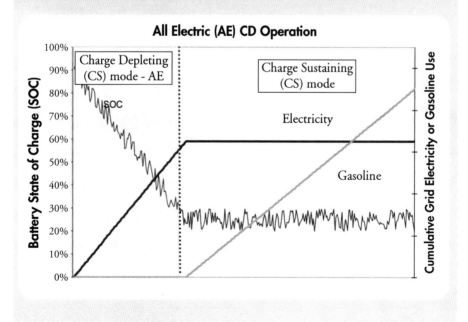

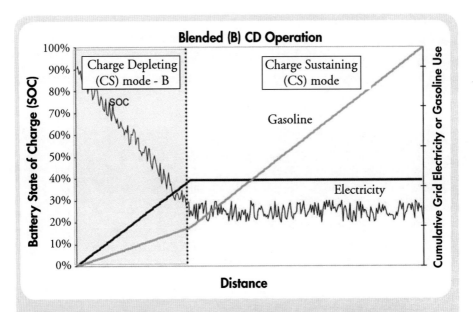

This figure compares the battery discharge patterns of two different PHEVs, one with a CD mode designed for all-electric (AE) operation (top graph) and one with a CD mode designed for blended (B) operation (bottom graph), measured as state of charge (SOC) on the left axis. Holding CD range constant, an AE-X design requires more battery energy and power capacity and is thus costlier than a B-X design (for the same X). On the other hand, at any distance cumulative gasoline use will be higher in the B-X design for any vehicle trips that include a portion of CD driving. Source: Adapted from M. A. Kromer and J. B. Heywood, Electric Powertrains: Opportunities and Challenges in the U.S. Light-Duty Vehicle Fleet, *LEFF 2007-02 RP (Sloan Automotive Laboratory, MIT Laboratory for Energy and the Environment, May 2007).*

The key requirements of PEV battery technology—power, energy capacity, durability, safety, and cost—depend on various assumptions about vehicle design. These factors include vehicle types (BEV versus PHEV), range in charge-depleting (CD) mode, and for PHEVs, type of CD operation (all-electric or blended), as well as use patterns.[4] The U.S. Advanced Battery Consortium (USABC) has set goals for batteries to be used in a PHEV with an all-electric range of 10 miles (AE-10) and one with an all-electric range of 40 miles (AE-40). Alternative targets have been suggested by the Sloan Automotive Laboratory at the Massachusetts Institute of Technology and the Electric Power Research Institute (EPRI). Here is a summary:[5]

Power: The rate of energy transfer is measured in kilowatts (kW) for automotive applications and typically portrayed as power density (W/kg) for batteries. Power goals range from 23 kW up to 99 kW, requiring densities between 380 and 830 W/kg.

Energy capacity: Battery storage capacity (kWh) relates to the size of the battery and its energy density (Wh/kg). (Note that there is an important distinction between available and total energy.

While a battery may have 10 kWh of battery storage capacity or total energy, only a portion of this capacity is available for vehicle operations. A battery with 10 kWh of total energy operating with a 65-percent depth of discharge would have only 6.5 kWh of available energy.) USABC goals range from 5.7 to 17 kWh of total energy, and from 100 to 140 Wh/kg.

Durability: With usage and time, battery performance—including power, energy capacity, and safety—can substantially degrade. Four measures are typically important: (1) calendar life, the ability to withstand degradation over time (15 years for USABC); (2) deep cycle life, the number of discharge-recharge cycles the battery can perform in CD mode (USABC's goal is 5000 cycles); (3) shallow cycles, state-of-charge variations of only a few percentage points, where the battery frequently takes in electric energy via a generator and from regenerative braking and passes energy to the electric motor as needed to power the vehicle (USABC targets 300,000 cycles); and (4) survival temperature range (USABC targets –46°C to +66°C).

Safety: Because batteries store energy and contain chemicals that can be dangerous if discharged in an uncontrolled manner, safety must be considered. Safety is typically measured through abuse tolerance tests, such as mechanical crushing, perforation, overcharging, and overheating.[6] USABC sets only the goal of "acceptability."

Cost: Although battery cost is thought to be one of the most critical factors in commercial PEV deployment, these costs are highly uncertain. USABC cost goals are $1,700 and $3,400 for AE-10 and AE-40 battery packs, respectively, under a scenario where battery production has reached 100,000 units per year, which equates to $200 to $300/kWh. The USABC estimates that in general, current advanced battery costs range from $800/kWh to $1000/kWh or higher. (See Chapter 4, Comparing Fuel Economies and Costs, for a look at how battery costs factor into the economic attractiveness of hybrid electric and plug-in hybrid electric vehicles in the future.)

There are inherent trade-offs among these attributes. Some existing battery technologies can achieve some of these goals, but meeting all goals simultaneously is far more challenging. For example, higher power can be achieved through the use of thinner electrodes, but these designs tend to reduce cycle life and safety while increasing material and manufacturing costs. In contrast, high-energy batteries use thicker electrodes that increase safety and life but reduce power density. Thus, it can be very difficult to meet ambitious targets for both power and energy density in the same battery, let alone also meet goals for longevity, safety, and cost. Understanding these trade-offs is key to understanding the requirements and challenges facing battery chemistries.

Currently, there are two main categories of battery chemistry that have been developed for electric drivetrains: nickel-metal hydride (NiMH) and lithium-ion (Li-ion). NiMH batteries are used for most HEVs currently sold in the United States, though some automatkers are starting to use Li-ion. The primary advantage of this chemistry is its proven longevity in calendar and cycle life, and overall history of safety, while drawbacks include limitations in energy and power density, and low likelihood of future cost reductions.[7] In contrast, Li-ion technology has the potential to meet the requirements of a broader variety of PEVs. Lithium is very attractive for high-energy batteries due to its light weight and potential for high voltage (while still falling short of the ambitious power targets of the USABC). Li-ion battery costs are predicted to fall as low as

$250–400/kWh with 100,000 units of production.[8] However, the high chemical reactivity of Li-ion provides a greater threat to calendar life, cycle life, and safety compared to NiMH batteries—thus, Li-ion batteries require a greater degree of control over cell voltage and temperature than do NiMH batteries.

POWER- AND ENERGY-DENSITY TRADE-OFFS FOR DIFFERENT BATTERY CHEMISTRIES

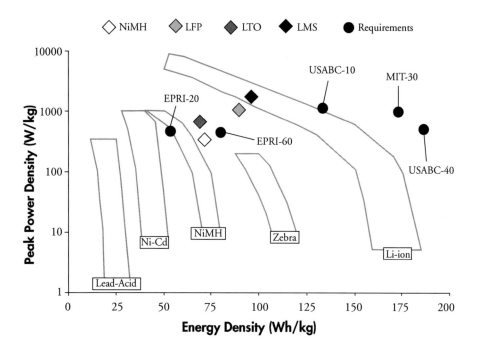

A Ragone plot represents the trade-offs between power density and energy density for a given battery chemistry. Power density (W/kg) is plotted on the vertical axis on a logarithmic scale. Energy density (Wh/kg) is presented on the horizontal axis for a specified discharge rate, say C/1 (complete discharge over 1 hour). Here the light gray bands represent the current power and energy capabilities of an individual battery cell of each of five different chemistries: lead-acid, nickel-cadmium, NiMH, ZEBRA, and Li-ion. The USABC, MIT, and EPRI battery requirements are plotted as black circles. The diamonds represent the performance of four PHEV batteries tested at UC Davis: one NiMH, and three Li-Ion.

Battery specifications assume a motor efficiency of 85 percent, a packaging factor of 0.75, and an 80-percent battery depth of discharge (DOD). The battery pack (or system) designed for a particular PHEV consists of many individual battery cells, plus a cooling system, inter-cell connectors, cell monitoring devices, and safety circuits. The added weight and volume of the additional components reduce the energy and power density of the pack relative to the cell. In addition, the inter-cell connectors and safety circuits of a battery pack can significantly increase resistance, decreasing the power rating from that achievable by a single cell. When applying cell-based ratings to a battery pack, and vice versa, a packaging factor conversion must be applied. There is typically a larger reduction for power density—and thus a smaller packaging factor—than energy density due to added resistance, in addition to the added weight. We assume an optimistic packaging factor of 0.75 for each conversion. Source: Ragone plots from Kromer and Heywood, Electric Powertrains. Figure adapted from J. Axsen, K. S. Kurani, and A. F. Burke, "Are Batteries Ready for Plug-in Hybrid Buyers?" Transport Policy 17 (2010): 173–82.

More important than this current snapshot are the long-term prospects for improvements to Li-ion batteries. Li-ion batteries can be constructed from a wide variety of materials and vary by electrolyte, packaging, structure, and shape. The main Li-ion cathode material used for consumer applications (such as laptop computers and cell phones) is lithium cobalt oxide (LCO). But there are safety concerns about using this chemistry for automotive applications, so several alternative chemistries are being piloted, developed, or researched for PEVs, including lithium nickel, cobalt, and aluminum (NCA); lithium iron phosphate (LFP); lithium nickel, cobalt, and manganese (NCM); lithium manganese spinel (LMS); lithium titanium (LTO); and manganese titanium (MNS). The attributes of any one of these chemistries may not represent Li-ion technology in general, and no single chemistry excels in all five requirement categories.

COMPARISON OF ALTERNATIVE CHEMISTRIES FOR PEV BATTERIES
A comparison of alternative chemistries for PEV batteries shows that no single chemistry excels in all five requirement categories. Trade-offs are necessary.

Name	Description	Automotive Status	Power	Energy	Safety	Life	Cost
NiMH	Nickel-metal hydride	Commercial production	Low	Low	High	High	Mid
LCO	Lithium cobalt oxide	Limited production	High	High	Low	Low	High
NCA	Lithium nickel, cobalt, and aluminum	Limited production	High	High	Low	Mid	Low
LFP	Lithium iron phosphate	Pilot	Mid-High	Mid	Mid-High	High	Low
NCM	Lithium nickel, cobalt,	Pilot	Mid	Mid-High	Mid	Low	High
LMS	Lithium manganese spinel	Development	Mid	Mid-High	Mid	Mid	Low-Mid
LTO	Lithium titanium	Development	High	Low	High	High	Mid
MNS	Manganese titanium	Research	High	Mid	High	?	Mid

Qualitative assessment by A. Burke at UC Davis, July 2010. Source: J. Axsen, A. Burke, and K. Kurani, "Batteries for PHEVs: Comparing Goals and the State of Technology," in Electric and Hybrid Vehicles: Power Sources, Models, Sustainability, Infrastructure and the Market, *ed. G. Pistoia (New York: Elsevier, 2010).*

Consumer-informed goals and the "battery problem"

Our summary of USABC goals and the capabilities of battery chemistries suggests there is a battery problem—that the inadequate performance and high cost of available battery technologies are the main barriers to the commercialization of electric passenger vehicles with plug-in capabilities. But how does the state of battery technology compare to what consumers actually want from PHEVs? STEPS researchers investigated this question in a web-based survey of a representative sample of new-vehicle-buying households in the United States in which consumers could create their own PHEV designs and thus set their own PHEV goals. [9]

ANTICIPATING THE PHEV MARKET WITH A CONSUMER SURVEY

To arrive at consumer-informed PHEV design goals and estimates of use behavior, STEPS researchers conducted a web-based survey in 2007 with a representative sample of 2,373 new-vehicle-buying households in the United States. The survey was implemented in three separate pieces, requiring multiple days for households to answer questions, conduct a review of their own driving and parking patterns, and then complete a sequence of PHEV design exercises.

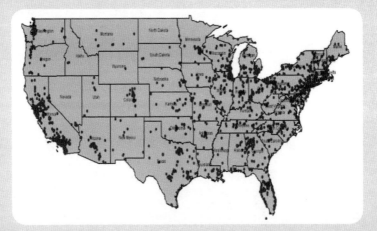

The sample was deemed representative of U.S. new car buyers according to geographic distribution, as well as income, age, education, and other sociodemographic variables. Source: J. Axsen and K. S. Kurani, "Early U.S. Market for Plug-in Hybrid Electric Vehicles: Anticipating Consumer Recharge Potential and Design Priorities," Transportation Research Record *2139 (2009): 64–72. Design choices presented to survey respondents included charge-depleting (CD) operation—all-electric or three levels of blended operation—and CD ranges of 10, 20, or 40 miles. The design space also offered respondents a choice of recharge times (8 hours, 4 hours, 2 hours, or 1 hour) and charge-sustaining fuel economy (+10, +20, or +30 MPG over a conventional vehicle). This offered respondents a choice of 144 possible combinations for cars and again for trucks. We focus here on results from the 33 percent of respondents we identify as "plausible early market respondents": those who currently have 110V recharge potential at home and who demonstrated interest in purchasing a PHEV even at a relatively high price.*

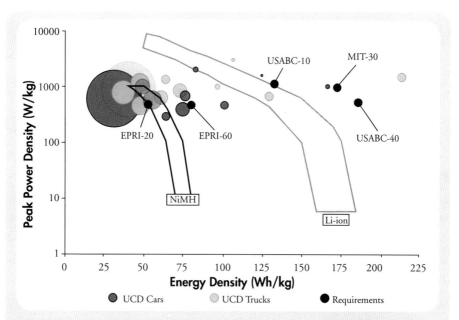

This Ragone plot summarizes the PHEV designs selected by our potential early-market PHEV buyers. The region bounded in black represents a range of NiMH capabilities and the region bounded in gray represents Li-ion chemistries. For comparison, we also plotted the battery cell requirements derived by USABC, MIT, and EPRI. The centers of the gray circles mark the location of the peak power density and energy density requirements derived from the respondents' designs; the sizes of the gray circles are proportional to the number of respondents who chose or designed the PHEV corresponding to those battery requirements. In contrast, the black circles marking the location of the USABC, MIT, and EPRI requirements have been sized solely to make them perceptible in the figure. What we see is that potential buyers have different requirements from those specified by USABC and MIT; they are closer to the EPRI goals, and especially the EPRI-20 goal. Source: Axsen et al., "Are Batteries Ready for Plug-in Hybrid Buyers?"

A substantial number of new-vehicle-buying households reported that they would like to buy vehicles with plug-in capabilities. The majority of these potential early market respondents selected the most basic PHEV design option: a B-10, requiring the lowest power and energy densities. Even including respondents who designed more demanding PHEVs, about 85 percent of the potential early buyers designed PHEVs that required peak power density and energy density within the current capabilities of NiMH batteries. In contrast, experts' projected PHEV designs all result in much higher peak power and energy density requirements, most of them seemingly beyond the present capabilities of Li-ion batteries.

The bottom line is that given consumers' preferred PHEV designs, the experts' aggressive battery technology goals may be unnecessary for near-term PHEV commercialization. To put it another way, the real battery problem may be better summarized as the challenge of aligning technological development with distribution of consumer interests in the near and long terms.

Other PEV components

The drivetrain configuration of a battery electric vehicle is relatively simple in that it consists of only a few components: electric motors, power electronics (a DC/AC inverter), a motor controller, and a battery pack. PHEVs are more complex in that they integrate an internal combustion engine and potentially a transmission into the drivetrain as well as the EV components. There is a great deal of design freedom for PHEVs in terms of the size and configuration of the various components (hardware) and the operation and control of these components during different types of driving (software).

While the individual component technologies beyond the batteries are relatively mature, the vehicle design, integration, and controls are the major areas for innovation and value added by the automakers. There will be a great deal of innovation in this arena over the next decade as we move from prototype vehicles in labs to commercial, mass-market vehicles that will attempt to appeal to regular drivers rather than just early adopters.

Infrastructure for PEVs

While the success of PEVs largely hinges on the development of robust, low-cost batteries that match consumer needs, the fueling and infrastructure side of the equation is also important. A key consideration is the present state and future prospects of recharge infrastructure to allow PEV recharging at home, work, and other locations. We must also consider the ability of the electrical grid to handle additional demand and anticipate the temporal and spatial distribution of charging behavior. This section discusses electricity demands for PEV charging and their potential interaction with the electricity grid and how costs and emissions depend on the quantity, location, and timing of vehicle electricity demands.

Charger technology

Electric vehicles need to be plugged in to recharge the vehicles' batteries. While current PEVs can plug into a conventional home 110V outlet, recharging this way takes a long time for BEVs and longer-range PHEVs (for many PHEVs, this will be sufficient). To recharge more quickly, it is necessary to use higher voltage (220V or higher) coupled with a PEV charger (also known as electric vehicle supply equipment or EVSE).

There are several categories of charging (Levels 1 through 3), depending on the voltage and power supplied to the vehicle. The EVSE designs that will allow for faster, higher-power charging will have higher costs, not only for the equipment but also for the electrical connection and installation. And batteries are charged with DC power, requiring conversion of AC to DC either onboard the vehicle or in the EVSE.

Public charging stations are expected to be relatively high power (Level 2) in order to allow for reasonable charging times for drivers. Very high power chargers (Level 3) will allow for significant recharging (perhaps 80 percent of battery capacity) in under half an hour.

The demand side: Anticipating recharge potential, timing, and location

To better understand the present state of charging infrastructure for PEVs, the 2007 PHEV survey also assessed respondent access to 110V electrical outlets over the course of one day of driving their conventional vehicle. As noted earlier, 110V outlets may be appropriate for PHEV recharging, while higher voltage will likely be necessary for most users of BEVs. However, 110V access may serve as a stand-in for proximity of access to circuits that may be upgraded to house 220V infrastructure or higher.

Survey results indicate that more new vehicle buyers may be pre-adapted for vehicle recharging than estimated in previous constraints analyses. Our study was different because it elicited reports of vehicle parking proximity to electrical outlets (not circuits), directly from respondents instead of via proxy data. About half of our U.S. new-vehicle-buying respondents have at least one viable 110V recharge location within 25 feet of their vehicle when parked at home. But this also means that approximately half do not have access to charging, perhaps because they park in an apartment parking lot or on the street, which is an important barrier to achieving high levels of PEV adoption. Only 4 percent of respondents found 110V outlets at work, and 9 percent found 110V outlets at other nonhome locations—for example, at a friend's home, school, and commercial sites. When we aggregated recharge potential across this sample, we found that total recharge potential ranges from more than 90 percent of respondents from 10:00 p.m. to 5:30 a.m., to less than 30 percent from 11:30 a.m. to 1:30 p.m. Throughout the day, home is by far the most frequent location of recharge opportunities for respondents.

ACCESS TO 110V RECHARGE SPOTS BY LOCATION AND OUTLET DISTANCE

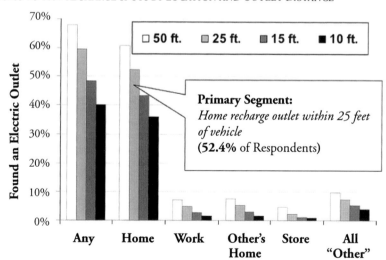

About half of our 2,373 U.S. new-vehicle-buying survey respondents have at least one viable 110V recharge location within 25 feet of their vehicle when parked at home. Only 4 percent of respondents found 110V outlets at work, and 9 percent found 110V outlets at other nonhome locations. Source: J. Axsen and K. S. Kurani, "Early U.S. Market for Plug-in Hybrid Electric Vehicles: Anticipating Consumer Recharge Potential and Design Priorities," Transportation Research Record *2139 (2009): 64–72.*

DRIVING AND RECHARGE POTENTIAL BY TIME OF DAY

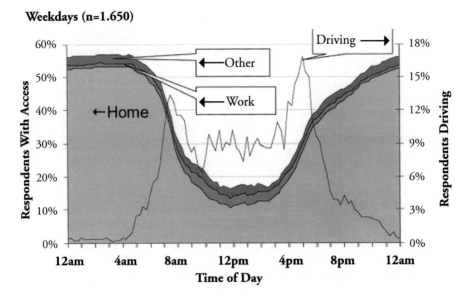

When we aggregated recharge potential across the 2,373 U.S. respondents, we found that total recharge potential ranges from more than 50 percent of respondents from 9:00 p.m. to 7:00 a.m., to fewer than 20 percent from 10:00 a.m. to 3:00 p.m. (weekdays only). Source: J. Axsen and K. S. Kurani, The Early U.S. Market for PHEVs: Anticipating Consumer Awareness, Recharge Potential, Design Priorities and Energy Impacts, *UCD-ITS-RR-08-22 (Institute of Transportation Studies, University of California, Davis, 2008).*

STEPS researchers integrated this consumer data to construct consumer-informed profiles representing the potential electrical demand of PEVs in California. Results suggest that the use of PHEV vehicles could halve gasoline use relative to conventional vehicles. Using three scenarios to represent plausible recharge patterns (immediate and unconstrained recharging, universal workplace access, and off-peak only), we assessed trade-offs between the magnitude and timing of PHEV electricity use. In the unconstrained recharge scenario, recharging peaks at 7:00 p.m., following a pattern throughout the day that is far more dispersed than anticipated by previous research. PHEV electricity use could be increased through policies that expand nonhome recharge opportunities (for example, the universal workplace access scenario), but most of this increase occurs during daytime hours and could contribute to peak electricity demand. Deferring all recharging to only off-peak hours (8:00 p.m. to 6:00 a.m.) could eliminate all additions to daytime electricity demand from PHEVs, although less electricity would be used and less gasoline displaced in this scenario.

CONSUMER-INFORMED PROFILES OF GASOLINE USE AND GRID ELECTRICITY RECHARGE

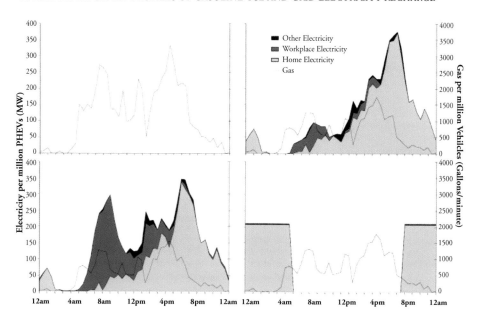

Based on the responses of 231 "early-market respondents" in California (consisting of those who identified an electrical outlet within 25 feet of where they park their vehicle at home and demonstrated interest in purchasing a PHEV in the survey described earlier in this chapter), STEPS researchers constructed four scenarios for weekday gasoline use and grid electricity recharging. A is "no PHEVs," B is "plug and play," C is "enhanced workplace access," and D is "off-peak only." Source: J. Axsen and K. S. Kurani, "Anticipating Plug-in Hybrid Vehicle Energy Impacts in California: Constructing Consumer-Informed Recharge Profiles," Transportation Research Part D 15 (2010): 212–19.

STEPS researchers also investigated PHEV recharge behavior by observing participants in a PHEV demonstration project in northern California. A total of 40 households took part in the project. Each household used a Toyota Prius converted to a PHEV (B-30) in place of one of their vehicles for a four-to-six-week trial. The resulting distribution of recharge potential and actual electricity use showed a broad weekday peak between 6:00 p.m. and midnight—during which period behavior varied substantially across respondents. The range of behaviors supports the contention that the success of PEVs in meeting energy and emission goals depends on PEV users' recharging and driving behavior as much as or more than on vehicle design.

OBSERVED HIGH AND LOW WEEKDAY ELECTRICITY AVAILABILITY AND POWER DEMAND

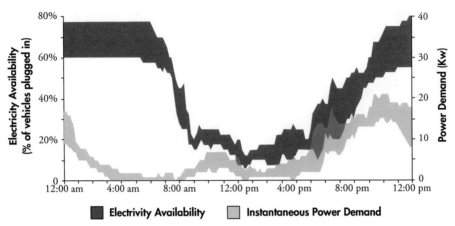

The power demand of the 40 households participating in a PHEV demonstration project in northern California showed a broad weekday peak between 6:00 p.m. and midnight—during which period behavior varied substantially across respondents. Source: J. Davis and K. Kurani, "Recharging Behavior of Households' Plug-In Hybrid Electric Vehicles: Observed Variation in Use of Conversions of 5-kW-h Blended Plug-In Hybrid Electric," Transportation Research Record 2191 (2010): 75–83.

Of course, recharge behavior concerning PEVs may differ substantially from that of actual and hypothetical PHEV drivers. As the market develops, utilities may offer incentives to motivate charging at off-peak, lower-cost rates as well as prevent charging that adds to peak demands and the need for additional power plants to be built. Recharging infrastructure availability will also play a role in where and when drivers charge their PEVs.

SPATIAL ANALYSIS OF EV ACTIVITY AND POTENTIAL FAST-CHARGE LOCATIONS

STEPS researchers conducted spatial research in an attempt to understand the limitations of electric vehicle range and potential for charging by comparing them to gasoline range and activity. This research explores questions such as: How important is range to the consumer? To what extent can an EV replace a gasoline vehicle? How would placement of fast chargers provide value to the customer?

The figure below represents the response of a single EV owner in San Diego to questions about where he drives his EV and where he drives his gasoline vehicle. The respondent never drove beyond the boundary of a small "activity space" near home in his EV. When asked which destinations he expected to be able to reach in his gasoline vehicle, he indicated a large area encompassing much of southern California as well as the

Lake Tahoe region and San Francisco. In response to the question of where he would like to place any number of fast chargers, he indicated only two locations, one to give him access to the Los Angeles area and one to help with range considerations within his existing EV activity space. It should be noted that this respondent had very limited access to charging away from home and had in fact never done it.

While conclusions cannot be drawn from a single response, this response highlights several themes surrounding electric vehicle range and charging. First, even though the EV activity space was significantly smaller than the gasoline activity space, the EV activity space represented a stated 90 percent of the respondent's driving. Further, meeting the need for this 90 percent of his driving resulted in the respondent being happy with the vehicle and satisfied with the range. The reason for not using the EV for the remainder of his driving could have been range limitations or cargo and passenger space.

The placement of fast chargers was also illustrative and highlights the concepts of intensification and extensification. In this context, *intensification* refers to the placement of chargers within a driver's EV activity space in order to recharge if the battery happens to be low. *Extensification* is the placement of chargers outside of the driver's primary EV activity space to enable travel outside of those boundaries. The respondent indicated that one of each type would be useful.

It is interesting to note that the respondent did not place fast chargers all along the highway to northern California. The implication is that an EV may not be seen as a viable substitute for a gasoline vehicle for long trips. The respondent only indicated that the immediately adjacent metropolitan area was a place he desired to go in his EV. Enabling travel along the corridor between adjacent metropolitan areas or metropolitan areas within 50 to 80 miles seems to be another oft-mentioned desire of EV owners.

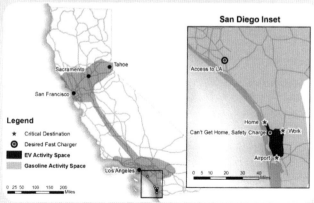

The response of one EV owner in San Diego to questions about where he drives his EV and where he drives his gasoline vehicle highlight themes surrounding EV range and charging. The respondent never drove beyond the boundary of the small black "activity space" near home in his EV. When asked which destinations he expected to be able to reach in his gasoline vehicle, he indicated the large gray area encompassing much of southern California as well as the Lake Tahoe region and San Francisco.

The supply side: Generating and delivering electricity for PEVs

Charging a PEV requires the grid to respond by providing more electricity. STEPS researchers have worked to better understand the electricity grid—the collection of power plants and transmission and distribution facilities that produces and delivers electricity to end users. The grid has evolved to meet continually changing electricity demands by using a suite of power plants that fulfill various roles in the grid network. Each type of power plant operates differently: baseload facilities (often large coal or nuclear plants) are designed to operate continuously and at low cost, and peaking power plants (often fired with natural gas or oil) are operated only a handful of hours per year when demand is highest and are more costly to operate. The mix of power plants that make up the grid varies significantly from one region to another—based on local demand profiles, resource availability and cost, and energy policy.

U.S. ELECTRICITY GENERATION BY RESOURCE TYPE, 2008

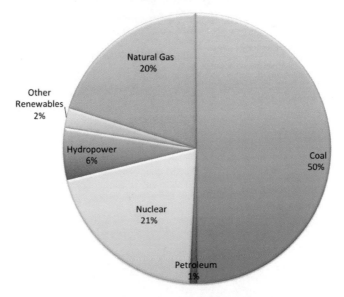

In 2008, 70 percent of the electricity in the United States was generated from fossil fuels (coal and natural gas). Hydropower and other renewables represented only 8 percent, but this percentage is growing. Source: Energy Information Agency, U.S. Department of Energy, Annual Energy Outlook 2010.

While fossil fuels (mainly coal and natural gas) account for 70 percent of U.S. electricity generation, the level of renewable generation is increasing. More than half of U.S. states and several European countries have a renewable portfolio standard (RPS), which mandates renewably based electricity generation. However, renewable resources are limited in quantity, temporal availability, and reliability. Intermittent renewables, such as solar and wind, can pose additional challenges to integration into the grid.

Vehicle recharging will impact the grid in both the immediate and long terms. In the near term, recharging vehicles will require additional electricity to be generated, although there is a large amount of excess capacity at night. A large number of PEVs will need to be driven in a region before power plants are operated differently or new ones are required. For example, adding 1 million PEVs in California (out of 26 million vehicles) increases total electricity consumption in the state by only about 1 percent.[12] Over time, as greater numbers of PEVs are introduced, their impact on the grid will increase. If each of the 240 million registered light-duty vehicles in the United States were charged at a rate of 5 to 10 kWh per day, an additional 12 to 23 percent of electricity generation would be required. However, if most PEVs were coordinated to charge overnight, additional capacity requirements could be much lower.

Typical U.S. households consumed approximately 11,000 kWh annually in 2001. If each household charged a PEV with 5 to 10 kWh of electricity once per day, this could add 21 to 43 percent (2200 to 4600 kWh) per year to the household electricity load, comparable to average central air conditioning and refrigeration loads.

Several studies show that existing grid capacity (including generation, transmission, and distribution) can fuel a significant number of PEVs in the U.S. light-duty vehicle fleet.[12] But specific points along some distribution lines may face congestion if local patterns of electricity demand change significantly because of vehicle recharging. At the substation and feeder levels, where demands are less aggregated—and as a result more variable and sensitive to the patterns of a few customers—distribution impacts are important. If many consumers in a given circuit recharged their plug-in vehicles simultaneously (for example, in the early evening after work), it could increase peak demand locally and require utilities to upgrade the distribution infrastructure.

The mix of power plants supplying a region is largely a function of peak demand and the hourly demand profile. Peak demand determines the total installed power plant capacity needed to supply a region, while the hourly demand profile determines the best mix of plants. Charging off-peak will flatten the demand profile, improving the economics of baseload and intermediate power plants and lowering average electricity costs. Charging at peak demand times will increase capacity requirements, while lowering the utilization of existing plants and increasing electricity costs. If charging could be controlled to occur when it was most optimal, PEV demand could respond to grid conditions. Given that cars are parked approximately 95 percent of the time[13] and potentially plugged in for a large fraction of the time they are parked, this is a real possibility.

One framework for understanding how PEVs can impact the electricity grid is based on the concept of passive and active grid elements (for example, generators and loads). Passive elements are imposed on the system and do not readily respond to grid conditions. Active elements can be controlled and utilized when optimal (i.e. "demand response" utility programs). Baseload and intermittent generators are passive, since they cannot easily turn on or off, or up or down, in response to changes in demand. Active generators can be operated to follow or match demand. Most electricity demand is passive, as it is imposed instantaneously on the electric system by millions of individual customers and not easily controlled. But electricity demand for some loads, including plug-in vehicles, can be active. The timing of recharging demand is controllable, because energy is stored onboard the vehicle in batteries, and vehicle travel is temporally separate from the time when recharging occurs.

The grid manages active and passive elements in real time to match supply and demand. Traditionally, the grid has consisted of passive electric demands, which require precise matching by active generation, such as dispatchable natural gas power plants. But active loads, such as those from plug-in vehicles, may be used to match passive elements, potentially reducing the need for active generation. Additionally, plug-in vehicles can enable the deployment of intermittent renewable generators, such as wind or solar. Since these passive generators are highly variable, they must be matched by standby active generation, typically natural gas-fired generators that are utilized when the renewable resource is unavailable. But aggregated active loads from plug-in vehicles could also be used, potentially reducing the required number of standby power plants and decreasing the costs associated with integrating intermittent power on the grid.

The smart grid, incorporating intelligence and communication between the supply and demand sides of the electricity equation, is needed in order to realize the full benefits of this vehicle charging flexibility. Managing vehicle recharging requires a smart charging system that enables communication between the customer and utilities. Consumers may give the utility greater control in exchange for lower rates. This type of charging interface can also permit vehicle charging emissions to be appropriately tracked and allocated, which will become increasingly important as states and countries adopt low-carbon fuel standards and impose caps on GHG emissions in different sectors.

While recharging vehicles during off-peak hours is preferable from a grid operations and cost perspective, off-peak recharging may not always be preferable to all stakeholders. For example, a consumer may be able to avoid a trip to the gas station by recharging during the day, and though this may be more costly than charging off-peak (the cost of peak electricity can be three times or more higher than off-peak power), it may still be cheaper and less polluting than operating the vehicle on gasoline. Some companies may even incentivize daytime recharging by offering recharging stations at the workplace or other public locations around town.

Environmental Impacts of PEV Use

The environmental impacts of PEVs need to be analyzed on a well-to-wheels (fuel production and end usage) basis to fully account for their operational differences. The generation of electricity accounts for the bulk of emissions from PEV use. Thus, characterizing the emissions associated with electricity generation and distribution is important in quantifying the environmental impacts of operating these plug-in vehicles. This requires an understanding of which power plants are operating during vehicle recharging that would not be generating power otherwise, also known as the marginal generation. Emissions from marginal power plants often differ significantly from the average emissions of all plants operating at a given time. STEPS researchers have developed a model of electricity dispatch (which determines which power plants are operating at any given hour and demand level) for the state of California in order to assess the environmental impacts of different timing profiles of PEV recharging.

Emissions attributable to PEVs depend on the regional characteristics of the grid and the magnitude and timing of demand. A commonly held assumption (which contrasts with the consumer-informed recharge profiles shown earlier) is that vehicle recharging is likely to occur at night, during off-peak hours. If coal power plants (~1000 gCO_2/kWh) provide marginal generation for off-peak vehicle demands, GHG emissions from plug-in vehicles could be *higher* than emissions from conventional hybrid electric vehicles. However, if natural gas-fired power plants (~400–600 gCO_2/kWh) operate on the margin, which is often the case, well-to-wheels GHG emissions from plug-in vehicles will likely be lower than those from conventional HEVs, and considerably lower than those from conventional vehicles. The exact emissions comparison will depend on the vehicle design (BEV versus PHEV), the efficiency of the conventional vehicle, and how the vehicles are driven and recharged.

MARGINAL GHG EMISSIONS BY TIME OF DAY AND MONTH OF YEAR FOR CALIFORNIA

Hour	Avg. recharging demand (MW)	Average hourly marginal generation GHG emissions rate (gCO$_2$-eq kWh^{-1})											
		J	F	M	A	M	J	J	A	S	O	N	D	Year
0	307	630	548	612	531	494	564	638	646	608	634	586	641	595
1	307	634	544	589	517	502	548	570	633	583	623	547	630	577
2	276	619	535	586	507	515	530	546	614	571	595	549	630	567
3	184	623	539	588	512	509	543	541	618	576	589	552	629	569
4	123	639	562	609	535	510	546	569	618	596	622	573	639	585
5	61	646	615	632	592	509	543	610	644	630	636	625	653	611
6	31	654	633	640	600	566	600	614	652	639	638	612	640	624
7	15	657	638	644	639	615	616	650	673	654	656	640	641	644
8	15	665	642	661	644	631	651	667	684	672	654	654	652	657
9	46	665	648	653	650	657	667	682	679	679	655	659	660	663
10	77	654	648	661	661	677	681	684	692	673	674	666	662	670
11	77	658	649	665	670	676	681	707	715	694	667	659	664	676
12	77	658	651	658	667	678	687	714	721	710	658	659	663	677
13	77	658	654	658	667	675	685	721	743	699	672	656	652	679
14	77	655	643	660	661	685	688	745	742	691	675	656	658	680
15	31	648	645	669	658	676	690	750	721	712	681	659	654	680
16	15	657	646	653	652	678	683	732	736	699	671	663	658	678
17	15	687	680	656	658	673	679	710	774	704	669	669	671	686
18	61	687	680	666	660	665	668	696	725	699	680	669	685	682
19	123	678	667	670	671	686	679	693	704	705	675	664	672	681
20	184	673	662	660	662	681	687	675	695	683	670	656	666	673
21	276	660	660	662	659	670	681	687	693	680	656	647	664	668
22	307	654	629	636	627	600	695	660	666	663	654	634	661	648
23	307	647	576	625	555	510	590	658	659	645	632	632	648	615
Demand-weighted avg.		647	601	629	590	580	617	639	665	640	640	613	650	626

This table compares the carbon intensity of marginal electricity (in gCO$_2$eq/kWh) by hour of day and month in California for 2010 as calculated by the Electricity Dispatch Model for Greenhouse Gas Emissions in California (EDGE-CA). It shows that the highest marginal emissions occur in the afternoon on summer days (when demand is highest due to high air-conditioning loads and when all power plants, including inefficient peaking plants, must be utilized) and lowest during the middle of the night in the spring (when demand is low and there is abundant hydro power available). Statewide average emissions for the entire year are calculated to be approximately 400 gCO$_2$eq/kWh. Source: R. McCarthy and C. Yang, "Determining Marginal Electricity for Near-Term Plug-in and Fuel Cell Vehicle Demands in California: Impact on Vehicle Greenhouse Gas Emissions," Journal of Power Sources 195 (2010): 2099–2109.

Studies of PEV environmental impacts rely on assumptions about vehicle designs, consumer values, driving and recharge behaviors, and the future electricity grid. Estimates of GHG reductions range from 32 percent to 65 percent relative to conventional vehicles.[14] But such analyses do not consider which designs PHEV buyers would want, or what design goals should be set. In short, most prior analyses of PHEV impacts assume a given PHEV design and that people will buy those PHEVs.

STEPS researchers sought to estimate potential PHEV GHG impacts in California by combining the consumer-informed recharge profiles described earlier with an electricity dispatch model representing the hourly GHG emissions associated with electricity demand across the year in California in 2010 and 2020. Results suggest that consumer-designed PHEVs can reduce well-to-wheels GHG emissions compared to conventional vehicles under all the recharge and energy conditions we simulated. Further, under present-day grid conditions, from a GHG perspective, these consumer-designed PHEVs may be more benign than the more ambitious AE-20 or AE-40 designs targeted by experts. However, as the carbon intensity of the California electricity grid falls in the future, more ambitious PEV designs will become increasingly advantageous.

POTENTIAL PHEV GHG IMPACTS IN CALIFORNIA, 2010 AND 2020

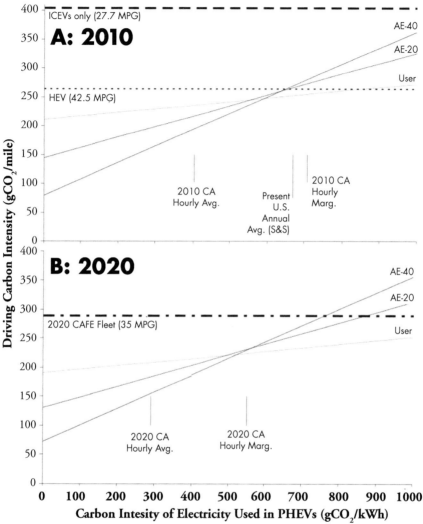

These graphs combine the consumer-informed recharge profiles described earlier with an electricity dispatch model representing the hourly GHG emissions associated with electricity demand across the year in California in 2010 and 2020. As the carbon intensity of the electricity used in PHEVs increases, so does the driving carbon intensity (the grams of CO_2 emitted per mile). Scenario A is for 2010; baselines include present conventional vehicles (CVs) and HEVs. Scenario B is for 2020; the baseline is the fuel economy stipulated by the 2016 CAFE standard. User is the distribution of respondent-designed PHEVs, while AE-20 and AE-40 map those vehicle technologies onto observed consumer driving behavior and recharge potential. Consumer-designed PHEVs can reduce well-to-wheels GHG emissions compared to conventional vehicles under all the recharge and energy conditions we simulated. Source: J. Axsen, K. Kurani, R. McCarthy, and C. Yang, "Plug-in Hybrid Vehicle GHG Impacts in California: Integrating Consumer-Informed Recharge Profiles with an Electricity-Dispatch Model," Energy Policy 39 (2011): 1617–29.

Over the longer term (out to 2050 and beyond), PEVs provide the potential for achieving the highest energy efficiencies of any technology for light-duty vehicles (LDVs). Also, given the move toward reducing the carbon intensity of electricity generation via the renewable portfolio standards (RPS) and other carbon policy measures, PEVs will have an increasingly clean source of low-carbon fuel to use. Further, PHEVs can use a combination of low-carbon biofuels and electricity, given their dual energy systems. However, the potential for GHG reduction that these vehicle technologies offer may be constrained by limitations such as high battery costs and the lack of universal home recharging. These limitations will need to be addressed with appropriate policies and business strategies if PEVs are to achieve their potential to reduce transportation GHG emissions.

Policies and Business Strategies Needed to Support the PEV Pathway

Perceptions of the "battery problem" hold important implications for policy and business strategy; it was the perceived gap between the capabilities of battery technology and the goals assumed by automakers for potential BEV buyers that convinced the California Air Resources Board to modify and reduce zero-emission vehicle sales requirements in the late 1990s. The commercialization potential for PEVs should be based on analysis of both the state of battery technology and the interests of consumers. As demonstrated in this chapter, there is a role for less ambitious PHEV designs with shorter CD ranges and blended CD operation in the near term. Such designs would meet the interests of many current vehicle buyers at relatively lower cost premiums while still significantly contributing to reductions in GHG emissions, air pollution, and petroleum use. Thus, it may not be necessary for battery technology to meet USABC's goals before PHEVs can be commercially viable, and business strategies should recognize this.

The successful commercialization of ambitious PHEV designs in the short term would likely require more aggressive policy actions—such as high financial incentives, large-scale vehicle demonstrations, and pervasive information campaigns—to overcome not just the higher cost of such added performance but also the lack of inherent interest in all-electric (versus blended) driving observed among a sample of potential PHEV consumers. Thus, while the PHEV performance assumed by the USABC and others provides a possibly useful benchmark for future targets for PHEV battery technology, a near-term focus on less aggressive goals may offset more petroleum and emissions in the long run.

Assumptions regarding future strategies for developing PHEVs should be continually reevaluated from a consumer standpoint to assure alignment with a developing market. By making incentives preferential for more aggressive PHEV goals, we risk stalling the market for PHEVs. For example, 90 percent of potential early PHEV buyers designed vehicles requiring less than 4 kWh of batteries—which are not eligible for the federal tax credit. Incentives should be designed to help develop the market for these vehicles even before they reach the most ambitious performance goals.

And attention should be paid to the importance of well-to-wheels emissions metrics. Although PEVs can reduce or eliminate tailpipe emissions, the emissions associated with electricity generation can be substantial. Such emissions are not easy to calculate given the wide regional and temporal variations in electricity carbon intensity by energy sources and power plants.

Vehicle policy, such as fuel economy standards, will need somehow to account for these upstream emissions. Further, efforts to commercialize PEVs for the sake of societal benefits should also be coordinated with efforts to integrate renewable energy sources into the electrical grid. The use of electricity is incentivized by the low-carbon fuel standard (LCFS), a policy that targets GHG emissions reductions from transportation by calculating and regulating the carbon intensity of all fuels used.

Summary and Conclusions

- Interest in PEVs is currently running high in industry, government and among consumers. Nearly every automaker is announcing vehicles that can plug in and run on electricity. The benefits of these vehicles stem from their high efficiency and their use of electricity that can be generated from numerous domestic low-carbon resources. But while PEVs offer significant potential for environmental benefits, they also present a radical departure from conventional vehicles in terms of efficiency, range, utility, flexibility, and the refueling experience. STEPS research on PEVs has attempted to enable better understanding of different vehicle designs, and their resource utilization and emissions impacts, especially when in the hands of consumers.

- STEPS analysis of battery technologies reveals trade-offs among different battery chemistries on key requirements—power, energy capacity, longevity, safety, and cost. Some existing battery technologies can meet some of the goals set by the USABC, but meeting all goals simultaneously is far more challenging. Our consumer research indicates that PHEV designs preferred by consumers are within the current capabilities of NiMH batteries, and thus the experts' aggressive battery technology goals may be unnecessary for near-term PHEV commercialization. Battery cost is thought to be one of the most critical factors in PHEV deployment.

- Our study of vehicle recharging behavior showed that more new vehicle buyers may be pre-adapted for vehicle recharging than estimated in previous analyses (about half have access to charging when parked at home) and that the success of PEVs in meeting energy and emission goals depends on PEV users' recharging and driving behavior as much as or more than on vehicle design. In terms of vehicle recharging and electricity supply, a large number of PEVs will need to be driven in a region before power plants are operated differently or new ones are required. A smart grid that enables communication between customer and utility will be the key to realizing the full benefits of vehicle charging flexibility.

- The generation of electricity accounts for the bulk of emissions from PEV use. Emissions attributable to PEVs depend on the regional characteristics of the grid and the magnitude and timing of demand. Consumer-designed PEVs can reduce well-to-wheels GHG emissions compared to conventional vehicles under all the energy and recharge conditions we simulated. Given the trend toward reducing the carbon intensity of electricity generation, PEVs will have an increasingly clean source of low-carbon fuel to use.

- This research has highlighted important challenges to mass adoption of PEVs but also laid out a potentially significant path forward that relies on lower battery capacity and cheaper blended PHEV designs rather than all-electric PHEV designs. Blended designs can potentially help reduce GHG emissions in the medium-to-long term relative to conventional and hybrid vehicles. This starting point of cheaper blended designs could set the stage for future commercialization of all-electric designs by increasing consumer experience with, and exposure to, PHEV technology, increasing consumer valuation of all-electric capabilities, and reducing battery and drivetrain costs due to increased manufacturing experience. Over time, with improvements in vehicle and battery technology and decarbonization of the electricity sector, a fleet with more all-electric driving could lead to deep long-term GHG reductions.

Notes

1. T. Turrentine and K. Kurani, *Advances in Electric Vehicle Technology from 1990 to 1995: The Role of California's Zero Emission Vehicle Mandate*, EPRI Report TR-106274 (Electric Power Research Institute, February 1996).
2. R. Service, "Hydrogen Cars: Fad or the Future?" *Science* 324 (2009): 1257–59.
3. A. C. Revkin, "The Obama Energy Speech, Annotated," *New York Times*, August 5, 2008.
4. See J. Axsen, A. Burke, and K. Kurani, "Batteries for PHEVs: Comparing Goals and the State of Technology," in *Electric and Hybrid Vehicles: Power Sources, Models, Sustainability, Infrastructure and the Market*, ed. G. Pistoia (New York: Elsevier, 2010).
5. A. Pesaran, T. Markel, H. S. Tataria, and D. Howell, "Battery Requirements for Plug-in Hybrid Electric Vehicles: Analysis and Rationale," presented at the 23rd International Electric Vehicle Symposium and Exposition (EVS-23), Anaheim, California, 2007; M. A. Kromer and J. B. Heywood, *Electric Powertrains: Opportunities and Challenges in the U.S. Light-Duty Vehicle Fleet*, LEFF 2007-02 RP (Sloan Automotive Laboratory, MIT Laboratory for Energy and the Environment, May 2007); R. Graham, *Comparing the Benefits and Impacts of Hybrid Electric Vehicle Options*, Report #1000349 (Electric Power Research Institute, 2001). These categories follow the California Air Resources Board's (CARB's) definition of PHEV-X, where X is the number of miles the vehicle can drive in all-electric mode during a particular drive cycle before the gasoline engine turns on. USABC goals are based on the Urban Dynamometer Driving Schedule (UDDS) to be consistent with CARB's testing methods. The USABC AE-10 goals are set for a "crossover utility vehicle" (an automobile-based SUV) weighing 1950 kg, and the AE-40 goals are set for a midsize sedan weighing 1600 kg.
6. D. Doughty and C. Crafts, *FreedomCAR Electrical Energy Storage System Abuse Test Manual for Electric and Hybrid Electric Vehicle Applications*, 2005-3123 (Sandia National Laboratories, 2005).
7. M. Anderman, "PHEV: A Step Forward or a Detour?" presented at SAE Hybrid Vehicle Technologies Symposium, San Diego, California, 2008; F. Kalhammer, B. Kopf, D. Swan, V. Roan, and M. Walsh, *Status and Prospects for Zero Emissions Vehicle Technology: Report of the ARB Independent Expert Panel 2007*, prepared for the State of California Air Resources Board, 2007, http://www.arb.ca.gov/msprog/zevprog/zevreview/zev_panel_report.pdf.
8. Kromer and Heywood, *Electric Powertrains*; Kalhammer et al., *Status and Prospects for Zero Emissions Vehicle Technology*.
9. Design games are summarized and portrayed in J. Axsen and K. S. Kurani, *The Early U.S. Market for PHEVs: Anticipating Consumer Awareness, Recharge Potential, Design Priorities and Energy Impacts*, UCD-ITS-RR-08-22 (Institute of Transportation Studies, University of California, Davis, 2008).
10. See, for example, B. D. Williams and K. S. Kurani, "Estimating the Early Household Market for Light-Duty Hydrogen-Fuel-Cell Vehicles and Other 'Mobile Energy' Innovations in California: A Constraints Analysis," *Journal of Power Sources* 160 (2006), 446–53; K. A. Nesbitt, K. S. Kurani, and M. A. Delucchi, "Home Recharging and Household Electric Vehicle Market: A Near-Term Constraints Analysis," *Transportation Research Record* 1366 (1992), 11–19.
11. One million PEVs each charging approximately 8 kWh/day (good for 25–35 all-electric miles) would require 2880 GWh/yr or 1 percent of 2010 California total electricity demand (288,000 GWh/yr).
12. See, for example, S. Hadley and A. Tsvetkova, *Potential Impacts of Plug-in Hybrid Electric Vehicles on Regional Power Generation* (Oak Ridge National Laboratory, 2008), and M. Duvall, E. Knipping, M. Alexander, L. Tonachel, and C. Clark, *Environmental Assessment of Plug-In Hybrid Electric Vehicles, Volume 1: Nationwide Greenhouse Gas Emissions* (Electric Power Research Institute, Palo Alto, CA, 2007).

13. Cars and light trucks typically are driven about 12,000 miles per year. At an average speed of 30 miles per hour, this translates into a vehicle's being driven 400 hours per year, or about 5 percent of the time.
14. M. Duvall, E. Knipping, M. Alexander, L. Tonachel, and C. Clark, *Environmental Assessment of Plug-in Hybrid Electric Vehicles, Volume 1: Nationwide Greenhouse Gas Emissions* (Electric Power Research Institute, Palo Alto, CA, 2007); C. Samaras and K. Meisterling, "Life-Cycle Assessment of Greenhouse Gas Emissions from Plug-in Hybrid Vehicles: Implications for Policy," *Environmental Science and Technology* 42 (2008): 3170–76.
15. See H.R. 1 *American Recovery and Reinvestment Act of 2009*, Section 1141 (U.S. Government Printing Office, Washington, DC, 2009), http://purl.access.gpo.gov/GPO/LPS111758.

Chapter 3: The Hydrogen Fuel Pathway

Joan Ogden, Christopher Yang, Joshua Cunningham, Nils Johnson, Xuping Li, Michael Nicholas, Nathan Parker, and Yongling Sun

We turn now from biofuels and electricity to a fuel pathway that holds out promise farther in the future. Hydrogen has been widely discussed as a long-term fuel option to address environmental and energy security problems posed by current transportation fuels. Hydrogen fuel cell cars are several times more efficient than today's conventional gasoline cars, and they produce zero tailpipe emissions. They offer good performance, a range of 270-430 miles,[1] and can be refueled in a few minutes. Hydrogen can be made with zero or near-zero emissions from widely available resources, including renewables (like biomass, solar, wind, hydropower, and geothermal), fossil fuels (such as natural gas or coal with carbon capture and sequestration), and nuclear energy. In principle, it should be possible to produce and use hydrogen transportation fuel with near-zero well-to-wheels emissions of greenhouse gases and greatly reduced emissions of air pollutants while simultaneously diversifying away from our current dependence on petroleum.[2]

To reach stringent long term goals for cutting greenhouse gas emissions from transportation, it appears likely that the light duty fleet will be largely electrified by 2050 (see Chapter 8). Hydrogen fuel cells are an important enabling technology for this vision. Automakers foresee a future electrified light duty fleet with batteries powering smaller, shorter range cars and hydrogen fuel cells powering larger vehicles with longer range. To electrify all segments of the light duty market, fuel cells are a necessary complement to batteries.

Recent assessments affirm the long-term potential of hydrogen to greatly reduce oil dependence as well as transportation emissions of greenhouse gases and air pollutants—far beyond what might be achieved by energy efficiency alone. They also highlight the complex technical and logistical challenges that must be addressed before a hydrogen-based transportation system can become a reality. This chapter discusses some of the major questions regarding future use of hydrogen in the transportation sector and highlights STEPS research on these issues.

- What is the technical outlook for hydrogen vehicles and hydrogen supply?
- What are the environmental impacts of hydrogen fuel compared to alternatives?
- What would a hydrogen infrastructure look like, and how could we make a transition to hydrogen?
- What policies and business strategies are needed to support hydrogen in both the near and long terms?

CHALLENGES ON THE HYDROGEN FUEL PATHWAY

These complex technical and logistical challenges must be addressed before a hydrogen-based transportation system can become widespread:

- **Technical challenges.** While many of the technologies exist to build a hydrogen energy system, further development is needed on key emerging technologies. In particular, further development is needed for proton exchange membrane (PEM) fuel cell cost and durability, hydrogen storage on vehicles, and technologies for zero-carbon hydrogen production.
- **Logistical challenges.** Full adoption of FCVs will require a widespread hydrogen infrastructure. The issue is not producing low-cost hydrogen at large scale but distributing hydrogen to many dispersed users at low cost, especially during the early stages of a transition.
- **Transition issues / coordination of stakeholders.** A hydrogen transition means many major changes at once: adoption of new types of cars, building a new fuel infrastructure, and development of new low-carbon primary energy resources. These changes will require coordination among diverse stakeholders with differing motivations (fuel suppliers, vehicle manufacturers, and policymakers), especially in the early stages when costs for vehicles are high and infrastructure is sparse. Factors that could ease transitions, like compatibility with the existing fuel infrastructure, are more problematic for hydrogen than for electricity or liquid synthetic fuels.
- **Policy challenges.** Finally, consistent policies that reflect the external costs of energy—such as global climate change and damage to health from air pollution, plus the costs of oil supply insecurity—are lacking. This is a barrier to introducing more-efficient, cleaner technologies, including hydrogen, and to assuring that hydrogen is made from low-carbon sources. It is almost certain that technology-specific policies will be needed to support a hydrogen transition.

Technology Status and Outlook

We start with the technology status and outlook for hydrogen vehicles and hydrogen supply. Technologies that use hydrogen, notably fuel cells, are making rapid and significant progress. But while many of the technologies to build a hydrogen-based transportation system already exist, further development is needed for key emerging technologies, especially proton exchange membrane (PEM) fuel cells for automotive use, hydrogen storage on vehicles, and technologies for zero-carbon hydrogen production.

Hydrogen vehicles

Although internal combustion engines can run on hydrogen, it is the higher-efficiency, zero-emission hydrogen fuel cell that has largely captured the attention of automakers. Several automakers have embraced fuel cells as a superior zero-emission technology and have large development and commercialization programs. Honda, Toyota, Daimler, GM and Hyundai have announced plans to commercialize FCVs sometime between 2015 and 2020.[3] Hydrogen and fuel cells represent a logical progression beyond efficiency and increasing electrification of cars with hybrid and electric drive trains. As noted above, many automakers see complementary roles for hydrogen fuel cells and battery electric vehicles and are pursuing both technologies.

Fuel cells are highly efficient electrochemical "engines" that combine hydrogen and oxygen in air to produce electricity to power the vehicle. Fuel cells operate without combustion or emissions of pollutants or greenhouse gases; the only tailpipe emission is water. Today's development FCVs have fuel economies twice that of comparable gasoline cars, and 35 to 65 percent higher than gasoline hybrids.[4] FCVs use electric drive trains but have a longer range, a faster refueling time, and the potential for lower cost than battery electric cars.[5] In addition to the fuel cell stack, other key components of a hydrogen FCVs include hydrogen storage, electric motors and power controllers, and batteries for hybrid operation and cold start support (most fuel cell vehicles today are hybrids).

A key technology for automotive applications is the proton exchange membrane or PEM fuel cell. Manufacturers have reduced the weight and volume of PEM fuel cell systems so that they easily fit under the hood of a compact car. Fuel cell systems have demonstrated good driving performance and meet goals for low-temperature operation and freeze tolerance. However, several issues remain. Current automotive PEM fuel cells still fall short of the 5,000-hour lifetime needed, lasting about 2,000 hours in on-road tests,[6] although durability is steadily increasing and researchers have reported new designs that might take fuel cells to 7,000 hours and beyond. Recently, 5,000 hours durability was demonstrated in laboratory cells under non-ideal conditions that resemble on-road operation.[7]

The U.S. Department of Energy (DOE) estimates that if today's automotive PEM fuel cell systems were mass-produced (at levels of 500,000 units per year), costs would drop to $51/kW (or about $4,000 for an 80-kW system), roughly twice the cost of a comparable internal combustion engine[8]. Fuel cell system costs are expected to continue declining toward the DOE goal of $30/kW because of improved materials, reductions in required platinum loading, and increased power density.

Storing enough hydrogen on a car for a reasonable traveling range (say 300 miles) is another key design issue. Storage requires high-pressure cylinders, liquid hydrogen at a super-cooled 20 K, or special materials such as metal hydrides that absorb hydrogen under pressure. Hydrogen storage systems are heavier and bulkier than those for gasoline, though less so than batteries, and compressing or liquefying hydrogen requires energy. Finding a better storage method is a major thrust of hydrogen R&D worldwide. In the absence of a breakthrough storage technology, most hydrogen vehicles today opt for the simplicity of compressed gas storage, which will be the system choice for early commercialization. Because of the low volumetric density of these systems, many FCV manufacturers have begun to design around the storage system in order to get adequate range without reducing passenger or cargo space in the vehicle. GM, Honda, Toyota, Daimler and Hyundai have all demonstrated light-duty fuel cell cars with a 270–400-mile range, using compressed hydrogen gas at 35–70 MPa (megapascals, a measure of pressure).[9] These vehicles meet the U.S. DOE goals for range.

Costs for mass-produced compressed storage tanks based upon current technology are estimated to be around $15–23/kWh (or about $2,500–3,700 for a compact FCV storing enough hydrogen for a 300-mile range).[10] Although these are substantially higher than the DOE's 2015 goals of $2–4/kWh, a recent National Academies study found acceptable overall vehicle costs with hydrogen storage tanks costing $10–15/kWh.[11]

H_2 TECHNOLOGIES: CURRENT STATUS VS 2015 DOE GOALS

	Today	**2015 Goals**
In-use durability (hrs)	2000	5000
Vehicle Range (miles/tank)	280-400	300
Fuel Economy (mi/kg H_2)	72	60
Fuel Cell Efficiency	53-58%	60%
Fuel Cell System Cost ($/kW)	51	30
H_2 Storage Cost ($/kWh)	15-23	10-15 (NRC)
		2-4 (USDOE)

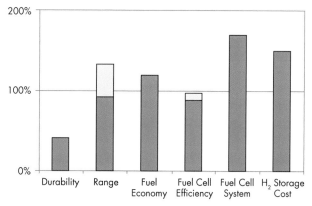

When we compare the status of current (2010) H_2 technologies in demonstration vehicles and goals set by the U.S. Department of Energy (DOE) for 2015, we can see that current technologies have farthest to go to reach durability, system-cost, and storage-cost goals. The figure at the bottom shows how close the current technology is to the 2015 goal.

While estimates of the price of mass-produced FCVs based upon projections for 2015 technology are within a few thousand dollars of conventional vehicles,[12] initial FCV models will not be produced in such high volumes and as a result will have a high price premium. At a scale of 50,000 FCVs being produced worldwide, estimated prices are around $75,000 per vehicle. Prices can drop quickly as manufacturing volume increases. Mass-produced, mature technology FCVs are estimated to have a retail price $3,600 to $6,000 higher than a comparable gasoline internal combustion engine vehicle (ICEV).[13]

ESTIMATED FCV RETAIL PRICE OVER TIME

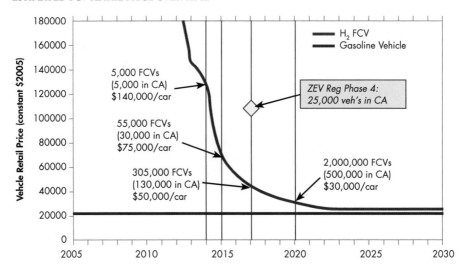

The estimated retail price for FCVs drops considerably as production scales from thousands (in 2012) to millions of vehicles per year (in 2025). The learned-out price difference between the FCV and the gasoline ICEV is about $3,600.[14]

Hydrogen production methods

Like electricity, hydrogen is an energy carrier that is produced from a primary energy resource. Almost any energy resource can be converted into hydrogen, although some pathways are superior to others in terms of cost, environmental impacts, efficiency, and technological maturity.

RESOURCES AND CONVERSION PATHWAYS FOR HYDROGEN

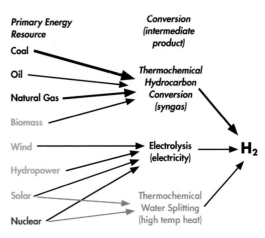

There are a multitude of potential primary energy resources and conversion pathways for producing hydrogen. Fossil resources are shown in black and renewable resources are shown in green. Pathways that are more technologically mature (for example, electrolysis and thermochemical conversion of hydrocarbons from coal and natural gas) are shown in bold, while the more speculative pathways (such as thermochemical water splitting) are in a lighter shade.

In the United States, about 9 million tonnes of hydrogen are produced each year (enough to fuel a fleet of about 35 million fuel cell cars). Steam reforming of natural gas is the most common method of hydrogen production today (mainly for industrial and refinery purposes), accounting for about 95 percent of hydrogen production in the United States.

In the near to medium term, fossil fuels (primarily natural gas) are likely to continue to be the least expensive and most energy-efficient resources from which to produce hydrogen. Conversion of these resources still emits some carbon into the atmosphere. However, future hydrogen production technologies could virtually eliminate GHG emissions. For large central plants producing hydrogen from natural gas or coal, it is technically feasible to capture the CO_2 and permanently sequester it in deep geological formations, although the widespread use of sequestration technology poses important challenges and will not happen until 2020 at the earliest.

Production of hydrogen from renewable biomass is a promising midterm option (post 2020) with very low net carbon emissions. In the longer term, vast carbon-free renewable resources such as wind and solar energy might be harnessed for hydrogen production via electrolysis of water. While this technology is still improving, high costs for electrolyzers and renewable electricity (in part because of the low capacity factors of intermittent renewable sources) suggest that renewable electrolytic hydrogen will likely cost more than hydrogen from fossil resources with carbon capture and sequestration (CCS) or biomass gasification.

DELIVERED COST OF HYDROGEN FROM VARIOUS PATHWAYS

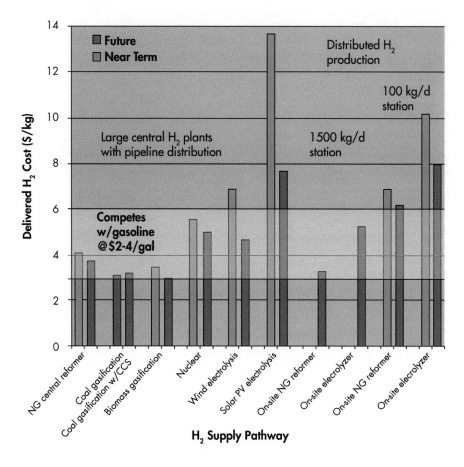

Here we compare the delivered cost of hydrogen transportation fuel produced via different pathways for "near term" (scaled up infrastructure with current technology) and "future" (full scale infrastructure with advanced technologies beyond 2015).[15] We see that costs will come down as technology advances, and that production from hydrocarbons generally costs less than electrolytic hydrogen production. All central alternatives assume hydrogen is deployed at a massive scale, which could happen beyond 2020. On-site alternatives use stations serving numbers of cars similar to today's gasoline stations (at 1,500 kg of H_2 per day). We also show estimated H_2 costs for smaller size stations (100 kg of H_2 per day) typical of near-term demonstration H_2 stations which serve a relatively small number of early FCVs. These small stations would have significantly higher hydrogen cost because of scale economies. The range for hydrogen fuel costs to compete with gasoline on a cents-per-mile basis is shown, based on an efficient gasoline hybrid competing with an FCV. If H_2 costs $3–6/kg, the fuel cost per mile for an FCV is about the same as for an efficient gasoline hybrid using gasoline at $2–4/gal, assuming that the fuel economy of a fuel cell vehicle is 1.5 times higher than that of a comparable gasoline hybrid.

In the United States, the lowest-cost low-carbon hydrogen supply pathways appear to be biomass gasification and hydrogen from coal with CCS. Each could contribute significantly to the long-term hydrogen supply. The lowest-cost option depends on the market penetration of FCVs, the local feedstock and energy prices, as well as geographic factors such as city size and density of demand. Detailed regional studies reveal possibilities for further optimizing the hydrogen supply system at the regional level. It appears that hydrogen could be delivered to consumers for about $3–4/kg, with near-zero emissions of greenhouse gases, on a well-to-wheels basis, which leads to a reduction in fuel cost per mile compared to gasoline vehicles, given the increased efficiency of FCVs.

H_2 AND ELECTRICITY AS PRIMARY ENERGY CARRIERS[16]

One compelling vision of a future decarbonized energy system involves the use of two primary energy carriers—hydrogen and electricity. H_2 and electricity are both decarbonized energy carriers that enable conversion, transport, and utilization of a wide variety of primary energy resources. In an integrated energy system, these two energy carriers could complement each other; they could be produced from the same primary energy resources and could in fact be co-produced and inter-converted. However, they have very different characteristics, which suggest specialized uses and applications for each.

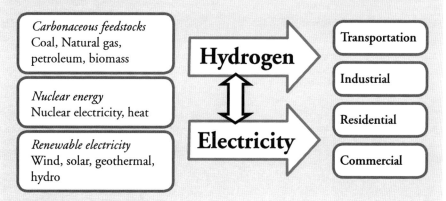

Given the benefits associated with electric-drive vehicles, hydrogen and electricity are in competition as the primary energy carrier for light-duty vehicles. However, many industry experts foresee a complementary role in the future light-duty sector dominated by electric-drive vehicles in which small, shorter-range vehicles are powered by batteries and longer-range, larger passenger vehicles are powered by fuel cells. The main technical challenges facing battery-powered vehicles stem from the energy density limitations and recharge times associated with batteries. Fuel cells appear to alleviate these issues with refueling speeds and vehicle ranges that approach those of gasoline vehicles, though these benefits are traded off for greater infrastructure requirements.

> Co-production is another area where these two energy carriers may interact. They can be made from the same primary resources and can be co-produced with higher efficiency and lower cost than producing either one separately. These can occur at the large scale (for example, thermochemical conversion from fossil fuels with carbon capture and sequestration) or the small scale (for instance, separate energy stations at one refueling station).
>
> Finally, hydrogen and electricity can be inter-converted via electrolyzers and fuel cells. While efficiency losses occur in converting one energy carrier to another, a number of circumstances may offer compelling reasons to do so. Such circumstances include electrolysis using cheap off-peak electricity, hydrogen production as a means of storing and leveling intermittent renewable electricity, and vehicle-to-grid electricity in a fuel cell vehicle.

Hydrogen delivery methods

Once hydrogen is produced, there are several ways to deliver it to vehicles. It can be produced regionally in large plants, stored as a compressed gas or cryogenic liquid (at −253° C), and distributed by truck or gas pipeline; or it can be produced on-site at refueling stations (or even homes) from natural gas, alcohols (methanol or ethanol), or electricity. No one hydrogen supply pathway is preferred in all situations.

TWO OPTIONS FOR SUPPLYING HYDROGEN

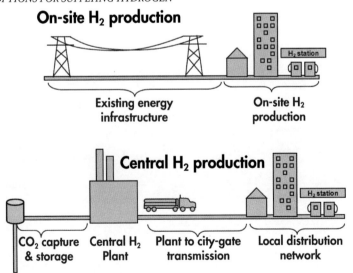

Options for producing and delivering hydrogen include on-site production and central production. Source: C. Yang and J. Ogden, "Determining the lowest-cost hydrogen delivery mode," International Journal of Hydrogen Energy 32 (2007): 268–86.

THE LOWEST-COST WAY TO DELIVER HYDROGEN

What is the least costly way to bring hydrogen to users? It all depends on how much hydrogen is needed (the hydrogen flow) and how far it needs to travel (distance). STEPS researchers developed models to find the lowest-cost delivery mode for hydrogen, given three choices: compressed gas hydrogen trucks, liquid hydrogen trucks, and hydrogen gas pipeline.

Hydrogen flow rate is an important factor determining delivery mode choice and cost. As the hydrogen flow rate goes up, costs come down, primarily because of scale economies in pipeline delivery. Pipeline delivery is the lowest-cost delivery option at high levels of hydrogen demand, while trucks dominate at smaller quantities of hydrogen. As distance increases, liquid trucks give a lower cost than compressed gas trucks because each truck carries more hydrogen. (If the gas pressure were increased allowing more hydrogen per truckload, compressed gas truck transport could become more competitive with liquid trucks, and the border between "L" and "G" might shift in the figure below). At a given distance, pipelines beat liquid trucks when the hydrogen flows are large enough. (For reference, 10 tonnes of hydrogen per day would fuel about 10,000 cars, and 100 tonnes per day about 100,000 cars. So pipeline transport is unlikely until large numbers of vehicles are present in a concentrated region.)

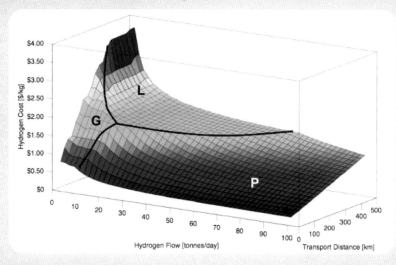

We compared three different hydrogen delivery modes to find the lowest-cost method. As hydrogen flow increases, delivery by hydrogen gas pipeline (P) starts to cost the least; as transport distance increases, liquid hydrogen trucks (L) win out. Compressed gas hydrogen trucks (G) cost least when both distance and flow are limited. Source: C. Yang and J. Ogden, "Determining the lowest-cost hydrogen delivery mode," International Journal of Hydrogen Energy *32 (2007): 268–86.*

Environmental Impacts of Hydrogen Fuel

The environmental impacts of hydrogen fuel vary with the production pathway. Most life-cycle analyses of alternative fuels have focused on emissions and energy use, but recently several authors have expanded their focus to estimate primary resource, land, water, and materials use associated with hydrogen energy systems as compared to other fuels.

GHG emissions, air pollution, and energy use

Most hydrogen production today is from fossil fuels, which releases CO_2, the major GHG linked to climate change. For the near term, FCVs using hydrogen produced from natural gas would reduce well-to-wheels GHG emissions by about half compared to current gasoline vehicles. For large central plants producing hydrogen from hydrocarbons (natural gas, coal or biomass), it is technically feasible to capture the CO_2 and permanently sequester it in deep geological formations, although sequestration technology will not be in widespread use before 2020 at the earliest. Production of hydrogen from renewable biomass is a promising midterm option with very low net carbon emissions. In the longer term, carbon-free renewables such as wind and solar energy might be harnessed for hydrogen production via electrolysis of water.

Air pollution reductions are significant with hydrogen pathways compared to gasoline, leading to better air quality[17] and lower social costs.[18] And petroleum use for hydrogen pathways is very small. The only oil use is associated with truck delivery and electricity generation for hydrogen compression or liquefaction, and this is much lower than with any gasoline pathway.

CHAPTER 3: THE HYDROGEN FUEL PATHWAY

COMPARISON OF EMISSIONS FROM DIFFERENT FUEL/VEHICLE PATHWAYS

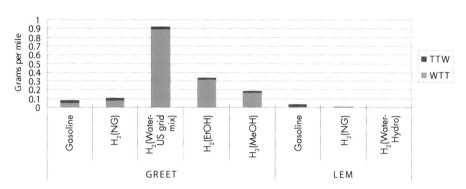

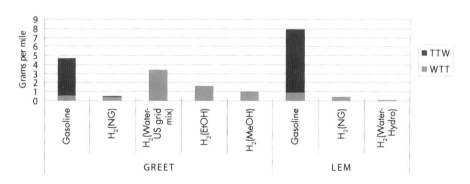

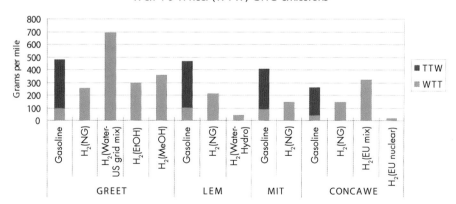

We compare the well-to-wheels (WTW) emissions of greenhouse gases, air pollutants, and particulate matter (PM) for a variety of hydrogen pathways, based on results from the Argonne National Laboratory GREET model, the UC Davis LEM model, MIT, and the European Union CONCAWE study. We break emissions down into phases: well-to-tank (WTT) and tank-to-wheels (TTW). Emissions are shown for $H_2(NG)$ = hydrogen from on-site natural gas reforming; $H_2(Water)$ = hydrogen from on-site water electrolysis; $H_2(EtOH)$ = hydrogen from ethanol at refueling stations; and $H_2(MeOH)$ = hydrogen from methanol at refueling stations.

Notes: The emission results from GREET V.1.8c.0 are for year 2010 using default input parameters in the GREET model. The LEM model assumes that electricity generation is from hydropower for hydrogen production from water electrolysis. PM emissions from GREET are larger than those from LEM as LEM considers emission reductions due to emission controls while GREET does not. GREET also includes the PM emissions from brake and tire wear. According to GREET, most of the PM emissions are from the WTT phase, about 1.8 percent of total air pollution is PM from gasoline, and about 20–27 percent of total air pollution is PM from H_2 pathways.

Sources: M. Wang, "Well-to-Wheels Analysis with the GREET Model," 2005 U.S. DOE Hydrogen Program Review, May 26, 2005; M. Wang, "Well to Wheels Analysis of Vehicle/Fuel Systems with GREET at Argonne National Lab," presentation at the U.S. DOE Hydrogen Analysis Deep Dive meeting, San Antonio, TX, March 22, 2007; M. A. Delucchi, "A Lifecycle Emissions Model (LEM): Lifecycle Emissions from Transportation Fuels, Motor Vehicles, Transportation Modes, Electricity Use, Heating and Cooking Fuels, and Materials," UCD-ITS-RR-03-17-MAIN (Institute of Transportation Studies, University of California, Davis, 2003); M. A. Kromer and J. B. Heywood, Electric Powertrains: Opportunities and Challenges in the U.S. Light-Duty Vehicle Fleet, *LEFF 2007-02 RP* (Sloan Automotive Laboratory, MIT Laboratory for Energy and the Environment, May 2007); EUCAR (European Council for Automotive Research and Development), CONCAWE, and ECJRC (European Commission Joint Research Centre), Well-to-Wheels Analysis of Future Automotive Fuels and Powertrains in the European Context, *Well-to-Wheels Report*, Version 2c, March 2007.

Use of primary energy resources, land, water, and other materials

With hydrogen fuel cells the amount of primary energy required is similar to that for gasoline hybrids and considerably less than for conventional gasoline cars. There are plentiful near-zero-carbon resources for hydrogen production in the United States. For example, a mix of low-carbon resources including natural gas, coal (with carbon sequestration), biomass, and wind power could supply ample hydrogen for vehicles. With 20 percent of the biomass resource, plus 15 percent of the wind resource, plus 25 percent added use of coal (with sequestration), 300 million hydrogen vehicles (approximately the entire U.S. fleet projected in 2030) could be served with near-zero GHG emissions.

ENERGY RESOURCES REQUIRED TO FUEL 100 MILLION CARS IN THE UNITED STATES

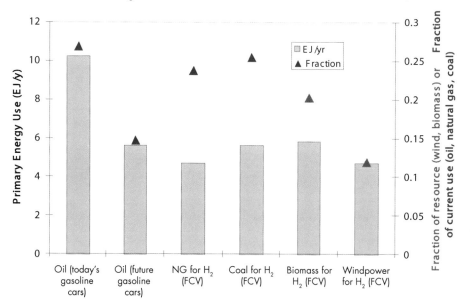

We did a sample calculation of the amount of primary energy needed to make hydrogen for 100 million FCVs in the United States (about 50 percent of the current U.S. fleet or 33 percent of the projected U.S. fleet in 2050). The amount of primary energy required is measured in exajoules (10^{18} joules) per year by the y-axis on the left. The fraction of the available annual resource (for biomass and wind) or the current use (for coal or natural gas) is measured by the y-axis on the right.

For reference, we also plot the energy use for 100 million current gasoline vehicles and 100 million gasoline hybrids. The biomass resource is assumed to be 800 million tonnes of biomass per year, and the wind resource is assumed to be 11,000 billion kWh of electricity per year. Source: J. Ogden and C. Yang, "Build-up of a Hydrogen Infrastructure in the U.S.," Chapter 15 in The Hydrogen Economy: Opportunities and Challenges, *ed. M. Ball and M. Wietschel (Cambridge, UK: Cambridge University Press, 2009), 454–82.*

The land and water requirements for producing hydrogen also depend on the production pathway. The table below shows the land requirements to produce hydrogen for a variety of renewable pathways. For comparison, the total U.S. land area is 9.1 million km² and the total cropland is 1.8 million km². The impacts of this level of land use have not been thoroughly examined in terms of competing uses. Regarding water use, hydrogen pathways relying on renewable electrolysis or steam methane reforming are estimated to use much less water than hydrogen pathways relying on synthetic fuels from coal or biomass, and somewhat less water than gasoline production.[19] Water could become an important constraint on future energy production.

LAND AREA REQUIRED TO PRODUCE RENEWABLE H_2

Hydrogen Pathway	Land Area (m²) to Produce 1 GJ H_2 per Year	Total Land Area (km²) to Produce H_2 for 100 Million Cars
Electrolytic H_2		
Solar PV	1.89	5,700
Solar thermal electric	5.71	17,000
Wind	6.3-33	19,000–99,000
Hydropower	11-500	
H_2 via biomass gasification	50	150,000

Materials availability could also become an issue for widespread use of hydrogen. For example, FCVs require use of a platinum catalyst in the fuel cell. If these vehicles come into widespread use in the future, significant quantities of platinum will be needed. However, studies by STEPS researchers and other have shown that there should be sufficient platinum for FCVs (see Chapter 7 for a full discussion).

Building a Hydrogen Infrastructure

Adoption of hydrogen will require a widespread hydrogen infrastructure to fuel vehicles. Unlike the case with gasoline and electricity, there is currently no large-scale infrastructure bringing hydrogen to consumers. Because there are many options for hydrogen production and delivery, and no one supply option is preferred in all cases, creating such an infrastructure is a complex design problem. The challenge is not so much producing low-cost hydrogen at large scale as it is distributing hydrogen to many dispersed users at low cost, especially during the early stages of the transition.

Recent studies (including those at UC Davis)[20] have found that the design of a hydrogen infrastructure depends on many factors, including these:

- **Scale**. Hydrogen production, storage, and delivery systems exhibit economies of scale, and costs generally decrease as demand grows.

- **Geography / regional factors**. The location, size, and density of demand, the location and size of resources for hydrogen production, the availability of sequestration sites, and the layout of existing infrastructure can all influence hydrogen infrastructure design.

- **Feedstocks**. The price and availability of feedstocks for hydrogen production, and energy prices for competing technologies (for example, gasoline prices), must be taken into account.

- **Technology status**. Assumptions about hydrogen technology cost and performance determine the best supply option.

- **Supply and demand**. The characteristics of the hydrogen demand and how well it matches supply must be considered. Time variations in demand (refueling tends to happen during the daytime, with peaks in the morning and early evening) and in the availability of supply (for example, wind power is intermittent) can help determine the best supply and how much hydrogen storage is needed in the system.

- **Policy**. Requirements for low-carbon or renewable hydrogen influence which hydrogen pathways are used.

In this section, we discuss a national rollout for the United States, early infrastructure and transition issues in southern California, and regional designs for two leading low-carbon options: biomass hydrogen and coal with CCS.

A scenario for hydrogen infrastructure build-up in the United States

Building a national hydrogen refueling infrastructure in a large, diverse country such as the United States is a complex design problem involving regional considerations. We developed the SSCHISM model to study this challenge. We use SSCHISM to determine the least-cost method for supplying hydrogen to a particular city at a given market penetration.

MODELING HYDROGEN INFRASTRUCTURE: THE UC DAVIS SSCHISM MODEL

To understand the design and economics of a hydrogen infrastructure, STEPS researchers developed the Steady-State City Hydrogen Infrastructure System Model (SSCHISM).[21] SSCHISM finds the lowest-cost infrastructure design based on regionally specific information (city population and physical size, energy prices, electricity grid characteristics), plus engineering/economic models of hydrogen infrastructure component costs, and market factors. We analyzed a wide variety of hydrogen supply pathways for each of 73 major U.S. urban areas. Outputs include the levelized cost of delivered hydrogen, the infrastructure capital cost, CO_2 emissions, and primary energy requirements.

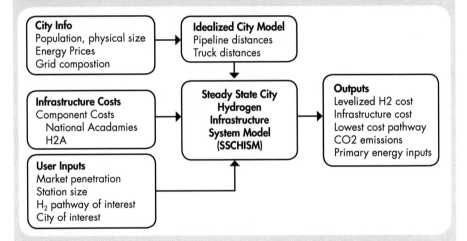

Our SSCHISM model finds the lowest-cost infrastructure design based on regionally specific information (city population and physical size, energy prices, electricity grid characteristics), plus hydrogen infrastructure component costs and market factors.

In the model, we assume that the first few thousand FCVs are successfully introduced in 2012, with tens of thousands of FCVs by 2015, 2 million in the fleet by 2020, 10 million by 2025, and about 200 million (60 percent of the fleet) by 2050. Because of the need to locate infrastructure and vehicles together, hydrogen is introduced in a succession of "lighthouse" cities, starting with the Los Angeles area. We assume that some minimum number of hydrogen stations is needed in each city to assure adequate coverage and consumer convenience and to help deal with the "chicken-or-egg" problem of assuring hydrogen fuel availability to early vehicle owners.

PROJECTED INTRODUCTION OF FCVS IN "LIGHTHOUSE" CITIES, 2012–2025

	2012	2013	2014	2015	2016	2017	2018	2019	2020	2021	2022	2023	2024	2025
Los Angeles	1	2	2	25	40	50	85	120	160	190	210	250	270	300
New York, Chicago				25	40	50	85	120	150	175	185	225	240	270
San Francisco, Washington/Baltimore					20	30	55	85	120	140	160	190	210	230
Boston, Philadelphia, Dallas						20	50	85	120	145	165	195	210	220
Detroit, Houston							25	50	80	120	140	160	190	210
Atlanta, Minneapolis, Miami								40	75	100	115	130	160	180
Cleaveland, Phoenix, Seattle									45	70	90	120	150	170
Denver, Pittsburgh, Portland, St.Louis, Cincinnati, Indianapolis, Kansas City										60	80	110	130	150
Milwaukee, Charlotte, Orlando, Columbus, Salt Lake City											55	80	110	130
Nashville, Buffalo, Raleigh												40	70	90
Nationwide													260	540

The number of light-duty FCVs sold annually in 27 "lighthouse" cities is given here in thousands of vehicles per year introduced between 2012 and 2025. The total number of hydrogen vehicles in 2025 is 10 million, and 2.5 million vehicles are sold that year. Source: S. Gronich, "Hydrogen and FCV Implementation Scenarios, 2010–2025," presented at the U.S. DOE Hydrogen Transition Analysis Workshop, Washington DC, August 9–10, 2006.

As new cities are phased in over time, hydrogen is initially costly because of the low demand in the new cities, but costs fall as demand grows. The phased introduction of hydrogen infrastructure and vehicles leads to differences in hydrogen market penetration and also contributes to differences in hydrogen cost for different cities. City size and density as well as local feedstock and energy prices also contribute to these cost differences.

PROJECTED HYDROGEN COSTS TO 2030

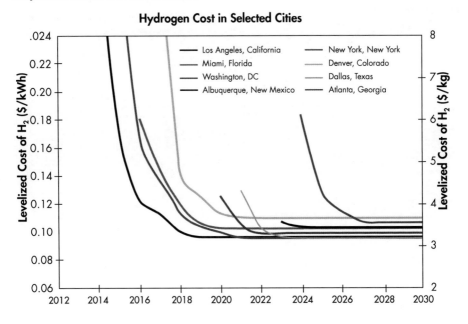

The range and progression of delivered hydrogen costs over time is shown for selected "lighthouse" cities. Cost differences are due to phased introduction of hydrogen cars as well as to city size and density and local feedstock and energy prices.

The choice of supply pathway also varies over time. At low demand, on-site steam methane reformers (SMRs) dominate because the large investments required for central production and hydrogen delivery are not yet justified. As hydrogen demand in a particular city grows, it makes sense to build central production plants and delivery systems when the economies of scale associated with large production plants overcome the additional cost associated with pipeline or truck delivery. This sequence is played out in each of the 73 urban areas in the model. However, the point at which this switch from distributed to central production occurs and the least-cost central pathway differ depending upon the size of the city, level of demand, demand density, and local energy and feedstock prices. On-site SMRs dominate until about 2025, and after that central biomass and coal plants with CCS come in along with pipeline distribution systems. The switch to central plants tends to occur at a lower market penetration for larger cities because the actual hydrogen demand is larger for these cities, while on-site SMRs tend to persist longer in smaller cities.

HYDROGEN SUPPLY PATHWAYS CONSIDERED IN OUR MODEL

Resource	H₂ Production Technology	H₂ Delivery Method
CENTRAL PRODUCTION		
Natural gas	Steam methane reforming (SMR)	Liquid H₂ truck Compressed gas truck H₂ gas pipeline
Coal	Coal gasification with carbon capture and sequestration	
Biomass (agricultural, forest and urban wastes)	Biomass gasification	
ON-SITE PRODUCTION (at refueling station)		
Natural gas	Steam methane reforming (SMR)	n/a
Electricity (from various electric generation resources)	Water electrolysis	

HYDROGEN INFRASTRUCTURE CAPITAL COSTS TO 2030

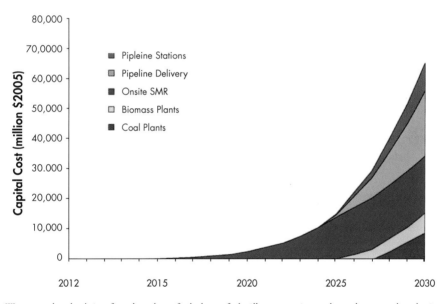

We assume that the choice of supply pathway for hydrogen fuel will vary over time as demand grows and production scales up. On-site SMRs dominate until about 2025, and after that central biomass and coal plants (with CCS) come in along with pipeline distribution systems. Source: J. Ogden and C. Yang, "Build-up of a Hydrogen Infrastructure in the U.S.," Chapter 15 in M. Ball and M. Wietschel, The Hydrogen Economy: Opportunities and Challenges *(Cambridge, UK: Cambridge University Press, 2009).*

Strategies for initiating a hydrogen infrastructure

We have just sketched how infrastructure might be built in the United States assuming that fuel cell vehicles are successful in the marketplace. But the early stages of infrastructure development are still a major hurdle. How can we begin a transition to hydrogen? Consumers will not buy the first hydrogen cars unless they can refuel them conveniently and travel to key destinations, and fuel providers will not build an early network of stations unless there are cars to use them. Major questions include how many stations to build, what type of stations to build, and where to locate them. Key concerns are cost, fuel accessibility, customer convenience, the quality of the refueling experience, network reliability, and technology choice.

Automakers seek a convenient, reliable refueling network, recognizing that a positive customer experience is largely dependent on making hydrogen refueling just as convenient as refueling gasoline vehicles. Energy suppliers are concerned about the cost of building the first stages of hydrogen infrastructure when stations are small and under-utilized. Installing a large number of stations for a small number of vehicles might solve the problem of convenience but would be prohibitively expensive. Energy suppliers are also concerned about how long it would take for hydrogen to reach competitive costs with gasoline and how to endure through the early phase of uncompetitive stations to a viable business case.

A series of studies by STEPS researchers[22] analyzed how many stations would be needed for consumer convenience (defined as travel time to the station), and used spatial analysis tools to estimate where stations would be located. Based on studies of four urban areas in California, Nicholas et al. found that a strategically sited hydrogen network could provide an acceptable level of convenience if only 10 to 30 percent of gas stations offered hydrogen.[23]

TRAVEL TIME TO REACH A HYDROGEN STATION AS A FUNCTION OF NUMBER OF STATIONS IN AN URBAN AREA

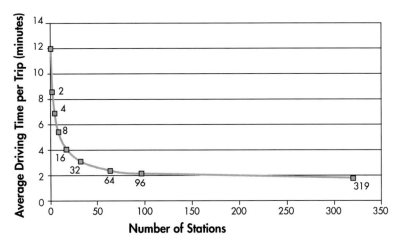

Not every existing fueling station in an urban area would need H_2 in order to provide convenience. Average driving time from home to an H_2 station goes down fast as H_2 becomes available at a relatively small fraction of existing stations. Source: M. Nicholas, S. Handy, and D. Sperling, "Usng Geographic Information Systems to Evaluate Siting and Networks of Hydrogen Stations," Transportation Research Record 1880 (2004): 126–34.

Later we explored a "cluster strategy" for introducing hydrogen vehicles and refueling infrastructure in southern California over the decade from 2010 to 2020 to satisfy California's zero-emission vehicle regulation. Clustering refers to coordinated introduction of hydrogen vehicles and refueling infrastructure in a few focused geographic areas such as smaller cities (like Santa Monica and Irvine) within a larger region (for instance, the Los Angeles Basin). We analyzed several transition scenarios for introducing hundreds to tens of thousands of vehicles and 8 to 40 stations, considering station placement, convenience of the refueling network (for both local—home to station—and regional travel), type of hydrogen supply, and economics (capital and operating costs of stations, hydrogen cost).

A cluster strategy provides good convenience and reliability with a small number of strategically placed stations, reducing infrastructure costs. (Clustering enables the average FCV driver to reach a hydrogen station in about 4 minutes, even with a sparse network of 16 stations. In rollout plans without clustering the average travel time for a 16-station network was 16 minutes.[24]) A cash flow analysis estimates infrastructure investments of $120–170 million might be needed to build a network of 42 stations serving the first 25,000 vehicles. As more vehicles are introduced, the network expands, larger stations are built, and the cost of hydrogen becomes competitive on a cents-per-mile basis with gasoline.

STRATEGIES FOR EARLY H_2 INFRASTRUCTURE

	2009-2011	**2012-2014**	**2015-2017**
	636 FCVs	3442 FCVs	25,000 FCVs
#Stations	8	20	42
#clusters	4 (2sta/cluster)	6 (3 sta/cluster)	12 (3 sta/cluster)
Connect.sta	0	2	6
Station Mix	4 Portable refuelers	8 Portable Refuelers	10 Portable Refuelers
	4 SMRs (100 kg/d)	12 SMRs (250 kg/d)	12 SMRs (250 kg/d)
			20 SMRs (1000 kg/d)
New Equip. Added	4 Portable refuelers	4 Portable Refuelers	2 Portable Refuelers
	4 SMRs (100 kg/d)	12 SMRs (250 kg/d)	20 SMRs (250 kg/d)
Capital Cost	$20 million	$52 million	$98 million
O&M Cost	3-5 $million/y	11-14 $million/y	30-40 $million/y
H_2 Cost $/kg	77	37	13
Ave travel time	3.9 minutes	2.9 minutes	2.6 minutes
Diversion time	5.6 minutes	4.5 minutes	3.6 minutes

Clustering is a good strategy for early H_2 infrastructure. Here is one plan for H_2 station build-out in southern California. Source: M. A. Nicholas and J. M. Ogden, "An Analysis of Near-Term Hydrogen Vehicle Rollout Scenarios for Southern California," UCD-ITS-RR-10-03 (Institute of Transportation Studies, University of California, Davis, 2010).

HOME REFUELING STRATEGIES FOR HYDROGEN VEHICLES

In contrast to the early infrastructure build-out strategies discussed above—which rely on a network of public stations, whose high cost and low utilization are discouraging to private investment—we have also explored the use of home and neighborhood refueling strategies as paths toward commercializing FCVs. In particular, we have assessed "tri-generation" systems, which are energy systems designed to meet the three energy needs of a typical household—electricity, heat, and transportation fuel. Current tri-generation technologies produce hydrogen by reforming natural gas. The economics of hydrogen refueling can be improved by co-producing electricity and heat. Home and neighborhood refueling both potentially offer convenience along with early availability of hydrogen fuel with less investment than a dedicated hydrogen station network.[25]

We developed an interdisciplinary framework and an engineering-economic model to evaluate the economic and environmental performance of tri-generation systems for home and neighborhood refueling. Based on near-term projections for system cost and performance, our model shows that residential tri-generation systems can become economically competitive, especially in regions with low natural gas prices and high electricity prices. In future work, we will examine neighborhood refueling concepts and tri-generation systems based on electrolyzers.

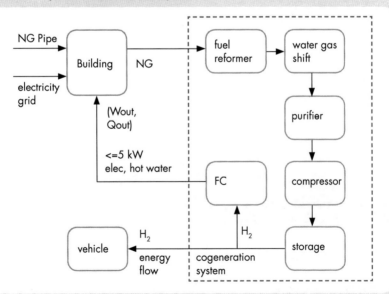

A typical tri-generation system simultaneously provides home electricity and heat along with hydrogen for a vehicle.

Regional hydrogen supply case studies

There are many options for hydrogen supply, and the lowest-cost design could vary by region. In this regard, hydrogen is more like electricity (which relies on regional primary energy sources) than like gasoline. To better understand the diversity of possible solutions for hydrogen supply in the United States, STEPS researchers have pioneered the use of engineering-economic models coupled with spatial information (GIS data) and optimization techniques. These models provide insight into the design, cost, and extent of regional hydrogen infrastructure. Unlike the SSCHISM model, which examines infrastructure for individual cities, these models let us evaluate whether economies of scale (and lower costs) can be achieved more quickly when infrastructure is designed for large regions encompassing multiple cities.

Coal with carbon capture and sequestration has been identified as one of the lowest-cost low-carbon, long-term hydrogen supply pathways.[26] To examine the regional deployment of centralized coal-based hydrogen infrastructure in the United States, we used regional spatial data to estimate the location and magnitude of demand and to identify potential locations for H_2 production facilities, CO_2 storage sites, and distribution networks for both H_2 and CO_2. We also used a network optimization tool to identify the lowest-cost infrastructure design for meeting demand at several market penetration levels. We evaluated both steady-state and dynamic deployment scenarios.[27]

In the steady-state scenarios, infrastructure is optimized independently for demand at different FCV market penetration levels ranging from 5 percent to 75 percent. Each design is independent of the others and represents a snapshot in time. A steady-state analysis for the state of Ohio, for example, indicates that a regional perspective lowers the levelized cost of hydrogen relative to models that examine individual cities.[28]

HYDROGEN INFRASTRUCTURE DESIGNS FOR OHIO BASED ON COAL WITH CCS

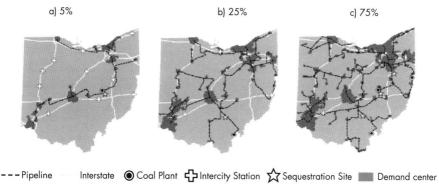

Our steady-state analysis for the state of Ohio came up with these optimal infrastructure designs at 5 percent, 25 percent, and 75 percent market penetration by FCVs. Source: N. Johnson, C. Yang, and J. Ogden, "A GIS-based Assessment of Coal-Based Hydrogen Infrastructure Deployment in the State of Ohio," International Journal of Hydrogen Energy *33 (2008): 5287–303.*

Modeling infrastructure deployment at the regional level allows for demand to be aggregated and economies of scale in production and distribution to be achieved at lower market penetration levels. At this level, pipeline delivery costs less than truck delivery, with the levelized cost of hydrogen delivered via pipeline ranging from $3.20/kg at 5-percent market penetration to $2.20/kg at 75-percent market penetration. However, the steady-state analysis assumes that the infrastructure is fully utilized and consequently does not account for the underutilization of capital that would occur during a transition. For this reason, steady-state models tend to underestimate the cost of hydrogen.

To address this issue, we conducted dynamic modeling of infrastructure deployment in which infrastructure is built over time to meet a growing demand and the timing of investments is tracked. This model accounts for underutilization of capital as large infrastructure investments are made to meet anticipated demand levels. For hydrogen infrastructure with pipeline distribution in Ohio, the levelized cost of hydrogen ranges from $4.30/kg at 5-percent market penetration to $2.70/kg at 75-percent market penetration. These costs represent a 20-percent to 35-percent increase in the cost of hydrogen compared with the results of the steady-state model.

We conducted similar dynamic modeling for California in which hydrogen infrastructure deployment with pipeline and liquid H_2 truck distribution was compared over a 30-year planning period.[29] This study found that truck distribution is competitive with pipelines in the first ten years (1-percent to 14-percent market penetration) since truck transport is less capital-intensive than pipelines and thus is impacted less by underutilization of capital. However, once the infrastructure becomes well utilized in later time periods, pipelines achieve better economies of scale since this mode is dominated by annual operating costs (for example, electricity for H_2 liquefaction and diesel for trucks). CCS represents a very small portion of the total infrastructure costs (less than 3 percent).

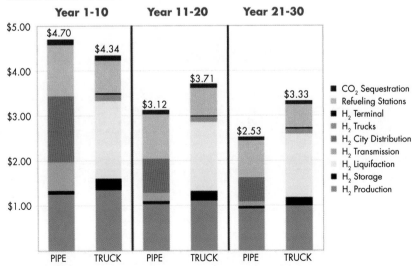

COSTS FOR HYDROGEN DELIVERY IN CALIFORNIA BASED ON COAL WITH CCS

Our dynamic modeling for California came up with these levelized costs for hydrogen delivered via pipeline and truck over a 30-year planning period.

Biomass hydrogen is another promising low-carbon pathway. STEPS researchers examined the possibility of using agricultural wastes to make biomass hydrogen in California, a region with an emphasis on renewable hydrogen.[30] He found that under certain circumstances it would be possible to reduce the costs of biomass hydrogen through optimal location of production plants and design of delivery systems. His best designs yielded delivered hydrogen costs of $3.5–4/kg, competitive with on-site natural gas reforming. The choice of delivery mode (pipeline vs. truck) depended on the market fraction and the type of waste (dense versus more dispersed).

A HYDROGEN INFRASTRUCTURE DESIGN FOR CALIFORNIA BASED ON BIOMASS

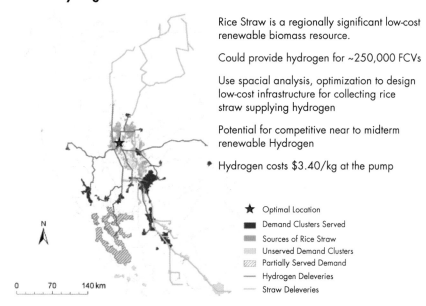

Parker et al. examined the possibility of using rice straw to make biomass hydrogen in California and found that under certain circumstances it would be possible to reduce the costs of biomass hydrogen through optimal location of production plants and design of delivery systems. Source: N. Parker, Y. Fan, and J. Ogden, "From Waste to Hydrogen: An Optimal Design of Energy Production and Distribution Network," Transportation Research E: Logistics 46 (2010): 534–45.

In summary, these regional case studies offer insight into what hydrogen infrastructure might look like in a specific region and illustrate the geographically specific nature of hydrogen supply design in the United States. As with the U.S. electricity system, it is likely that hydrogen will be produced from a variety of feedstocks. Regional case studies allow decision makers to assess the magnitude of required infrastructure and quantify the investments required to make it happen. These case studies can also be used to explore how and why these investments might differ between geographic regions.

Policies and Business Strategies Needed to Support Hydrogen

The results presented in this chapter (and in Chapter 9), as well as those of several recent studies,[31] indicate that the costs to buy down FCVs to market-clearing levels (through technological learning and mass production) and build the associated infrastructure might be tens of billions of dollars, spent over the course of one to two decades. The majority of the cost will be associated with early FCVs, and a lesser amount with early infrastructure. It is almost certain that government policy will be needed to bring these technologies to cost-competitive levels.

How might policy and business strategy support the future of hydrogen in the energy system? Since the start of the 21st century, the vision of hydrogen-fueled transportation has received attention from policymakers and industry worldwide, with investments of billions of dollars in public and private funds.[32] Eighteen countries have national programs to develop hydrogen energy; in North America, more than 30 U.S. states and several Canadian provinces have announced regional "roadmaps" or "hydrogen highways."[33] Automakers and energy companies like Shell and Total are working with governments to introduce the first fleets of hydrogen vehicles and refueling mini-networks in Europe (notably Germany and Norway), Japan, Korea, and the United States (notably California, Hawaii and New York[34]).

However, while there is a growing imperative for alternative fuels driven by concerns about oil supply, rising fuel costs, and climate change, and the search by politicians for a quick technical fix, the context for considering future alternative fuels is dynamic and uncertain. In the early 2000s, hydrogen and fuel cells were widely seen as the endgame. Over the past few years, though, it has become apparent that hydrogen infrastructure will take more time to develop and implement than was previously assumed. Meanwhile, technical progress continues in a variety of other alternative-fuel and efficient-vehicle technologies that are nearer term and/or more compatible with the existing energy system, especially liquid biofuels and plug-in hybrid electric vehicles. Still hydrogen fuel cell vehicles are moving forward rapidly, and several automakers plan to commercialize hydrogen fuel cell vehicles around 2015, just a few years after battery cars, which are making their initial appearance now. Hydrogen and fuel cells are part of a technical progression, building on efficiency and increasing electrification of cars that encompasses hybrid electric drive trains, plug-in hybrids, and improved batteries.

Hydrogen should be seen as one aspect of a broad move toward lower-carbon energy. To realize hydrogen's full benefits will require making hydrogen from domestic and widely available zero-carbon or decarbonized primary energy supplies. Hydrogen can benefit from ongoing efforts to develop biomass and coal gasification with carbon sequestration for electric power, as well as renewable energy sources such as wind and solar.

Finally, public policy is needed to move toward a goal of zero-emission, low-carbon transportation with diversification away from oil-derived transportation fuels. This calls for a comprehensive strategy, based on developing and encouraging the use of clean, efficient internal combustion engine vehicles in the near term, coupled with a long-term strategy supporting the introduction and scale-up of advanced transportation technologies including hydrogen and fuel cells, advanced batteries, and biofuels.

Summary and Conclusions

- Hydrogen fuel cell vehicles are making rapid progress; it appears likely that they will meet their technical and cost goals and could be commercially ready by 2015. Hydrogen infrastructure technologies are also progressing, and the technology to produce natural-gas-based hydrogen is commercial today. In the near term (up to 2025), hydrogen fuel will likely be produced from natural gas, via distributed production at refueling stations, or, where available, excess industrial or refinery hydrogen. Beyond 2025, central production plants with pipeline delivery will become economically viable in urban areas and regionally, and low carbon hydrogen sources such as renewables and fossil with CCS will be phased in.

- The environmental impacts of hydrogen fuel vary with the production pathway. For the near term, FCVs using hydrogen made from natural gas would reduce well-to-wheels GHG emissions by about half compared to current gasoline vehicles. Future hydrogen production technologies could virtually eliminate GHG emissions. On the other hand, important constraints on use of land, water, and materials required by the hydrogen pathway are not well understood. This is a key area for future work under the STEPS program.

- Building a hydrogen infrastructure will be a decades-long process in concert with growing vehicle markets. We have modeled infrastructure deployment in individual "lighthouse" cities as well as at the regional level. Since it is likely that hydrogen will be produced from a variety of feedstocks, optimal supply strategies will differ between geographic regions.

- When FCVs are mass marketed and sold to consumers in 2015 or soon after, hydrogen must make a major leap to a commercial fuel available initially at a small network of refueling stations and must be offered at a competitive price. The first steps are providing hydrogen to test fleets and demonstrating refueling technologies in mini-networks. Several such projects are now underway in Germany, Japan, and North Anerica. Learning from these programs will include development of safety codes and standards. If strategically placed, these early sparse networks could provide good fuel accessibility for early users, while forming a seedbed for a large scale hydrogen infrastructure rollout after 2015.

- Getting through the transition to hydrogen will involve significant costs and some technological and investment risks. Concentrating hydrogen projects in key regions like southern California will focus efforts, lower investment costs to make refueling available to consumers, and hasten infrastructure cost reductions through faster market growth and economies of scale.

- Even under optimistic assumptions, it will be several decades before FCV technologies can significantly reduce emissions and oil use globally, because of the time needed for new vehicle technology to gain major fleet share. Beyond this, hydrogen can yield significant benefits, greater than those possible with efficiency alone. This underscores the importance of providing consistent support for hydrogen and fuel cell vehicle technologies as they approach commercial introduction, so they can progress more quickly to scale, yielding competitive costs and greater societal benefits.

Notes

1. K. Wipke, D. Anton, S. Sprik, Evaluation of Range Estimates for Toyota FCHV-adv Under Open Road Driving Conditions, Savannah River National Laboratory Report, SRNS-STI-2009-00446, August 10, 2009, available from: http://www.cleancaroptions.com/Toyota_431_mile_range.pdf
2. See National Research Council, National Academy of Engineering, Committee on Alternatives and Strategies for Future Hydrogen Production and Use, *The Hydrogen Economy: Opportunities, Costs, Barriers, and R&D Needs* (Washington, DC: National Academies Press, 2004), available from http://www.nap.edu/catalog.php?record_id=10922; National Research Council, Committee on Assessment of Resource Needs for Fuel Cell and Hydrogen Technologies, *Transitions to Alternative Transportation Technologies: A Focus on Hydrogen* (Washington, DC: National Academies Press, 2008), available from http://www.nap.edu/catalog.php?record_id=12222; D. Gielen and G. Simbolotti, *Prospects for Hydrogen and Fuel Cells* (Paris, France: OECD/IEA, International Energy Agency Publications, 2005); M. Ball and M. Wietschel, *The Hydrogen Economy: Opportunities and Challenges* (Cambridge, UK: Cambridge University Press, 2009); A. Rousseau and P. Sharer, "Comparing Apples to Apples: Well-to-Wheel Analysis of Current ICE and Fuel Cell Vehicle Technologies," 2004-01-1015 (Argonne National Laboratory, 2004); M. A. Weiss, J. B. Heywood, A. Schafer, and V. K. Natarajan, "Comparative Assessment of Fuel Cell Cars," MIT LFEE 2003-001 RP (MIT Laboratory for Energy and the Environment, 2003); M. A. Kromer and J. B. Heywood, *Electric Powertrains: Opportunities and Challenges in the U.S. Light-Duty Vehicle Fleet*, LEFF 2007-02 RP (Sloan Automotive Laboratory, MIT Laboratory for Energy and the Environment, May 2007), http://web.mit.edu/sloan-auto-lab/research/beforeh2/files/kromer_electric_powertrains.pdf; A. Bandivadekar, K. Bodek, L. Cheah, C. Evans, T. Groode, J. Heywood, E. Kasseris, M. Kromer, and M. Weiss, *On the Road in 2035: Reducing Transportation's Petroleum Consumption and GHG Emissions* (MIT Laboratory for Energy and the Environment, 2008); S. Plotkin and M. Singh, "Multi-Path Transportation Futures Study: Vehicle Characterization and Scenario Analyses (DRAFT)" (Argonne National Laboratory, June 24, 2009); EUCAR (European Council for Automotive Research and Development), CONCAWE, and ECJRC (European Commission Joint Research Centre), *Well-to-Wheels Analysis of Future Automotive Fuels and Powertrains in the European Context*, Well-to-Wheels Report, Version 2c, March 2007, available at http://ies.jrc.ec.europa.eu/our-activities/support-for-eu-policies/well-to-wheels-analysis/WTW.html.
3. For statements from automakers about their HFCV commercialization timelines, see http://www.fuelcells.org/automaker_quotes.pdf. See also the 2009 USCAR (United States Council for Automotive Research) white paper "Hydrogen Research for Transportation: The USCAR Perspective," available at http://www.uscar.org/guest/article_view.php?articles_id=312.
4. M. A. Weiss, J. B. Heywood, A. Schafer, and V. K. Natarajan, "Comparative Assessment of Fuel Cell Cars," MIT LFEE 2003-001 RP (MIT Laboratory for Energy and the Environment, 2003); M. A. Kromer and J. B. Heywood, *Electric Powertrains: Opportunities and Challenges in the U.S. Light-Duty Vehicle Fleet*, LEFF 2007-02 RP (Sloan Automotive Laboratory, MIT Laboratory for Energy and the Environment, May 2007), http://web.mit.edu/sloan-auto-lab/research/beforeh2/files/kromer_electric_powertrains.pdf.
5. Kromer and Heywood, *Electric Powertrains*.
6. K. Wipke, S. Sprik, J. Kurtz, and T. Ramsden, "Controlled Hydrogen Fleet and Infrastructure Analysis," 2010 DOE Annual Merit Review and Peer Evaluation Meeting, Washington, DC, June 10, 2010, http://www.nrel.gov/hydrogen/pdfs/tv001_wipke_2010_o_an8.pdf.
7. D. Papageorgopoulos, "Fuel Cells," 2010 Annual Merit Review and Peer Evaluation Meeting, Washington, DC, June 8, 2010, http://www.hydrogen.energy.gov/pdfs/review10/fc00a_papageorgopoulos_2010_o_web.pdf.
8. S. Satyapal Hydrogen Fuel Cells Program Overview2011 Annual Merit Review and Peer Evaluation Meeting May 9, 2011. available at: http://www.hydrogen.energy.gov/pdfs/review11/pl003_satyapal_joint_plenary_2011_o.pdf
9. Wipke et al., "Controlled Hydrogen Fleet and Infrastructure Analysis."
10. N. Stetson, "Hydrogen Storage, 2010 Annual Merit Review and Peer Evaluation Meeting (8 June 2010) available at: http://www.hydrogen.energy.gov/pdfs/review10/st00a_stetson_2010_o_web.pdf.
11. National Research Council, *Transitions to Alternative Transportation Technologies*.
12. Kromer and Heywood, *Electric Powertrains; National Research Council, Transitions to Alternative Transportation Technologies*.
13. National Research Council, Committee on Assessment of Resource Needs for Fuel Cell and Hydrogen Technologies, *Transitions to Alternative Transportation Technologies: A Focus on Hydrogen* (Washington, DC: National Academies Press, 2008), available from http://www.nap.edu/catalog.php?record_id=12222.
14. Ibid.

15. For comparison see hydrogen cost estimates from the U.S. Department of Energy, http://www.hydrogen.energy.gov/pdfs/progress10/vii_10_ruth.pdf.
16. Adapted from C. Yang, "Hydrogen and Electricity: Parallels, Interactions, and Convergence," *International Journal of Hydrogen Energy* 33 (2008): 1977–94.
17. G. Wang, J. Ogden, and D. Sperling, "Comparing Air Quality Impacts of Hydrogen and Gasoline," *Transportation Research Part D: Transport and Environment* 13 (2008): 436–48.
18. Y. Sun, J. Ogden, and M. Delucchi, "Societal Life-cycle Buy-down Cost of Hydrogen Fuel Cell Vehicles," accepted for publication in the *International Journal of Hydrogen Energy*, 2010.
19. C. W. King and M. E. Webber, "Water Use and Transportation," *Environmental Science and Technology* 42 (2008): 7866–872.
20. M. Mintz, "Hydrogen Delivery Infrastructure Analysis," U.S. Department of Energy Hydrogen Program *FY 2008 Annual Progress Report*, 368–71; J. Schindler, "E3 Database—A Tool for the Evaluation of Hydrogen Chains," L-B-Systemtechnik GmbH Ottobrunn, presentation at the IEA Workshop, March 22, 2005; C. Yang and J. Ogden, "U.S. Urban Hydrogen Infrastructure Costs Using the Steady State City Hydrogen Infrastructure System Model (SSCHISM)," presented at the 2007 National Hydrogen Association Meeting, San Antonio, TX, March 18–22, 2007; C. Yang and J. Ogden, "Determining the Lowest-cost Hydrogen Delivery Mode," *International Journal of Hydrogen Energy* 32 (2007): 268–86; B. D. James and J. Perez, "Hydrogen Infrastructure Pathways Analysis Using HYPRO," poster presented at the NHA Meeting, San Antonio, TX, March 18–22, 2007; D. L. Greene, P. N. Leiby, B. James, J. Perez, M. Melendez, A. Milbrandt, S. Unnasch, M. Hooks, "Analysis of the Transition to Hydrogen Fuel Cell Vehicles and the Potential Hydrogen Energy Infrastructure Requirements," ORNL/TM-2008/30 (Oak Ridge National Laboratory, March 2008), http://www-cta.ornl.gov/cta/publications/reports/ornl_tm_2008_30.pdf.
21. C. Yang and J. Ogden, "U.S. Urban Hydrogen Infrastructure Costs Using the Steady State City Hydrogen Infrastructure System Model (SSCHISM)," presented at the 2007 National Hydrogen Association Meeting, San Antonio, TX, March 18–22, 2007. A beta copy of the model is posted on Christopher Yang's website at UC Davis Institute of Transportation Studies, available at www.its.ucdavis.edu/people.
22. M. Nicholas, S. Handy, and D. Sperling, "Using Geographic Information Systems to Evaluate Siting and Networks of Hydrogen Stations," *Transportation Research Record* 1880 (2004), 126–34; M. A. Nicholas and J. M. Ogden, "Detailed Analysis of Urban Station Siting for California Hydrogen Highway Network," *Transportation Research Record* 1983 (2007): 121–28; M. Nicholas, "The Importance of Interregional Refueling Availability to the Purchase Decision," UCD-ITS-WP-09-01 (Institute of Transportation Studies, University of California, Davis, 2009), http://pubs.its.ucdavis.edu/publication_detail.php?id=1269; M. Nicholas, "Driving Demand: What Can Gasoline Refueling Patterns Tell Us About Planning an Alternative Fuel Network?" *Journal of Transport Geography*, in press, 2010; M. A. Nicholas and J. M. Ogden, "An Analysis of Near-Term Hydrogen Vehicle Rollout Scenarios for Southern California," UCD-ITS-RR-10-03 (Institute of Transportation Studies, University of California, Davis, 2010).
23. Nicholas, Handy, and Sperling, "Using Geographic Information Systems to Evaluate Siting and Networks of Hydrogen Stations"; Nicholas and Ogden, "Detailed Analysis of Urban Station Siting for California Hydrogen Highway Network."
24. Ibid.
25. X. Li and J. Ogden, "Understanding the Design and Economics of Distributed Tri-generation Systems for Home and Neighborhood Refueling," presented at the 2010 National Hydrogen Association Meeting, Long Beach, CA, May 3–6, 2010; X. Li, J. Ogden, and K. Kurani, "An Overview of Automotive Home and Neighborhood Refueling," *Proceedings of the 24th Electric Vehicle Symposium*, Stavanger, Norway, May 13–16, 2009.
26. National Research Council, *The Hydrogen Economy*; National Research Council, *Transitions to Alternative Transportation Technologies*.
27. N. Johnson and J. Ogden, "Moving Towards a National Assessment of Coal-Based Hydrogen Infrastructure Deployment with Carbon Capture and Sequestration," presented at the Seventh Annual Conference on Carbon Capture and Sequestration, Pittsburgh, PA, May 5–8, 2008; N. Johnson, C. Yang, and J. Ogden, "Build-Out Scenarios for Implementing a Regional Hydrogen Infrastructure," presented to the National Hydrogen Association, Long Beach, CA, March 11–16, 2006; N. Johnson, C. Yang, and J. Ogden, "A Blueprint for the Long-Term Deployment of Hydrogen Infrastructure in California," presented to the National Hydrogen Association, Sacramento, CA, March 30–April 4, 2008; N. Johnson, C. Yang, and J. Ogden, "A GIS-Based Assessment of Coal-Based Hydrogen Infrastructure Deployment in the State of Ohio," *International Journal of Hydrogen Energy* 33 (2008): 5287–303; N. Johnson, C. Yang, and J. Ogden, "A Regional Model of Coal-Based Hydrogen Infrastructure Deployment with Carbon Capture and Storage (CCS)," presented at the Pittsburgh Coal Conference, Pittsburgh, PA, September 29–October 2, 2008.

28. Johnson, Yang, and Ogden, "GIS-based Assessment of Coal-Based Hydrogen Infrastructure Deployment."
29. Ibid.
30. N. Parker, Y. Fan, and J. Ogden, "From Waste to Hydrogen: An Optimal Design of Energy Production and Distribution Network," *Transportation Research Part E: Logistics* 46 (2010): 534–45.
31. D. Greene, P. Leiby, and D. Bowman, *Integrated Analysis of Market Transformation Scenarios with HyTrans*, ORNL/TM-2007/094 (Oak Ridge National Laboratory, June 2007); S. Gronich, "Hydrogen and FCV Implementation Scenarios, 2010–2025," U.S. DOE Hydrogen Transition Analysis Workshop, Washington, DC, August 9–10, 2006, http://www1.eere.energy.gov/hydrogenandfuelcells/analysis/scenario_analysis_mtg.html; Z. Lin, C.-W. Chen, J. Ogden, and Y. Fan, "The Least-Cost Hydrogen for Southern California," *International Journal of Hydrogen Energy* 33 (2008): 3009–3014; Gielen and Simbolotti, Prospects for Hydrogen and Fuel Cells.
32. See National Research Council, Committee on Assessment of Resource Needs for Fuel Cell and Hydrogen Technologies, *Transitions to Alternative Transportation Technologies: A Focus on Hydrogen* (Washington, DC: National Academies Press, 2008), available from http://www.nap.edu/catalog.php?record_id=12222. This report estimates a current RD&D funding level for hydrogen and fuel cells in the U.S. private sector of about $700 million per year. Globally, the total is several times this amount.
33. See the International Partnership for Hydrogen and Fuel Cells in the Economy, http://www.iphe.net, and *State Activities That Promote Fuel Cells and Hydrogen Infrastructure Development* (Washington, DC: Breakthrough Technologies Institute, 2006), http://www.fuelcells.org/info/StateActivity.pdf.
34. See these documents from the California Fuel Cell Partnership: "Hydrogen Fuel Cell Vehicle and Station Deployment Plan, Action Plan" (February 2009), available at http://www.cafcp.org/sites/files/Action%20Plan%20FINAL.pdf, and "Hydrogen Fuel Cell Vehicle and Station Deployment Plan, Progress and Next Steps" (April 2010), available at http://www.cafcp.org/sites/files/FINALProgressReport.pdf.

Part 2: Pathway Comparisons

Analyzing single-fuel pathways has given us a basis for comparing these pathways in terms of how well they promise to meet important objectives for the transportation system of the future. The four chapters in this section take a comparative approach to fuel economy and cost, fuel infrastructure requirements, and environmental impacts.

- **Chapter 4** focuses on the question of how much each pathway promises to trim fuel consumption relative to today's conventional vehicles. It also asks how the cost to the consumer of various types of vehicles will compare some number of years in the future, particularly when the price of gasoline is factored in. To begin to answer these questions, the researchers ran computer simulations of the operation of a midsize passenger car and a small/compact SUV at three points in the future: 2015, 2030, and 2045. They compared advanced higher-efficiency engines, hybrid-electric vehicles, and all-electric vehicles with a conventional vehicle marketed in 2007.

- **Chapter 5** looks at the infrastructure development required if biofuels, electricity, and/or hydrogen are to assume major roles as transportation fuels over the next several decades. It also examines the challenges—given that today's transportation system is 97-percent dependent on petroleum-based liquid fuels. For each fuel pathway, the chapter considers and compares system design, resources, technology status, cost, reliability, transition barriers, and policies that might be needed to provide incentives for new infrastructure development.

- **Chapter 6** considers the matter of reducing greenhouse gas (GHG) emissions from vehicles and fuels, one key to lessening transportation's contribution to the climate change problem. This chapter presents much of what is known about the relative emissions of GHGs from battery, fuel cell, and plug-in hybrid electric vehicles versus conventional internal combustion engine vehicles. It provides background on the issue of GHG emissions and their climate impact, reviews and compares recent estimates of GHG emissions from the fuel cycles of various types of electric vehicles, and examines the potential for electric vehicles to rapidly scale up to meet the climate challenge.

- **Chapter 7** considers the environmental impact of transportation fuels and vehicles beyond GHG emissions—impacts on land, water, and materials. Biofuel and oil production in particular can result in land-use impacts that must be acknowledged and weighed. Production of fossil fuels, biofuels, electricity, and hydrogen all have water footprints that must be considered. And advanced vehicle technologies use materials that might become a barrier to development if they are either scarce or else concentrated in a few countries. This chapter focuses on work comparing the sustainability of different fuel/vehicle pathways along these lines.

Chapter 4:
Comparing Fuel Economies and Costs of Advanced vs. Conventional Vehicles

Andrew Burke, Hengbing Zhao, and Marshall Miller

A key question in comparing advanced and conventional vehicles is how much of a reduction in fuel consumption we can expect from new technologies. One approach to answering this question is to run computer simulations of the operation of advanced vehicles on different driving cycles using the best component models available and control strategies intended to maximize the driveline efficiency. In these simulations we can vary the vehicle and component characteristics to reflect projected improvements in technologies in the future.

This chapter describes simulations run for a midsize passenger car and a small/compact SUV for the time period 2015 to 2045. The baseline vehicle is a conventional vehicle marketed in 2007. Technologies we compared are advanced, higher-efficiency engines, hybrid-electric vehicles, and electric-drive battery and fuel cell-powered vehicles. We present the results of our simulations in terms of the equivalent gasoline consumption of the various vehicle designs and the projected reductions in fuel usage, and we compare our results with those presented in previous studies at MIT, the U.S. Department of Energy (DOE), and the National Research Council (NRC). We also compare the alternative advanced vehicle technologies in terms of their costs relative to conventional and advanced engine/transmission power trains that would be available in the same time periods.

THE SIMULATION TOOLS WE USED

Studies directed toward projecting the performance of vehicles using various advanced power train technologies have been performed at the UC Davis Institute of Transportation Studies since about 2000.[1] A number of computer models have been developed to simulate advanced vehicles.[2] For this chapter, we performed the conventional and hybrid (HEV and PHEV) vehicle simulations using the UC Davis version of ADVISOR,[3] which includes special power train schematic and control strategy files.

The computer program we used to simulate fuel cell vehicles is a modification of the program developed previously at UC Davis[4] that permits scaling of the fuel cell stack and accessories and improves the treatment of system transients, particularly those due to the compressed air system. In addition, we added control strategies using batteries or ultracapacitors that permit operation of the fuel cell in either the load-leveled or power-assist mode. These simulation tools allow us to calculate the fuel consumption of advanced vehicles on various driving cycles.

To the extent possible, the results of the simulation programs have been validated by comparing simulation results for vehicles currently being marketed with EPA dynamometer test data[5] for vehicles using the same power trains / engines. In all cases, the comparisons are reasonable, as shown in the following table.

Model/Year	Engine	Driveline Type	City mpg	Highway mpg
Ford Focus/2010 simulation	Focus	conventional	28	44
EPA test 2007/ Ford Focus	Focus	conventional	30	44
Honda Civic simulation	i-VTEC	conventional	33	45
EPA test 2007/ Honda Civic	i-VTEC	conventional	33	50
Honda Civic simulation	i-VTEC	hybrid	56.5	62.5
EPA test 2007/ Honda Civic	i-VTEC	hybrid	54.4	65.4
Toyota Prius simulation	Atkinson	hybrid	68	67.5
EPA test 2007/Toyota Prius	Atkinson	hybrid	66.6	65.4
EPA test 2007/Honda Accord	4 cyl. 140 kW	conventional	26.6	43.6
EPA test 2007/Toyota Camry	4 cyl. 140 kW	conventional	26.6	42.3

Note: EPA test results from U.S. Department of Energy and U.S. Environmental Protection Agency, "Fuel Economy Guide—2007," are corrected by 1/.9 for the Federal Urban Driving Schedule (FUDS) and 1/.78 for the Federal Highway Driving Schedule (FHWDS) to obtain the dynamometer test data. The .9 and .78 values are the factors used by EPA to relate their test data for the vehicles on the city and highway cycles, respectively, to the fuel economy values given in the Fuel Economy Guide for those test cycles.

Fuel Economy and Energy Savings Simulation Inputs

The primary challenge in simulating vehicle operation is to come up with the vehicle and power train inputs to be used in the simulations. If the inputs are realistic, the simulation results should be a reliable estimate of the performance and fuel consumption of vehicles in the future. There is, of course, considerable uncertainty in the inputs used in any study, particularly regarding when specific improvements in the technologies will be achieved. Thus, results can also be interpreted as representing the fuel savings that would result if vehicles are marketed having the vehicle and power train characteristics assumed in the inputs. This makes the simulation results useful in setting design targets for future development programs for advanced vehicle technologies. Similarly, component costs assumed in our economic estimates are useful as targets for future pricing.

Power train configurations and component characteristics

We compared three types of power trains—conventional internal combustion engine/transmission (ICE), hybrid-electric (HEV and PHEV), and all-electric powered by batteries alone or by a hydrogen fuel cell. The ICE vehicles we studied used an automatically shifted multi-speed transmission with increasing mechanical efficiency; we made no attempt to optimize the transmission gearing or shifting strategy. The efficiency of the transmission was assumed to be a constant value varying from 92 percent in 2015 to 95 percent in 2045.

All the vehicle simulations were performed using gasoline, spark-ignition (SI) engines. The engine characteristics (efficiency maps as a function of torque and RPM) used in the simulations are based on those available in ADVISOR and PSAT (vehicle system modeling tools developed and supported by the National Renewable Energy Laboratory and Argonne National Laboratory, respectively). This included engines currently in passenger cars (such as the Ford Focus engine and the Honda i-VTEC engine) and more advanced engines like those employing an Atkinson cycle (Prius 2004), variable valve timing (An_iVTEC), and direct injection (An_GDi). We increased the maximum engine efficiencies in the simulations for future years based on expected significant improvements in engine efficiencies using upcoming technologies.[6] Modifying the engine maps in this way does not include the effects of changes in the basic shape of the contours of constant efficiency, which would likely show even more drastic increases in efficiency at low engine torque/power. The uncertainty in the engine maps is one of the largest uncertainties in the inputs needed to perform the simulations.

MAP OF THE ADVANCED VTEC ENGINE USED IN THE ICE VEHICLE SIMULATIONS

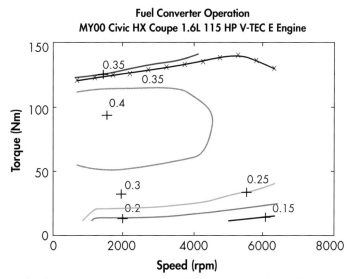

The engines used in the ICE vehicle simulations were scaled from the four-cylinder Honda VTEC engine, for a maximum efficiency of 40 percent (value for 2030).

The electric motor/controller efficiency maps were scaled from the map for the 15 kW permanent magnet AC motor in the hybrid Honda Civic and Accord. The maximum efficiency of these motors is presently quite high—in the 92 to 96 percent range—so large improvements are not expected in future years.

The power trains for all the hybrid vehicles (HEVs and PHEVs) used a single-shaft, parallel arrangement with clutches that permit on/off engine operation at any vehicle speed[7] and the engine to be decoupled and coupled in an optimum manner. The same engine maps and maximum efficiencies were used for the hybrids as for the ICE vehicles. The HEVs operated in the charge-sustaining mode and utilized the "sawtooth" control strategy[8] for splitting the power demand between the engine and the electric motor. This strategy results in the vehicle operating in the electric mode when the power demand is low; when the vehicle power demand is higher, the engine is turned on, providing power to meet the vehicle demand and to recharge the batteries or ultracapacitors. It is likely that engines designed to operate primarily at the high torque conditions, such as the Atkinson cycle engines, will have higher efficiency than the standard designs used in ICE vehicles. The effects of engine redesign have not been included in the present study.

Characteristics of the batteries used in the simulations are shown in the table below. The battery models for the various battery chemistries were based on test data taken in the battery laboratory at UC Davis.[9] Modest improvements in both energy density and resistance are projected in future years.[10] These improvements will result in lower vehicle weight and more efficient power train operation but should not significantly affect the fuel economy projections, as all the batteries used in the simulations have high power capability and thus high round-trip efficiency.

CHARACTERISTICS OF THE BATTERIES USED IN OUR SIMULATIONS

Vehicle Configuration	2015				2030–2045			
	Battery Type	Ah	Wh/kg	Resist. mOhm	Battery Type	Ah	Wh/kg	Resist. mOhm
HEV	Li Titanate	4	35	1.1	Li Titanate	4	42	.9
PHEV-20	Ni MnO2	15	120	1.5	Ni MnO2	15	135	1.3
PHEV-40	Ni MnO2	50	140	.8	Ni MnO2	50	170	.65
FCHEV	Li Titanate	4	35	1.1	Li Titanate	4	42	.9

Notes: Ah = ampere-hour; Wh/kg = watt hours per kilogram; Resist. mOhm = electrical resistance in milliohms.

For the PHEVs, the batteries were sized (in terms of useable kWh) for either a 10–20 mile or a 40–60 mile range with all-electric operation on the Federal Urban Driving Schedule (FUDS) and Federal Highway Driving Schedule (FHWDS) driving cycles in the charge-depleting mode. After the batteries were depleted to their minimum state-of-charge, the PHEVs operated in the charge-sustaining mode using the same sawtooth strategy used for the HEVs. The same single-shaft, parallel hybrid power train arrangement used in the HEVs was used in the PHEVs with the larger battery.

The power train arrangement for the fuel cell-powered vehicles (FCHEVs) consisted of a PEM fuel cell and a lithium-ion battery. The battery is connected to the DC bus by a DC/DC converter that controls the output power of the battery such that the output power of the fuel cell is load leveled.[11] This control strategy greatly reduces the voltage fluctuations of the fuel cell and should significantly increase its life expectancy. The peak efficiency of the fuel cell is increased in future years. The batteries used in the FCHEVs are the same as those used in the HEVs.

The batteries used in the all-electric battery powered vehicles were the same as those used in the PHEV-40. The range of BEVs was about 100 miles (160 km). The characteristics of the mid-size passenger car were selected to give performance similar to the Nissan Leaf. The BEVs with a range of 100 miles are not all-purpose vehicles unless the batteries have fast charge capability of 10 minutes or less.

Vehicle weight and road load characteristics

The most important and uncertain inputs used in the simulations are the vehicle characteristics—weight and road load characteristics (drag coefficient frontal area, and tire rolling resistance). The weight and drag reductions assumed for the future are aggressive. The weights were reduced about 20 percent compared to 2007 models and the drag coefficients were reduced about 25 percent in 2030; hence the fuel consumption reduction projections should be considered to be reasonably optimistic. The tire rolling resistance was assumed to decrease only slightly from a baseline value of .007 due to the need to maintain traction for driving safety. The frontal area of the vehicles was not changed in future years. There is a marked difference in the drag characteristic, C_DA, between the passenger car and the SUV, which will have significant effects on the projected fuel consumption of the two classes of vehicles.

VEHICLE WEIGHT AND DRAG REDUCTIONS PROJECTED FOR ADVANCED ICE VEHICLES

Significant reductions in vehicle weight and drag are assumed for both the passenger car and the SUV. The values used are the same as assumed in S. Plotkin and M. Singh, "Multi-Path Transportation Futures Study: Vehicle Characterization and Scenarios," Argonne Lab and DOE Report (draft), March 5, 2009, and are not much different from those used for 2030 in E. Kasseris and J. Heywood, "Comparative Analysis of Automotive Powertrain Choices for the Next 25 Years," SAE paper 2007-01-1605, 2007. Nevertheless, whether the vehicles in the future will meet these targets for weight and drag reduction remains to be seen.

Year	Midsize Passenger Car		Small/Compact SUV	
	Test weight (kg)	Drag coef. CD	Test weight (kg)	Drag coef. CD
2007–10	1615	.30	1750	.40
2015	1403	.25	1629	.37
2030	1299	.22	1497	.35
2045	1299	.20	1497	.33

Note: Vehicle test weight = curb weight + 136 kg

SUMMARY OF INPUTS USED IN THE VEHICLE SIMULATIONS

Midsize passenger cars
Acceleration performance for all vehicles: 0–60 mph in 9–10 seconds, 0–30 mph in 3–4 seconds

Vehicle Configuration	Parameter	2015	2030	2045
	CD	.25	.22	.20
	AF m2	2.2	2.2	2.2
	Fr	.007	.006	.006
Advanced ICE	Engine kW	105	97	97
	Max. engine efficiency %	39	40	41
	Vehicle test weight (kg)	1403	1299	1299
	DOE mpg FUDS/FHWDS	29/47	33/54	34/57
HEV	Engine kW	73	67	67
	Max. engine efficiency %	39	40	41
	Motor kW	26	24	24
	Battery kWh	1.0	.9	.9
	Vehicle test weight (kg)	1434	1324	1324
	DOE mpg FUDS/FHWDS	73/61	84/82	89/88
PHEV-20	Engine kW	75	69	68
	Motor kW	61	57	57
	Battery kWh	4.0	3.6	3.6
	Vehicle test weight (kg)	1475	1361	1354
PHEV-40	Engine kW	77	71	67
	Motor kW	63	59	59
	Battery kWh	11.1	9.8	9.4
	Vehicle test weight (kg)	1535	1415	1407
FCHEV	Fuel cell efficiency %	60	62	65
	Fuel cell kW	83	76	72
	Motor kW	103	100	99
	Battery kWh	.93	.85	.85
	Vehicle test weight (kg)	1516	1383	1366
	DOE mpg FUDS/FHWDS	70/79	102/114	114/130
BEV	Motor kW	80	72	70
	Battery kWh	24	28	32
	Vehicle curb weight kg	1521	1400	1350

Small/compact SUVs

Acceleration performance for all vehicles: 0–60 mph in 10–11 seconds, 0–30 mph in 3–4 seconds

Vehicle Configuration	Parameter	2015	2030	2045
	CD	.37	.35	.33
	AF m2	2.9	2.94	2.94
	Fr	.0075	.007	.007
Advanced ICE	Engine kW	122	112	112
	Max. engine efficiency %	39	40	41
	Vehicle test weight (kg)	1629	1497	1497
	DOE mpg FUDS/FHWDS	24/34	27/38	28/39
HEV	Engine kW	89	81	81
	Max. engine efficiency %	39	40	41
	Motor kW	31	28	28
	Battery kWh	1.2	1.1	1.1
	Vehicle test weight (kg)	1669	1532	1530
	DOE mpg FUDS/FHWDS	55/46	61/51	63/54
PHEV-20	Engine kW	96	90	89
	Motor kW	66	62	61
	Battery kWh	5.6	5.1	5.0
	Vehicle test weight (kg)	1719	1576	1570
PHEV-40	Engine kW	99	93	91
	Motor kW	69	64	64
	Battery kWh	15.2	14.0	13.5
	Vehicle test weight (kg)	1802	1654	1644
FCHEV	Fuel cell efficiency %	60	62	65
	Fuel cell kW	104	95	92
	Motor kW	129	119	116
	Battery kWh	1.2	1.1	1.1
	Vehicle test weight (kg)	1875	1705	1683
	DOE mpg FUDS/FHWDS	62/59	73/68	82/77

Notes: The first three rows of each table show the road load characteristics: drag coefficient C_D, frontal area A_F in meters squared, and tire rolling resistance F_r.

Vehicle test weight = curb weight + 136 kg

FUDS = Federal Urban Driving Schedule (a driving cycle that simulates city driving) and FHWDS = Federal Highway Driving Schedule (a driving cycle that simulates highway driving); mpg ratings arrived at by the U.S. Department of Energy (DOE) are shown here.

The PHEV-20 has a small battery (25–33 kg, all-electric range or AER of 10–20 mi); the PHEV-40 has a large battery (55–80 kg, AER 40–60 mi); batteries are assumed to be discharged to a 30-percent state-of-charge. Battery kWh refers to the total energy stored in the battery.

Fuel Economy and Energy Savings Simulation Results

The simulation results are shown in the following tables for midsize passenger cars and small/compact size SUVs in 2015, 2030, and 2045, with the corresponding fuel savings (as a percentage) compared to 2007 vehicles for each case. Also shown when they are available are simulation results previously published by the DOE,[12] MIT,[13] and the NRC.[14] In all cases the fuel saving comparisons are made based on the simulation results. It is thus assumed that on a percentage basis, the fuel savings would be the same for actual on-road driving.

The results for vehicles using each type of advanced technology are discussed separately in the following sections.

FUEL ECONOMY SIMULATION RESULTS FOR VARIOUS DRIVING CYCLES

Midsize passenger cars

% Fuel Saved = $(1-(mpg)_0/mpg)) \times 100$, $(mpg)_0 = 34.5$, which is the average of the city and highway dynamometer fuel economy of the 2007 baseline vehicle.

Year	Study By	FUDS mpg	FHWDS mpg	% Fuel Saved	US06 mpg	Accel. 0–30/0–60
Baseline 2007		26	42	0		
Adv. ICE						
2015	UCD	41.4	62.3	33.5	37.5	4.3/9.7
	DOE	29	47	9		
	NRC			29		
2030	UCD	47.4	73.3	42.8	44.0	4.7/10.3
	DOE	33*	54*	20.7		
	MIT	42	68	37.3	44	
2045	UCD	48.9	77.1	45.2	46.1	4.6/10.3
	DOE	34*	57*			
HEV						
2015	UCD	73.3	74.1	53.1	46.5	4.3/9.7
	DOE	73	61	48.5		
	NRC			44		
2030	UCD	85.7	84	59.3	53.7	4.7/10.3
	DOE	84	82	41.6		
	MIT	95	88	62.2	58	
2045	UCD	87.9	89.2	61.0	55.8	4.6/10.3
	DOE	89	88	61.0		
FCHEV						
2015	UCD	82.6	90.8	60.2	61.3	
	DOE	70	79	53.7		
2030	UCD	102.8	111.5	67.8	76.2	
	DOE	102	114	68.1		
2045	UCD	108.9	119.5	69.8	82.3	
	DOE	114	130	71.7		
Battery Electric (BEV)		FUDS Wh/mi/ range	FHWDS Wh/mi/ range	% Fuel Saved (1)	US06 Wh/mi/ range	Accel. 0–30/0–60 mph
2015	UCD	220/ 75mi	206/ 82mi	76.1/40.1	400/ 45mi	3.4/11.1
2030	UCD	198/ 97mi	184/ 104mi	78.6/46.3	365/ 54mi	3.2/10.5
2045	UCD	194/ 122mi	176/ 122mi	79.3/48.0	352/ 63mi	3.1/10.2

(1) gasoline energy / powerplant source energy; 90% charger effic., 40% powerplt. effic.

* The DOE fuel economy values for the Adv. ICEV in 2030 and 2045 do not properly reflect improvements in engine technology and as a result are too low.

Small/compact SUVs

% Fuel Saved = $(1-(mpg)_0/mpg)) \times 100$, $(mpg)_0 = 30$, which is the average of the city and highway dynamometer fuel economy of the 2007 baseline vehicle.

Year	Study By	FUDS mpg	FHWDS mpg	% Fuel Saved	US06 mpg	Accel. 0–30/0–60
Baseline 2007		25	34	0		
Adv. ICE						
2015	UCD	34	44.4	23	27.3	
	DOE	24	34			
2030	UCD	38.9	50.3	33	30.8	
	DOE	27*	38*	8		
2045	UCD	40.2	53	36	32.5	
	DOE	28*	39*	10		
HEV						
2015	UCD	52.7	44.7	39	29.7	
	DOE	54.6	46.4	41		
2030	UCD	58.7	51	45	34	
	DOE	61	51	46		
2045	UCD	61	54.1	48	34.9	
	DOE	63	54	49		
FCHEV						
2015	UCD	61	60	50	40.5	
	DOE	62	59	50		
2030	UCD	74.7	73	59	48.8	
	DOE	73	68	57		
2045	UCD	80.8	78.7	62	52.9	
	DOE	82	77	62		

* *The DOE fuel economy values for the Adv. ICEV in 2030 and 2045 do not properly reflect improvements in engine technology and as a result are too low.*

Notes: FUDS mpg = Federal Urban Driving Schedule mpg; FHWDS mpg = Federal Highway Driving Schedule mpg; US06 mpg = US06 Driving Schedule mpg

Fuel consumption in L/100 km = 238/mpg

Conventional engine/transmission vehicles

The simulation results indicate that large improvements in the fuel economy of conventional midsize passenger cars and compact SUVs can be expected in 2015 to 2020. Further improvements are projected for 2030 and 2045. These improvements relative to 2007 models for midsize cars are 50 percent (2015) to 70 percent (2030) for fuel economy and 33 percent (2015) to 43 percent (2030) for fuel savings. For conventional compact SUVs, the projected improvements in fuel economy are 30 percent (2015) to 49 percent (2030) with fuel savings of

23 percent (2015) to 33 percent (2030). These improvements result from the combined effects of decreases in weight and drag coefficient and increases in engine efficiency. In the table below, it is shown that projected increases in engine efficiency have a considerably larger effect than reductions in weight and drag for both vehicle types.

CHANGES IN FUEL ECONOMY FROM TECH IMPROVEMENTS, ICE VEHICLES

Midsize passenger cars

Technology	2015		2030	
	FUDS mpg	FHWDS mpg	FUDS mpg	FHWDS mpg
2007 engine (baseline)	27	42	28	43
Engine efficiency improvements, but no weight and CD reduction	39	56	42	61
All improvements	43	63	48	72

Small/compact SUVs

Technology	2015		2030	
	FUDS mpg	FHWDS mpg	FUDS mpg	FHWDS mpg
2007 engine (baseline)	22	31	24	32
Engine efficiency improvements, but no weight and CD reduction	30.1	37.4	34.7	43.2
All improvements	34.8	44	38.1	48.1

Hybrid vehicles (HEVs and PHEVs)

This category of advanced technology includes HEVs (gasoline fueled) and PHEVs (wall plug-in electricity and gasoline). Large improvements in the fuel economy of HEVs are projected for both midsize passenger cars and small/compact SUVs, resulting in fuel savings of 50–60 percent for the cars and 40–50 percent for the SUVs compared to the 2007 baseline vehicles. Relatively large fuel economy improvements are projected for HEVs compared to advanced conventional vehicles using the same engine technologies.

IMPROVEMENTS (AS RATIOS) IN THE FUEL ECONOMY OF HEVS COMPARED TO ADVANCED ICE VEHICLES

Technology	2015		2030	
	FUDS	FHWDS	FUDS	FHWDS
Midsize passenger car	1.65	1.15	1.79	1.21
Small/compactSUV	1.55	1.05	1.56	1.06

Two types of PHEVs were simulated—one with a small battery and an all-electric range of 10–20 miles and one with a larger battery and a range of 40–50 miles. There is not a large reduction (only about 15 percent) in electrical energy usage (Wh/mi) in the all-electric mode projected for 2015 to 2045, and the fuel economy of the various vehicle designs in the charge-sustaining mode is similar to the corresponding HEV. As a result, one would expect the energy usage (electricity plus gasoline) of the 10–20 mile PHEV would decrease by a greater fraction in the future than the 40–50 mile PHEV, which would travel a greater fraction of miles on electricity. The split between electricity and gasoline for either vehicle will depend on its usage pattern (average miles driven per day and number of long trips taken).

Assuming for the PHEV-20 and PHEV-40 mid-size car that 20% and 65% of the total annual miles (city plus highway), respectively, are driven on electricity, one can calculate the wall-plug electricity and gasoline used and the total energy (gasoline plus energy needed to generate the electricity) savings. Assuming 15,000 annual miles, a battery charger efficiency of 90%, and a powerplant efficiency of 40%, one calculates the following results for the two PHEVs in 2030 compared to an advanced ICE vehicle. For the PHEV-20, one finds a gasoline saving of 40% and a total energy saving of 26% for the 40% efficient powerplant. The PHEV-20 would use 480 kWh of electricity from the wall-plug. The corresponding values for the PHEV-40 are 75% gasoline savings, 30% total energy savings, and 1538 kWh electricity used from the wall-plug. Note that the total energy savings (gasoline plus that to generate the electricity) are about the same for a 40% efficient powerplant. For a 50% efficient powerplant, the difference in total energy savings is larger being 29% for the PHEV-20 and 39% for the PHEV-40.

PHEV FUEL ECONOMY AND ELECTRICITY USAGE SIMULATION RESULTS

Midsize passenger cars

Year	Driving Cycle	Electric Range mi	Charge-depleting mpg	Charge-depleting Wh/mi (at battery)	Charge-sustaining mpg
PHEV-20					
2015	FUDS	17	All-elec	163	70.0
	FHWDS	17	All-elec	165	69.6
	US06	10	1570	280	45
2030	FUDS	17	3333	143	77
	FHWDS	17	7500	145	84
	US06	11	1500	234	53
2045	FUDS	18	All-elec	140	85.6
	FHWDS	19	All-elec	134	87.8
	US06	11	1400	233	52.8
PHEV-40					
2015	FUDS	46	All-elec	167	69.1
	FHWDS	45	All-elec	171	71.7
	US06	31	800	251	46.2
2030	FUDS	49	All-elec	141	84.6
	FHWDS	48	All-elec	143	86.0
	US06	32	1495	218	54.5
2045	FUDS	49	All-elec	135	87.8
	FHWDS	49	All-elec	134	92.5
	US06	32	1731	205	59

Small/compact SUVs

Year	Driving Cycle	Electric Range mi	Charge-depleting mpg	Charge-depleting Wh/mi (at battery)	Charge-sustaining mpg
PHEV-20					
2015	FUDS	19	All-elec.	213	51.9
	FHWDS	16	All-elec.	257	45.4
	US06	12	379	384	30.6
2030	FUDS	19	All-elec.	192	57.9
	FHWDS	14	All-elec.	255	50.6
	US06	10	525	360	34
2045	FUDS	19	All-elec.	188	62.0
	FHWDS	16	All-elec.	226	53.8
	US06	10	576	348	36.3
PHEV-40					
2015	FUDS	49	All-elec.	218	54.6
	FHWDS	40	All-elec.	266	46.1
	US06	28	547	385	30.7
2030	FUDS	51	All-elec.	192	60.4
	FHWDS	41	All-elec.	239	51.4
	US06	28	781	351	33.9
2045	FUDS	50	All-elec.	188	62.6
	FHWDS	41	All-elec.	230	55.2
	US06	28	879	338	36.5

Notes: FUDS = Federal Urban Driving Schedule; FHWDS = Federal Highway Driving Schedule; US06 = US06 Driving Schedule; Wh/mi = watt hours per mile.

The PHEV-20 has a small battery (25–33 kg, all-electric range or AER of 10–20 mi); the PHEV-40 has a large battery (55–80 kg, AER 40–60 mi).

Electric vehicles (Fuel cell–powered and battery vehicles)

Fuel cell-powered vehicles use hydrogen as the fuel. As with gasoline-fueled hybrids, the batteries are recharged onboard the vehicle from the fuel cell and not from the wall plug. The fuel economies calculated in our simulation for FCHEVs are gasoline equivalent values but are easily interpreted as mi/kg H_2 since the energy in a kilogram of hydrogen is close to that in a gallon of gasoline. Hence the fuel savings shown for the fuel cell vehicles can be interpreted as the fraction of energy saved relative to that in the gasoline used in the baseline 2007 conventional vehicle. Fuel cell technology would thus reduce energy use by 60 percent (2015) to 72 percent (2030) for the midsize passenger car and by 40 percent (2015) to 53 percent (2030) for the compact SUV. This reduction in energy use of the fuel cell vehicles compared to the baseline gasoline vehicle is for tank-to-wheels (TtW). The energy use reduction from the hydrogen production plant-to-wheels (the so-called well-to-wheels reduction) would be less depending on the relative efficiencies of production and distribution of hydrogen and gasoline.

Battery-powered vehicles are recharged with electricity from the wall-plug. The energy use of the BEVs is given as Wh/mi from the battery. The gasoline equivalent can be calculated from $(gal/mi)_{gas.equiv.} = (kWh/mi)/33.7$. The energy saved depends on the battery charging efficiency and the efficiency of the powerplant generating the electricity. For 2030 BEV, the gasoline energy equivalent saved is 79% from the wall-plug and 45% at a 40% efficient powerplant compared to the 2007 baseline ICE mid-size car. Compared to a 2030 HEV, the gasoline equivalent saved is only 47% from the wall-plug and there are no savings at the powerplant until the efficiency of the powerplant exceeds about 55%.

SUMMARY: FUEL SAVINGS FOR THE VARIOUS TECHNOLOGIES

The fuel savings results for the midsize passenger car and the compact SUV compared to the baseline 2007 conventional vehicle are summarized in the table below.

Technology	Percentage fuel savings, 2015–2045	
	Midsize passenger car	Compact SUV
Advanced ICE vehicle	33–45 (tank)	23–36 (tank)
HEV	53–61 (tank)	39–48 (tank)
PHEV-20	62% (wall-plug, tank))	—
PHEV-40	75% (wall-plug, tank))	—
FCHEV	60–72 (tank)	50–62 (tank)
BEV	79% (wall-plug) 45% (powerplant)	—

As expected, the magnitude of the fuel/energy savings is greatest for the fuel cell technology. However, the differences between the fuel savings achieved by the different technologies are not as large as we might have expected. Fuel cell vehicles achieve only about twice the fuel savings of the improved conventional engine/transmission power trains and only about 15 percent better savings compared to the HEV (charge-sustaining) power trains. This does not include a consideration of the differences in the efficiencies

> of producing gasoline from petroleum and hydrogen from natural gas or coal, however. The battery-powered vehicle (BEV) has a high energy savings (79%) from the wall-plug, but only modest savings (40%) when the power generation losses at the powerplant are included.
>
> In terms of saving petroleum, the BEV and PHEV offer the greatest opportunity for fuel savings, especially the 40–50 mile PHEV design. It is difficult to quantify the savings of the PHEV because they depend on the usage pattern of the vehicle and the energy source used to generate the electricity. In any case, gasoline-only fuel economy of the PHEV will be significantly greater than for the HEV.

Comparisons of the simulation results from the various studies

The UC Davis simulation results are close to the DOE results except for advanced conventional vehicles (as noted previously the DOE projections are known to be low). However, the UC Davis and MIT fuel economy projections for the midsize passenger car for 2030 are in good agreement for both the advanced ICE and HEV technologies. In addition, the percentage fuel savings projected by the NRC for the advanced ICE vehicle in the near term is close to that projected in the UC Davis simulation (29 percent compared to 33 percent in 2015). In the case of the HEV technology, the NRC projects a fuel saving of 44 percent and UC Davis projects 53 percent in 2015. For the HEV and FCHEV technologies, the DOE and UC Davis results are in good agreement over the complete time period of the simulations, with the agreement being closest in the 2030–2045 time periods. It should be noted that the vehicle characteristics used in the UC Davis simulations were selected to match those used in the DOE study. Hence the agreement between the two studies indicates consistency in the modeling approaches in the two studies for the HEV and FCHEV technologies.

Cost Projections

The second part of our advanced vehicle study involved projecting costs for each of the power train combinations simulated. We did this using a spreadsheet cost model[15] that permits the quick analysis of the economics of hybrid vehicle designs for vehicles of various sizes operated in North America, Europe, and Japan. We analyzed the economics as a function of fuel price, usage pattern (driving cycle and miles/year), and discount rate.

Methodology and cost inputs

The key inputs to the cost analysis are the fuel economy projections for each of the vehicle/driveline combinations and the unit costs of the driveline components. The costs of the engine/transmission and electric motor/electronics are calculated from the maximum power rating of the components and their unit cost ($/kW). The component power (kW) and energy storage (kWh) ratings for the calculations of the component costs were taken from the earlier "Summary of Inputs Used in the Vehicle Simulations" tables. In all cases, the values for 2030 were used in the cost projections. The input values for the fuel economy projections were taken from the earlier

"Fuel Economy Simulation Results for Various Driving Cycles" tables. The fuel economy values shown in the tables correspond to the EPA chassis dynamometer test data and have been corrected to obtain real-world fuel economy using the .9 and .78 factors used by EPA to obtain the fuel economy values given in their Fuel Economy Guide. The real-world fuel economy values are used in all the economic study calculations.

Considerable uncertainty currently surrounds the costs of electric driveline components—the electric motor, power electronics, batteries, and fuel cell. This is especially true of the cost of the batteries and the fuel cell. For this reason, we estimated a range of values for the unit costs of those components. There is a smaller uncertainty about the costs of advanced conventional engine components, so we used single unit cost values for those components. The values we used were based on information in Kromer and Heywood (2007) and Lipman and Delucchi (2003).[16] In all cases, we assumed that the vehicles and driveline components are manufactured in large volume for a mass market.

The inputs to the spreadsheet were selected to model the specific vehicle designs analyzed in this study. In the case of PHEVs, the fuel economy used was the equivalent value based on the sum of the electricity and gasoline usage for the usage pattern (fraction of miles driven in the all-electric, charge-depletion mode). We assumed that this value of equivalent fuel economy was applicable to both the urban (FUDS) and highway (FHWDS) driving cycles. In the case of FCHEVs, the gasoline equivalent of the hydrogen consumption (kgH2/mi) was used to determine the equivalent gasoline break-even price. In the case of the BEVs, the electrical energy cost for the operation of the vehicle was determined using the Wh/mi value from the simulations assuming an electricity price of 8 cents/kWh.

In estimating the retail or showroom cost of vehicles, we used a markup factor of 1.5—that is, the retail price is 1.5 times the OEM (original equipment manufacturer) cost of the component. The cost of reducing the weight and the drag of the vehicle is included as a fixed cost based on values given in the NRC's 2010 "Assessment of Fuel Economy Technologies for Light-duty Vehicles" report. Additional input values to the cost model include the price of the fuel, the annual mileage use of the vehicles, the years over which the analysis is to be done, and the discount rate. Values of all the input parameters can be changed by the user from the keyboard as part of setting up the economic analysis run. Key output parameters are the average composite fuel economy for the vehicle in real world use, differential driveline cost, fraction of fuel saved, and actual and discounted breakeven fuel price ($/gal). All vehicle costs and fuel prices are in 2007–2010 dollars.

Discussion of the cost projection results

We show the results of the economic analysis of the various advanced vehicle cases for a midsize passenger car in 2030. The energy saved and cost differentials are relative to the 2007 baseline vehicle using a port fuel-injected (PFI) engine. The break-even gasoline price is calculated for a vehicle use of 12,000 miles per year and time periods of 5 or 10 years. The 5-year period is used for the ICE vehicles and the HEVs because it is commonly assumed that new car buyers would desire to recover their additional purchase cost in that period of time. Both the 5-year and 10-year periods are used for the PHEVs, BEVs, and FCHEVs since the lifetimes of the batteries and the fuel cells are uncertain at the present time and it seems reasonable to recover the high cost of those components over their lifetimes. Discount rates of 4 and 10 percent are used for the 5- and

10-year periods, respectively. These discount rates are likely more appropriate for society as a whole than for individual vehicle buyers. The economic calculations were made for ranges of battery and fuel cell costs because those costs are particularly uncertain and sure to change significantly over the next 10 to 20 years.

First consider the economic results for the ICE and HEV vehicles. The fractional energy savings are .43 and .62 for the ICE vehicle using an advanced engine and the HEV using the same engine technology, respectively. The corresponding discounted break-even gasoline prices ($/gal) are $3.62 for the ICE vehicle and $2.30–$2.60 for the HEV. The gasoline price is lower for the HEV than for the ICE vehicle because the fuel economy of the HEV is significantly higher. These results indicate the economic attractiveness of the HEV even at battery costs of $1000/kWh. It appears that both the advanced ICE and the HEV will make economic sense even at the gasoline prices in 2010 and with a 5-year payback period.

Next consider the economic results for the PHEVs. The fractional energy savings are .65 and .79 for the PHEV-20 (small battery, AER =10–20 miles) and PHEV-40 (large battery, 40–50 miles), respectively. The energy used by the PHEVs includes both gasoline fuel and the gasoline equivalent of the electrical energy from the battery. The cost differentials of the PHEVs are relatively high compared to those of the HEVs and depend markedly on the cost of the batteries. As would be expected, the differential costs and break-even gasoline prices are significantly higher for the large-battery PHEV than for the small-battery PHEV, which is significantly higher than for the HEV with about the same energy savings. In the case of the PHEV with the small battery, the break-even gasoline price is in the same range as that of the HEV only when the retail battery cost is about $400/kWh and the time period of the calculation is 10 years, the assumed lifetime of the battery. For the PHEV with the large battery, a retail battery cost of $300/kWh and at least a 10-year life is needed to make the vehicle cost competitive with either the small-battery PHEV or the HEV. However, the fuel and energy savings using the large-battery PHEV are the highest among the advanced vehicles considered.

The break-even gasoline prices do not include the effect of possible battery replacement. We assumed the batteries will last through at least the time period of the calculation (5 years or 10 years). Results for the PHEVs are shown for 5 years at a 4-percent discount rate and 10 years at a 10-percent discount rate. The break-even gasoline prices are lower for the longer time period, even using the higher discount rate. The short discount period (5 years) corresponds to the time we expected the first owner of the vehicle to own the car, and the 10-year period corresponds to the expected lifetime of the batteries. In all cases, the economics are more attractive for the longer time period, indicating a leasing arrangement for the batteries seems to make sense. The cost of the electricity to recharge the batteries was included in the calculations using the equivalent fuel economy, which was determined by adding the gasoline equivalent of the electricity (kWh) used in the all-electric charge-depleting mode to the gasoline used in the charge-sustaining mode. This approximation is almost exact for electricity costs of 6–10 cents/kWh.

The economic calculations for the FCHEVs were done for a range of fuel cell unit costs ($30–75/kW). An intermediate battery cost ($800/kWh) was used for all the calculations. The break-even fuel cost (hydrogen equivalent) becomes comparable to that of the HEV when the fuel cell unit cost is less than $50/kW. This is especially the case when the time period of the analysis is 10 years. The energy savings of the fuel cell vehicles (70 percent) are intermediate between those of the HEV and the large-battery PHEV. The break-even fuel cost represents the gasoline ($/gal) and hydrogen ($/kg) prices for which the vehicle owner would recover the differential vehicle cost in the time period of the calculation. If the price of the hydrogen is lower than the breakeven gasoline price, the vehicle owner would recover more than the vehicle price differential from fuel cost savings compared to the baseline ICE vehicle. These economic results for the FCHEVs indicate that target fuel cell costs of $30–50/kW, 10-year life, and hydrogen prices in the $2.50–$3.00/kgH$_2$ range should make fuel cell vehicles cost competitive with HEVs and ICE vehicles using advanced engines.

We have also analysed the economics of battery-powered vehicles with a range of 100 miles for battery costs between $300–700/kWh. The differential costs of the BEVs are greater than any of the other vehicle designs being $20294 for batteries costing $700/kWh and $9094 for $300/kWh. The breakeven gasoline prices for the BEVs are also higher than for the other advanced vehicles being $4–5/gal even for the $300/kWh batteries. Based on the energy equivalent of the wall-plug electricity to recharge the batteries, the BEVs have an energy savings of 77 %, but much less savings if the powerplant efficiency is included. In that case, the energy savings are only 40%.

All the breakeven gasoline prices considered thus far were determined for differential costs and fuel savings relative to the 2007 baseline vehicle. It is of interest to consider the breakeven gasoline prices of the BEV, PHEV-40, and FCHEV using the Advanced ICE and HEV vehicles as the baseline. These comparisons indicate that none of the electric drive vehicles with large batteries, even at the lowest battery cost of $300/kWh, are economically attractive relative to the Adv. ICE and HEV vehicles. This is especially true of the BEVs. As expected the breakeven gasoline prices are highest when the HEV is used as the baseline. The FCHEV is the most attractive of the electric drive vehicles when compared to the HEV.

SUMMARY OF COST RESULTS FOR A MIDSIZE PASSENGER CAR IN 2030

Component cost assumptions (changes in retail price of the vehicle):
Added vehicle cost to reduce drag and weight, $1,600
Advanced engine/transmission, $45/kW
Standard engine/transmission, $32/kW
Electric motor and electronics, $467 + $27.6/kW
Batteries $/kg = $/kWh x Wh/kg /1000
Fuel cell, $30/kW–$75/kW

Notes:
1. 5 years and 4% discount rate, 12,000 miles/yr
2. 10 years and 10% discount rate, 12,000 miles/yr
3. 10 years and 6% discount rate, 12,000 miles/yr
4. Equivalent (includes gallon equivalent of gasoline for electricity used in the all- electric operation) including electricity, 20% of vehicle miles on electricity
5. Equivalent (includes gallon equivalent of gasoline for electricity used in the all- electric operation) including electricity, 65% of vehicle miles on electricity
6. Hydrogen equivalent kg/mi
The PHEV-20 has a small battery (25–33 kg, all-electric range or AER of 10–20 mi); the PHEV-40 has a large battery (55–80 kg, AER 40–60 mi).

Vehicle Configuration	Real-World mpg	Battery Inputs			Energy Saved	Vehicle Cost Differential	Discounted Break-even Gas Price
		$/kWh	Wh/kg	$/kg			
Baseline vehicle 2007	27.1						
Adv. ICE	47.8				.43	$3095	$3.62/gal1
HEV	71.1	1000	70	70	.62	$3204	$2.61/gal1
		800	70	56		$3003	$2.45/gal1
		600	70	42		$2802	$2.29/gal1
PHEV-20	75.34	800	100	80	.65	$6409	$5.03/gal1
							$3.64/gal2
		600	100	60		$5605	$4.40/gal1
							$3.19/gal2
		400	100	40		$4801	$3.77/gal1
							$2.73/gal2
PHEV-40	1275	700	150	105	.79	$10,228	$6.58/gal1
							$4.77/gal2
		500	150	75		$8218	$5.29/gal1
							$3.83/gal2
		300	150	45		$6208	$3.99/gal1
							$2.89/gal2
FCHEV	89.8						
$75/kW FC		800	70	56	.70	$7549	$5.47/gal1
							$3.31/gal3
$50/kW FC		800	70	56		$5549	$4.02/gal1
							$2.43/gal3
$30/kW FC		800	70	56		$3949	$2.86/gal1
							$1.73/gal3
Battery electric BEV	Equiv. 176						
Range 100 mi.		$700	170	119	.77 wallplug	20294	10.72 (1)
							8.09 (3)
		$500	170	85		14694	7.90 (1)
							6.04 (3)
		$300	170	47		9094	5.06 (1)
							3.99 (3)

2030 BREAKEVEN FUEL PRICE $/GAL GASOLINE EQUIV.

Vehicle design	2007 ICE baseline		Adv. ICE baseline		HEV baseline	
Battery electric *						
5 yr at 4% disc						
• battery cost $/kWh	w/o disc	with disc	w/o disc	with disc	w/o disc	with disc
700	9.57	10.72	14.43	16.16	21.50	24.08
500	7.05	7.90	9.97	11.17	14.91	16.70
300	4.52	5.06	5.50	6.17	8.28	9.27
10 yr at 10% disc						
• battery cost $/kWh	w/o disc	with disc	w/o disc	with disc	w/o disc	with disc
700	4.99	8.09	7.58	12.28	11.31	18.30
500	3.72	6.04	5.35	8.67	7.99	12.94
300	2.46	3.99	3.12	5.05	4.63	7.50
PHEV large battery **						
5 yr at 4% disc						
• battery cost $/kWh	w/o disc	with disc	w/o disc	with disc	w/o disc	with disc
700	5.6	6.27	8.07	9.04	14.1	15.79
500	4.55	5.10	6.0	6.72	10.45	11.70
300	3.51	3.93	3.9	4.37	6.8	7.62
10 yr at 10% disc						
• battery cost $/kWh	w/o disc	with disc	w/o disc	with disc	w/o disc	with disc
700	2.94	4.76	4.32	7.00	7.54	12.22
500	2.42	3.92	3.27	5.30	5.71	9.25
300	1.89	3.06	2.22	3.60	3.88	6.29
Fuel cell HEV***						
5 yr at 4% disc						
fuel cell cost	w/o disc	with disc	w/o disc	with disc	w/o disc	with disc
75	5.07	5.68	6.48	7.26	9.62	10.77
50	4.16	4.66	4.88	5.47	7.25	8.12
30	3.44	3.85	3.61	4.04	5.36	6.00
10 yr at 10% disc						
fuel cell cost$/kW	w/o disc	with disc	w/o disc	with disc	w/o disc	with disc
75	3.06	4.96	4.17	6.76	6.19	10.02
50	2.61	4.23	3.37	5.46	5.00	8.10
30	2.25	3.64	2.73	4.42	4.06	6.58

* *electric cost 8¢/kWh; 12000 miles/yr.*
** *65% of miles on electricity, 12,000 miles/yr.*
*** *fuel cell cost includes hydrogen storage at $10/kWh, 4 kg H2; $3.5/kg H_2*

Summary and Conclusions

- To determine how much of a reduction in fuel consumption we can expect from new vehicle technologies, we ran simulations for a midsize passenger car and a small/compact SUV for 2015, 2030, and 2045. We compared fuel economy (mpg) and fractional energy saved by advanced, higher-efficiency engines, hybrid-electric vehicles (HEVs and PHEVs), and electric-drive vehicles (BEVs and FCVs) in relation to a conventional vehicle marketed in 2007.

- According to our simulation results, large improvements in the fuel economy of conventional midsize passenger cars (50–70 percent) and compact SUVs (30–49 percent) relative to 2007 models can be expected in the next ten to twenty years even without large changes in the basic power train technology. These improvements will result from the combined effects of decreases in weight, vehicle drag, and tire rolling resistance and increases in engine efficiency.

- We found that gasoline/energy savings of about 40 percent can be expected due to vehicle and engine improvements and up to 60% when the powertrain is hybridized. A fuel/gasoline savings of nearly 80 percent is projected for a PHEV with a large battery (40- to 50-mile all-electric range). The corresponding total energy saving is about 40% for a 50% efficient electricity powerplant. The fuel cell vehicle has a projected energy savings (tank-to-wheels) of 72 percent in 2030 and an equivalent fuel economy of more than 100 mpg.

- For 2030 BEV, the gasoline energy equivalent saved is 79% from the wall-plug and 57% at a 50% efficient powerplant compared to the 2007 baseline ICE mid-size car. Compared to a 2030 HEV, the gasoline equivalent saved is only 47% from the wall-plug and there are no savings at the powerplant until the efficiency of the powerplant exceeds about 55%.

- Although we did expect that the magnitude of the fuel/energy savings would be greatest for the fuel cell technology, the differences between the fuel savings achieved by the different technologies are not as large as we might have expected. FCVs achieve only about twice the fuel economy of the improved conventional engine/transmission power trains and only about 15 percent better savings compared to the HEV (charge-sustaining) power trains. This does not include a consideration of the differences in the efficiencies of producing gasoline from petroleum and hydrogen from natural gas or coal, however. The BEV has a high energy savings (79 percent) from the wall plug, but more modest savings (40–55 percent) when the power generation losses at the power plant are considered.

- In terms of saving petroleum, the BEV and the PHEV offer the greatest opportunity for fuel/gasoline savings, especially the 40–50 mile PHEV design. It is difficult to quantify the real-world savings because they depend on the detailed usage pattern of the vehicle and the energy source used to generate the electricity. In any case, the gasoline-only fuel economy of the PHEV will be significantly greater than for the HEV.

- Our cost studies indicate that both the advanced ICEV and HEVs using advanced high-efficiency engines would be cost competitive with the baseline vehicle in 2015 to 2030, with a break-even gasoline price of $2.50–$3.50/gal calculated for a five-year performance period (12,000 mi/yr) and a 4-percent discount rate.

- The PHEV with the small battery (all-electric range of about 20 miles) becomes competitive with the HEV when the retail battery cost is $400/kWh and the performance period is ten years. The PHEV with the large battery (all-electric range of about 50 miles) becomes cost competitive at a battery cost of $300/kWh. The cycle life of the batteries was assumed to be ten years. The FCV becomes cost competitive with the HEV when the retail fuel cell cost is $30–$50/kW and the price of hydrogen is about $3/kg. BEVs are not cost competitive with advanced ICEVs and HEVs even at a battery cost of $300/kWh.

Notes

1. See, for example, A. F. Burke, "Saving Petroleum with Cost-Effective Hybrids," SAE Paper 2003-01-3279, presented at the Powertrain and Fluids Conference, Pittsburgh, PA, October 2003; and A. F. Burke and A. Abeles, "Feasible CAFÉ Standard Increases Using Emerging Diesel and Hybrid-electric Technologies for Light-duty Vehicles in the United States," *World Resource Review* 16 (2004).
2. See Burke, Zhao, and Van Gelder, "Simulated Performance of Alternative Hybrid-Electric Powertrains"; Burke and Van Gelder, "Plug-in Hybrid-Electric Vehicle Powertrain Design and Control Strategy Options"; Burke, "Batteries and Ultracapacitors for Electric, Hybrid, and Fuel Cell Vehicles"; Burke, "Ultracapacitor Technologies and Application in Hybrid and Electric Vehicles"; Burke, "Saving Petroleum with Cost-Effective Hybrids"; and Burke and Abeles, "Feasible CAFÉ Standard Increases."
3. K. Wipke, M. Cuddy, D. Bharathan, S. Burch, V. Johnson, A. Markel, and S. Sprik, "Advisor 2.0: A Second-Generation Advanced Vehicle Simulator for Systems Analysis," NREL/TP-540-25928, March 1999.
4. K.-H. Hauer, A. Eggert, R. M. Moore, and S. Ramaswamy, "The Hybridized Fuel Cell Vehicle Model of the University of California, Davis," SAE paper 2001-01-05432001, 2001.
5. U.S. Department of Energy and U.S. Environmental Protection Agency, "Fuel Economy Guide—2007," DOE/EE-0314.
6. The upcoming technologies are those discussed in S. Plotkin and M. Singh, "Multi-Path Transportation Futures Study: Vehicle Characterization and Scenarios," Argonne Lab and DOE Report (draft), March 5, 2009; National Research Council, *Assessment of Fuel Economy Technologies for Light-duty Vehicles* (National Academies Press, 2010); K. G. Duleep, "Technologies to Reduce Greenhouse Emissions for Light-duty Vehicles," prepared for the Heywood, "Comparative Analysis of Automotive Powertrain Choices for the Next 25 Years," SAE paper 2007-01-1605, 2007.
7. A. F. Burke, H. Zhao, and E. Van Gelder, "Simulated Performance of Alternative Hybrid-Electric Powertrains in Vehicles on Various Driving Cycles," EVS-24, Stavanger, Norway, May 2009 (paper on the CD of the meeting); A. F. Burke and E. Van Gelder, "Plug-in Hybrid-Electric Vehicle Powertrain Design and Control Strategy Options and Simulation Results with Lithium-ion Batteries," paper presented at EET-2008 European Ele-Drive Conference, Geneva, Switzerland, March 12, 2008 (paper on CD of proceedings).
8. A. F. Burke, "Batteries and Ultracapacitors for Electric, Hybrid, and Fuel Cell Vehicles," *Proceedings of the IEEE* 95 (April 2007): 806–820; A. F. Burke, "Ultracapacitor Technologies and Application in Hybrid and Electric Vehicles," *International Journal of Energy Research* 34 (February 2010): 133–51.
9. A. F. Burke and M. Miller, "Performance Characteristics of Lithium-ion Batteries of Various Chemistries for Plug-in Hybrid Vehicles," EVS-24, Stavanger, Norway, May 2009 (paper on the CD of the meeting).
10. See Burke and Miller, "Performance Characteristics of Lithium-ion Batteries," and A. F. Burke and H. Zhao, "Simulations of Plug-in Hybrid Vehicles using Advanced Lithium Batteries and Ultracapacitors on Various Driving Cycles," IAAMF Conference, Geneva, Switzerland, March 2010.

11. H. Zhao and A. F. Burke, "Fuel Cell Powered Vehicles Using Supercapacitors—Device Characteristics, Control Strategies, and Simulation Results," *Fuel Cells* 10 (2010): 879–96; H. Zhao and A. F. Burke, "Optimization of Fuel Cell System Operating Conditions for Fuel Cell Vehicles," *Journal of Power Sources* 186 (2009), 408–16.
12. Plotkin and Singh, "Multi-Path Transportation Futures Study."
13. Kasseris and Heywood, "Comparative Analysis of Automotive Powertrain Choices for the Next 25 Years"; A. Schafer, J. Heywood, D. Jacoby, and I. A. Waitz, "Transportation in a Climate-Constrained World," Chapter 4 of Road Vehicle Technology (MIT Press, 2009).
14. National Research Council, *Assessment of Fuel Economy Technologies for Light-duty Vehicles* (National Academies Press, 2010).
15. The model was developed in A. F. Burke, "Saving Petroleum with Cost–Effective Hybrids," SAE Paper 2003-01-3279, 2003.
16. M. Kromer and J. B. Heywood, Electric Powertrains: Opportunities and Challenges in the U.S. Light-duty Vehicle Fleet, MIT report LFEE 2007-03- RP, May 2007; T. Lipman and M. A. Delucchi, Hybrid-Electric Vehicle Design Retail and Lifecycle Cost Analysis, UCD_ITS Report No. UCD-ITS-RR-03-01, April 2003.

Chapter 5: Comparing Infrastructure Requirements

Joan Ogden, Christopher Yang, Yueyue Fan, and Nathan Parker

For biofuels, electricity, and hydrogen to assume major roles as transportation fuels over the next several decades—as they must if we are to meet future goals for low-carbon transportation—one or more new fuel infrastructures will have to be developed. We define a fuel infrastructure as all of the components of the physical system needed to provide transportation fuels to the end user, including extracting primary resources, transporting them to a fuel production plant, processing them to produce transportation fuels, providing refueling sites, and delivering fuels to these refueling locations. In some cases one fuel infrastructure can depend on another (for example, the electricity system depends on other infrastructures that deliver coal or natural gas). In this chapter we will focus mainly on infrastructure issues for the particular fuel supply chain in question and less on the underlying infrastructures (for example, more on the hydrogen infrastructure itself and less on the natural gas or electricity infrastructure supplying energy to make hydrogen).

Today's transportation system is 97-percent dependent on petroleum-based liquid fuels. A vast petroleum infrastructure has developed over a century, encompassing worldwide oil exploration and production, long-distance transport of crude oil to hundreds of refineries, and an extensive network of pipelines and trucks delivering gasoline and diesel to terminals and refueling stations. Since 1980, the global capital expenditure to maintain and expand this massive infrastructure has averaged hundreds of billions of dollars per year, about 80 percent of which is devoted to finding and extracting crude oil, and the remainder to refineries, storage, and pipelines. Infrastructure costs are rising: the investment in petroleum fuel infrastructure between 2007 and 2030 is projected to be about $1 trillion in North America alone and $6 trillion globally.

In this chapter we discuss general considerations for building transportation fuel infrastructures and compare infrastructure challenges for biofuels, electricity, and hydrogen with respect to system design, resources, technology status, cost, reliability, and transition barriers such as compatibility with existing infrastructures. Finally, we discuss policies that might be needed to provide incentives for new infrastructure development.

Infrastructure Design and Deployment

A transportation fuel infrastructure needs to satisfy certain requirements: it must bring adequate supplies of fuel to consumers at a competitive and stable cost, it must be reliable and robust enough to resist disruptions (natural or human), and ideally it should impose minimal environmental costs and security risks. The infrastructures for each of the fuels we consider (biofuels, hydrogen, and electricity) will need to meet these requirements while at the same time placing different emphases on key infrastructure components. These different emphases result from the fact that the costs, technical challenges, and/or other important considerations and barriers are different for each type of infrastructure.

A GENERIC FUEL SUPPLY INFRASTRUCTURE

A generic fuel supply infrastructure has the components shown here.

Fuel infrastructures involve large capital and investment costs, and long-lived assets. Most are complex networks rather than single chains, with a wide resource base and feedstock transport, multiple conversion facilities, an extensive delivery system, and numerous points of use. For example, the electricity system uses diverse primary sources (fossil, renewable, and nuclear), numerous conversion power plants, an extensive transmission and distribution system, and potentially chargers in every garage.

Because of the high cost of building a major new infrastructure, it would be desirable to utilize existing infrastructure where possible. For instance, new "drop-in" liquid biofuels might be developed that would be compatible with the petroleum system and could use existing assets such as petroleum refineries, pipelines, trucks, and stations. Even if existing infrastructure could not be used directly with a new fuel, piggybacking on today's systems could reduce costs—for example, adding chargers for electric vehicles to homes that already have electricity service, or making hydrogen from natural gas already available at refueling stations.

Infrastructure for biofuels

An extensive infrastructure is required to supply liquid biofuels to a refueling station. It begins with feedstock production. The energy, material (fertilizer and water), and capital inputs required for this step can vary depending on the type of biomass used. First-generation biofuels are made primarily from food crops that produce sugars/starch or vegetable oil. Future generations of biofuels will be made from cellulosic materials as well, including agricultural and forestry wastes and also dedicated energy crops. Using waste products limits the input requirements because these inputs were already being used to produce the primary crop (food, fiber, or forest products); growing crops specifically for use as a transportation fuel feedstock requires more inputs.

In either case, this biomass must be collected and transported to a biofuel production facility, commonly in trains and trucks. Because of the low energy and spatial density of biomass, feedstock transport can account for a significant portion of the energy input to biofuel production.

The conversion of biomass into a biofuel can be a complex process and differs widely depending upon the type of biomass being converted and the technologies being employed. Currently, the major processes for biofuel production involve the production of alcohol from sugars/starch via biological fermentation, and the production of biodiesel via transesterification.

Conversion of sugar or cellulosic-based biomass to biofuels requires processing and separating the materials to yield sugars that can be fermented. Next-generation fuel production could also involve thermal treatment (for example, gasification or pyrolysis).

The transport of a biofuel is similar to the transport of petroleum products like gasoline and diesel. Some forms of biofuel might be transported in the existing gasoline and diesel distribution infrastructure, but some forms cannot. For example, pure ethanol is a gasoline substitute but cannot be transported in gasoline pipelines because of its tendency to absorb water and its corrosiveness. By contrast, ethanol blended with gasoline at concentrations of 10 to 20 percent can be transported without infrastructure changes, and if "drop-in" biofuels were produced they could be co-transported with existing fuels.

Infrastructure for hydrogen

Hydrogen can be produced either on-site or in a centralized facility. The infrastructure required for on-site production is much less extensive than that required for centralized production. For on-site production, the infrastructure is confined primarily to the refueling station, where energy resources (natural gas or electricity) are delivered using existing infrastructure. At the refueling station, these energy resources can be converted to hydrogen using steam reforming or electrolysis and then compressed, stored, and dispensed to fuel cell vehicles.

Centralized production of hydrogen requires more capital-intensive infrastructure investments and is justified only with large demands for hydrogen. Large central plants for producing hydrogen can use many different resources, including fossil fuels (natural gas or coal), biomass, or electricity. The choice of energy resource will dictate the type and scale of the first stages of fuel infrastructure. For example, hydrogen plants can be built at sites for renewable electricity generation (wind or solar farms) or have energy resources delivered to them (biomass, coal, natural gas, or electricity via the grid). In either case, these primary energy resources are not unique to hydrogen but are developed for other purposes as well, such as electricity production.

Once hydrogen is produced, the remainder of the supply chain infrastructure is unique to the hydrogen fuel pathway. Because hydrogen is a gas, storage and delivery are more energy intensive and costly than for a liquid fuel. There are several different methods for delivering hydrogen to the refueling station, including compressed gas truck, liquid hydrogen truck, and pipeline. The choice of method depends upon many factors, including demand density, transport distances, and size of refueling stations (see Chapter 3). Refueling stations make up the last piece of the hydrogen fuel supply chain; they have equipment for compression, storage, and fuel dispensing to vehicles.

Infrastructure for electric vehicles

Similar to other fuels, electricity for use in vehicles requires an infrastructure that consists of the entire system of extracting primary energy resources, converting those resources into electricity, distributing it to the point of use, and then providing a way to recharge batteries. Unlike the other fuels, electricity is already in widespread use for a variety of purposes, so much of this infrastructure already exists and can be used to provide electricity to vehicles as well. As a result, analysis of infrastructure for providing fuel to electric vehicles is largely focused on the point of refueling—the vehicle charger.

It is expected that most drivers of plug-in electric vehicles (PEVs) will refuel primarily at home, so much of this recharging infrastructure will be concentrated there. However, there is also significant activity in the development of public charging infrastructure. The thinking is that some level of public access to charging away from the home is needed to overcome the range limitations of pure battery electric vehicles (BEVs), though the appropriate balance between private and public charging equipment will depend on the mix of BEVs and plug-in hybrid electric vehicles (PHEVs), and the needs and preferences of their drivers. There is some evidence that public recharging stations may be needed to reassure drivers (that is, to ease "range anxiety") without being used very often. Aside from that concern, public charging will need to be ubiquitous if electricity is to displace most petroleum fuel usage because many drivers do not have access to overnight off-street parking.

Widespread infrastructure for electricity generation and distribution already exists, and this infrastructure has quite a bit of underused capacity. Even with significant penetration of BEVs and PHEVs in the next few decades, electricity demand for recharging these vehicles will make only a minor contribution to total electricity demands. Thus, there may not be a need for additional generating capacity to meet this additional demand. If PEV adoption is concentrated in certain neighborhoods, this could require some upgrades to distribution infrastructure. STEPS researchers have analyzed how the addition of PEV recharging will change the pattern of electricity generation and affect emissions from electricity generation.[1] Over the long term as PEV demands grow and affect the overall timing of electricity demands, this could induce changes in the mix of generation capacity that would be used to meet all demands.

Resource Issues

Fuel infrastructure supply chains begin with primary resource extraction. Each fuel faces different resource challenges, especially given the imperative to adopt a low-carbon primary supply over time.

Many different kinds of biomass resources could be converted to biofuels, each of which has different environmental impacts (see Chapters 1 and 12). The total biomass resource available for biofuel production is constrained by a variety of economic and environmental factors that can vary regionally. Competing uses for biomass—for example, to generate renewable power and heat—could further reduce the biomass resource base available for transportation fuel production. Global estimates suggest that 10 to 30 percent of transportation fuel needs could be met with biofuels, with biofuels playing a larger role if vehicles are made more efficient and biomass productivity is increased.

Hydrogen and electricity could access a much wider primary resource base, including low-carbon options such as fossil fuels with carbon capture and sequestration (CCS), renewables (solar, wind, biomass, hydro, geothermal), and nuclear. In theory, the availability of low-carbon resources should not be a limiting factor for either electricity or hydrogen, although the higher cost of zero-carbon pathways could increase fuel costs.

Technology Status

For biofuel infrastructure, the largest technology gap is the need to develop low-cost, low-net-carbon, advanced biofuel production methods for cellulosic ethanol and Fischer-Tropsch (F-T) liquids (diesel-like liquid fuels compatible with the existing petroleum infrastructure). The technologies to harvest, store, and transport biomass feedstocks, and to store and deliver biofuels, are mature, although scale-up is needed for biofuel transport systems to reach low costs. Technical improvements in crop yields and productivity could also be very important to the overall role of biomass in the energy system.

For electricity, one of the major technical issues is development of a low-carbon supply. As shown in Chapters 6 and 9, electric vehicles do not represent much of an improvement over gasoline hybrids in terms of greenhouse gas emissions unless the grid is substantially decarbonized. Another technology gap is the implementation of a "smart grid" that can manage the time-changing demands for charging a fleet of electric vehicles, and time-variable renewable energy sources. Finally, bulk storage for electricity could play a role in a future grid heavily dependent on variable renewable electricity sources such as wind and solar, and in serving time-varying vehicle-charging demands. Improved batteries that could accommodate fast charging could influence the relative role of fast charging in the electricity infrastructure.

Commercial technologies to produce hydrogen from fossil fuels and to deliver and store it are already in use in the chemical industry today. However, there is a need to develop technologies for cost-effective low-carbon production. Hydrogen from coal or natural gas with CCS, hydrogen from biomass gasification, and electrolytic hydrogen powered by low-carbon electricity are all low-carbon options. In general, production options based on thermochemical processing of hydrocarbons will offer lower costs than electrolytic hydrogen. Hydrogen storage is another area where technical breakthroughs could transform the design and cost of infrastructure.

Cost Considerations

Some factors that influence infrastructure cost are technology maturity, scale economies in both fuel production plants and delivery systems, geography-specific factors including location and costs of feedstocks for fuel production, geographic density of demand, and compatibility with existing energy systems.

For biofuels, technology advancement and scale-up of biorefineries are the most important factors in reaching competitive costs. As shown in Chapter 1, the cost of biorefineries is the largest single cost in the supply chain (about 85 percent of the investment), with fuel delivery costs playing a much smaller role. The capital investment for mature biorefineries is expected to be about $3–5 per gallon gasoline equivalent (gge) per year. Studies by Parker et al. suggest that to produce between 12 and 46 billion gge per year in the United States in 2018, the investment in biorefineries would total between $100 billion and $360 billion (see Chapter 1). This could supply enough biofuels to meet between 5 and 21 percent of the projected U.S. demand for transportation fuel in 2018 at an average infrastructure capital investment cost of several thousand dollars per car. Biofuel delivery systems would add another 15 to 20 percent to the cost.

For hydrogen, recent studies by the National Academies and others suggest that the capital investment for mature infrastructure would be $1,400–2,000 per light-duty vehicle served, depending on the pathway. The National Academy of Sciences (NAS) found that building a fully developed hydrogen infrastructure serving 220 million vehicles in the United States in 2050 would cost about $400 billion over a period of about 40 years.[2] (The NAS scenario is based mostly on fossil-fueled hydrogen with CCS and biomass hydrogen. Electrolysis-based pathways could cost more to build.) Early infrastructure investment costs per car (to serve the first million vehicles) would be higher ($5,000–10,000 per car).

The investment cost for electric vehicle infrastructure is difficult to estimate because the electric generation, transmission, and distribution system is shared by multiple end users. Moreover, the grid will undergo a transformation toward lower-carbon resources independent of the introduction of PEVs. A recent study by the U.S. Department of Energy suggested that it would cost between $800 and $2,100 to install a charger in a typical home, in part because of circuit upgrades to accommodate "level 2" charging, and in part for metering and utility interface.[3] And some fraction of costs for smart grid upgrades would be borne by PEVs among the other demands. If charging of PEVs were primarily confined to off-peak hours, this would increase the utilization of existing power plants and reduce the average cost of supplying electricity. There could be benefits with regard to system reliability, depending on how smart-grid technologies are implemented. Thus, benefits as well as costs to the electricity system might accompany the large-scale use of electric vehicles.

Reliability and Resilience

Because energy is an important part of lifeline systems that touch almost every aspect of modern society, an energy supply chain is considered a critical infrastructure system. In view of the extreme vulnerability of such systems to disasters and disruptions (as evidenced by the World Trade Center terrorist attack in 2001, the tsunamis in 2004 and 2011, and hurricanes Katrina and Rita in 2005), infrastructure security, especially reliability and resilience, has become an important issue to be addressed in renewable energy infrastructure system design. More specifically, it is important to plan for potential disruptions caused by feedstock fluctuation, demand and price spikes, and unexpected facility failures caused by natural disasters and human errors.

Strategic supply chain management aims at finding the best supply chain configuration—including location setup, procurement, production, storage, and distribution—to support efficient operation of the whole supply chain. On the other hand, reducing redundancy and buffers, which improves the system efficiency under normal conditions, may make the supply chain more vulnerable to unexpected events such as supply shortage, demand spike, technological failure, or attacks and disasters. Because different components of the supply chain are so interdependent, failure of one component might reverberate through the entire supply chain.

From the viewpoint of the physical structure of an energy supply chain, storage facilities hedge against disruptions in two important ways: (1) by storing energy, they provide a buffer for the system to adjust to fluctuations in supply and demand, and (2) by redistributing energy over space and time, they increase the self-healing ability of the system. Simulating these systems using advanced stochastic modeling approaches (approaches that estimate probability distributions of potential outcomes by allowing for random variation in one or more inputs over time) considering a wide range of future possibilities may produce results that hedge better against future uncertainties.

Compatibility with Existing Infrastructure

Current liquid biofuels such as ethanol are at least partly compatible with existing petroleum infrastructure in that they can be blended with petroleum-based fuels at concentrations of up to 10 to 20 percent without infrastructure changes. The main issues with ethanol transport in existing infrastructure have to do with its water absorption and corrosiveness. Transporting neat ethanol or E85 is also feasible but requires its own infrastructure. Future "drop-in" biofuels produced via gasification and Fischer-Tropsche synthesis might be able to use the petroleum storage and pipeline system. An interesting question is whether the existing petroleum delivery system is located in the right places for ready access by future large biorefineries. In the United States, for example, a majority of biorefineries would be sited in the Midwest and Southeast, but the petroleum pipeline system is focused in the Gulf Coast area. Given this geographic mismatch, some new biofuel infrastructure might be required anyway to bring "drop-in" biofuels to existing gasoline and diesel terminals.

The infrastructure for PEVs will likely be based on home recharging plus a network of public "fast charge" stations to facilitate long-distance travel. The electricity system reaches most homes, and about 50 percent of these households appear to be well adapted for private recharging (see Chapter 2). Changes to the electric transmission, distribution, and generation systems will take place over a long time, and with a trend toward low-carbon sources and smart-grid technologies to manage time-variable renewable sources and demands. These developments should be synergistic with adoption of PEVs.

There is little opportunity to use hydrogen directly in existing energy systems, and a new dedicated infrastructure would be needed (see Chapter 3). In the early stages of infrastructure development, hydrogen might rely on truck delivery of small quantities of "merchant" hydrogen produced from natural gas, moving toward on-site hydrogen production at stations and eventually toward centralized production of low-carbon hydrogen with pipeline delivery. It has been suggested that hydrogen could be blended at up to 15 percent by volume with natural gas without infrastructure changes, but there would be only a modest environmental benefit to this approach. For large quantities of pure hydrogen, a new dedicated production and delivery system would be needed.

Both electricity and hydrogen rely on other underlying infrastructures that deliver feedstocks to production plants. Expanding use of either carrier could require an expansion of the underlying feedstock infrastructure as well. (For example, to make large quantities of hydrogen from coal would require extra rail and barge capacity to deliver coal to hydrogen production plants.)

REQUIREMENTS FOR BUILDING INFRASTRUCTURE SUPPLY CHAINS

Infrastructure requirements are summarized here for each fuel along the entire supply chain. Opportunities to use existing infrastructure are highlighted in boldface.

	Hydrogen	**Electricity**	**Biofuels**
Resource extraction and collection	**Use existing infrastructure for fossil resources (natural gas, coal).** New infrastructure may be needed for expanded use of renewables, CCS.	**Use existing infrastructure for fossil resources (natural gas, coal).** New infrastructure may be needed for expanded use of renewables, CCS.	Use some food crops, and wastes produced as part of existing agriculture, forestry, urban systems. Wastes require collection; energy crops require dedicated operation.
Resource transport	**Use existing infrastructure for fossil resources or electricity.** New transport system may be needed for biomass-to-H2 plant.	**Use existing infrastructure for current resources.** New transport system may be needed for biomass-to-power plant.	Transport by truck or rail. Low biomass energy density limits transport distances.
Conversion facility	**Initial supply from merchant H2 system.** New on-site reformers or electrolyzers, or large-scale central reformers, gasifiers (w/CCS), or electrolyzers needed.	**Use existing electric generation infrastructure.** New power plants or retrofits may be needed for renewable electric production, CCS.	New biorefineries (including feedstock processing and conversion) needed.
Fuel transport	Use trucks or pipelines (for central H2 production).	**Use existing distribution infrastructure.** May require upgrades in some places.	**Use existing infrastructure for "drop-in" biofuels.** New transport system needed for pure ethanol/E85.
Fuel dispensing/charging	Hydrogen refueling station network needed. Early options could include in-home or neighbourhood refueling.	In-home chargers and some public chargers needed.	**Use existing stations for "drop-in" biofuels.** New liquid fuel stations needed for pure ethanol/E85.

Transition Issues and Timing

Biofuel internal combustion engine vehicles could be introduced rapidly. The rate of biofuel adoption will be determined by investments in biorefineries, and to a lesser extent associated biofuel delivery infrastructure. The Renewable Fuel Standard in the United States requires production of 36 billion gallons of biofuel per year by 2022, which will require a tripling of current biofuel production capacity. (Reaching even higher levels of biofuel production in the longer term would require increased biomass productivity and breakthroughs in biofuel conversion technologies, or both.) The enticing possibility of future "drop-in" biofuels could potentially delay investments in nearer-term biofuels like ethanol that are less compatible with the petroleum system.

For electricity, vehicle adoption rates will be the main factor determining the transition time. For PEVs to capture major market share, battery costs must come down by a factor of 3 to 5 through technology advances and manufacturing scale-up (see Chapters 4 and 9). Early infrastructure availability should not be a major issue for battery cars. Electric infrastructure is ubiquitous and many consumers could readily adopt home charging. The existing electricity grid (generation, transmission, and distribution) should have enough underutilized capacity to handle millions of PEVs without major changes (see Chapter 2). In the longer term, the evolution of a smart grid should help enable wider use of electric vehicles. A low-carbon grid will be required for PEVs to achieve deep cuts in well-to-wheels carbon emissions.

The rate of hydrogen vehicle adoption will strongly influence the transition rate. As with electric vehicles, there is a need to buy down the cost of hydrogen vehicles through technical advancement and manufacturing scale-up. Hydrogen faces an additional transition barrier: the "chicken and egg" problem. Early adopters of hydrogen vehicles must be sure of a convenient, cost-effective fueling network, while early fuel suppliers must be sure that there are enough vehicles to use their stations. To assure adequate fuel supply, it will be important to collocate the first vehicles and early infrastructure in "lighthouse cities." STEPS researchers have developed placement strategies for early vehicles and infrastructure that could achieve good fuel accessibility with a very sparse station network, but implementing these will require close coordination among automakers and fuel suppliers, and strong policy support.

FACTORS LIMITING THE RATE OF DEPLOYMENT OF NEW FUEL INFRASTRUCTURES

The rate-limiting factors for infrastructure deployment are summarized here for each fuel.

	Hydrogen	**Electricity**	**Biofuels**
Resources	No major resource limitations, due to diversity of resources available for hydrogen production.	No major resource limitations, due to diversity of resources available for electricity production.	Limits on providing enough low-carbon biomass for all transportation.
Technology gaps	No major technology limitations for delivery infrastructure. Low-cost, low-C hydrogen production needed (renewable, CCS). Fuel cells are critical for vehicles.	No major technology limitations for delivery infrastructure. Low-cost, low-C electricity production needed (renewable, CCS). Batteries are critical for vehicles.	No major technology limitations for delivery infrastructure or vehicles. Biorefineries are critical technology.
Costs	High initial costs for small, underutilized stations until number of hydrogen vehicles rises. As hydrogen demand increases, hydrogen costs decrease, because of scale economies associated with central hydrogen production, delivery systems, and hydrogen stations.	Initial infrastructure costs should not be a limiting factor for PEVs: home chargers have relatively low initial investment costs because they can be added one at a time. Need for public charging and distribution upgrades could raise costs.	Biorefineries are primary infrastructure cost. Need to build large-scale biorefineries for low fuel costs.
Transitions	Need for coordinated, geographically focused deployment of vehicles and infrastructure.	Rate of vehicle adoption, which will determine the rate of infrastructure deployment.	No vehicle-related limitations. Rate of deployment of biofuels and biorefineries in next few decades (RFS) will determine transition rate.

INFRASTRUCTURE INVESTMENTS NEEDED TO SUPPORT 10 PERCENT OF U.S. LDVS USING HYDROGEN, ELECTRICITY, OR BIOFUELS

As an example of the infrastructure investments needed to support introduction of new vehicles, we sketch in the table below the infrastructure that would be needed to support about 10 percent of U.S. light-duty vehicles (20 million vehicles) using hydrogen, electricity, or biofuels. Even at this relatively modest level, which might be reached by 2025, tens of billions of dollars would be needed to build infrastructure. For hydrogen, investments would occur primarily in production and delivery; for biofuels, biorefineries are the major capital cost; and for electricity, in-home chargers make up the majority of the infrastructure cost. Building a larger-scale infrastructure (serving 50 percent of U.S. vehicles) would cost at least five times as much.

	Hydrogen	**Electricity**	**Biofuels**
Fuel consumption (assumed vehicle fuel economy)	5 billion kg H2/yr 0.6 EJ/yr (60 kg H2/mi)	90 billion kWh/y 0.33 EJ/y (300 Wh/mi)	12 billion gge/yr 1.4 EJ/yr (25 mi/gge)
Primary resources required (EJ/y)	To supply all hydrogen from natural gas would require about 0.8 EJ/y, about 3% of total natural gas use in the United States today.	In the near term (2020), there will be a growing use of renewable electricity. Future grid scenarios imply a mix of low-carbon sources by 2050.	Corn (about 30% of 2008 supply) Forest wastes, 0.4 EJ/y (24 million tons of estimated 61 million tons available) Ag. residues, 0.5 EJ/y (33 million tons of estimated 238 million tons available) Municipal solid waste, 0.4 EJ/y (29.5 million tons of estimated 135 million tons available)
Fuel production plants (number of plants, average size [bbl oil equiv/day or GJ/d])	24 central biomass H2 plants (30–200 tonnes H2/day); most H2 production via 1–2 tonne/day on-site natural gas reformers at refueling stations	28 GW at 35% capacity factor = nighttime electricity from 28 1000-MW coal or nuclear plants or 10,000 3-MW wind turbines (~ total installed wind capacity today) (28 GW < 5% of U.S. electricity generation capacity)	150 corn ethanol plants 76 cellulosic biorefineries 16 biodiesel plants

	Hydrogen	**Electricity**	**Biofuels**
Fuel distribution network (type, extent in miles, compatibility w/existing system)	9,000 miles hydrogen pipeline in urban areas; most production on-site	Use electricity transmission and distribution system. May need "smart grid" upgrades.	Additional 7,000 rail tank cars; rail receiving yards at 25% of fuel terminals
Vehicle refueling or recharging interface (number of stations)	18,000 stations total: 14,000 on-site SMR, 4,000 pipeline stations	Home recharging + fast-charge stations on interstates	None if cellulosic biofuels are "drop-in"; 20,000 E85 stations if all ethanol
Cost breakdown for infrastructure capital investment	$38 billion total: $4 B biomass plants, $9 B pipelines, $21B on-site SMRs, $4B pipeline stations	$16–42 billion total: $800–2,100 per vehicle for in-home chargers	$50–70 billion; more than 80% of investment is for biorefineries, rest is for biofuel delivery system

Policies to Encourage Infrastructure Development

A variety of policies, listed in this book's introduction, are driving toward lower-carbon fuels and zero-emission vehicles. These are covered in more detail in Part 4. Realistic policies should recognize the large capital investments and long planning horizon to build the new fuel supply infrastructures required to enable new types of vehicles. In some cases, new vehicle types (such as battery electric fuel cell cars) are mandated while the corresponding energy infrastructure is not. Increasingly, policy should seek to encourage the whole pathway (vehicle and fuel) with incentives so that different stakeholders are motivated strongly enough to participate and coordinate in infrastructure transitions.

Summary and Conclusions

- Each fuel type (hydrogen, electricity, and biofuels) faces infrastructure deployment challenges, which differ by pathway.

- Liquid biofuels are relatively easy to store and transport, and require few vehicle changes to implement compared to the fundamentally new drive trains needed for electric vehicles and hydrogen fuel cells. Some biofuels may be at least partly compatible with the existing petroleum delivery and refueling infrastructure. The main technology gap is development of low-cost advanced biorefineries. The rate of biorefinery deployment will determine the rate of biofuel adoption. Ultimately, availability of biomass is the factor that will limit how much biofuel (and biofuel infrastructure) will be deployed.

- Hydrogen requires the biggest infrastructure changes: new hydrogen production and delivery systems and a network of refueling stations. The main technology gap is development of low-cost, low-carbon hydrogen production technology. Successful introduction will require coordination of vehicle and infrastructure deployments in carefully chosen geographic areas.

- Electricity is already being produced and delivered to users, so the main near-term infrastructure needs are new home chargers. In the longer term, integration of charging demands will occur as part of the larger electric power system, and a low-carbon electricity supply will be needed. The infrastructure build-out rate will be paced by the rate of market penetration of electric vehicles.

- Both electricity and hydrogen could utilize large low-carbon resources. Continued development of low-cost, low-carbon supplies is needed for both electricity and hydrogen.

Notes

1. C. Yang and R. W. McCarthy, "Electricity Grid: Impacts of Plug-In Electric Vehicle Charging," *Environmental Management* 43 (June 2009), 16–20.
2. National Research Council, Committee on Assessment of Resource Needs for Fuel Cell and Hydrogen Technologies, *Transitions to Alternative Transportation Technologies: A Focus on Hydrogen* (Washington, DC: National Academies Press, 2008), available from http://www.nap.edu/catalog.php?record_id=12222.
3. K. Morrow, D. Karner, and J. Francfort, "Plug-in Hybrid Electric Vehicle Charging Infrastructure Review," INL/EXT-08-15058 (Idaho National Laboratory, November 2008).

Chapter 6: Comparing Greenhouse Gas Emissions

Timothy Lipman and Mark A. Delucchi

We turn now to comparing the environmental impacts of our alternative fuel / advanced vehicle pathways. Reducing greenhouse gas (GHG) emissions from vehicles and fuels is one key to lessening transportation's contribution to the climate change problem. This chapter presents much of what is known about the relative emissions of GHGs from battery, fuel cell, and plug-in hybrid electric vehicles versus conventional internal combustion engine vehicles.

We first give some background on the issue of GHG emissions and their climate impact, and review previous research. We then discuss how GHG emissions from electric vehicle (EV) fuel cycles are estimated, before reviewing and comparing recent estimates of GHG emissions from the fuel cycles of various types of EVs. (Note that researchers generally distinguish emissions related to the life cycle of fuels and energy used to power the vehicle—the fuel cycle—from emissions related to the life cycle of the vehicle and the materials it is made from—the vehicle life cycle. In this chapter we focus mainly but not exclusively on fuel-cycle emissions, because there has been relatively little work on vehicle life-cycle emissions.) We next examine the potential for EVs to rapidly scale up to meet the climate challenge, and finally we discuss key uncertainties, areas for further research, and conclusions.

Background and Previous Research

GHGs are a number of different gases and aerosols that have climatic impacts. For EVs of various types that are fueled with electricity and/or hydrogen, the GHGs of greatest interest are carbon dioxide (CO_2), methane (CH_4), nitrous oxide (N_2O), nitrogen oxides (NO_X), the latest automotive refrigerants (HFC-134a, HFO-1234yf, and so on), ozone (O_3), and direct and secondary particulates from power production. Some other gases with apparently lesser significance (due in part to their relatively weak global warming potentials) but that also contribute are carbon monoxide (CO) and various nonmethane hydrocarbons (NMHCs).

Scientists compare the climatic impact of these various gases in terms of what is called radiative forcing. Radiative forcing is a direct measure of the imbalance between the energy flowing into the earth's atmosphere from the sun and the energy being reflected and radiated back out into space; if there is more energy coming into than leaving the atmosphere, the earth is going to heat up. The year 1750, before world industrialization began, is used by many scientists and the Intergovernmental Panel on Climate Change as the baseline or zero point in relation to which radiative forcing is computed. When we look at radiative forcing, CO_2 has had the single largest effect, but various other gases and atmospheric species are significant as well. For example, ozone and aerosols—which are omitted from most analyses of GHG emissions from EVs—have had a greater absolute radiative forcing effect than nitrous oxide.

RADIATIVE FORCING 1750–2005 FROM GREENHOUSE GASES CAUSED BY HUMAN AND NATURAL ACTIVITIES

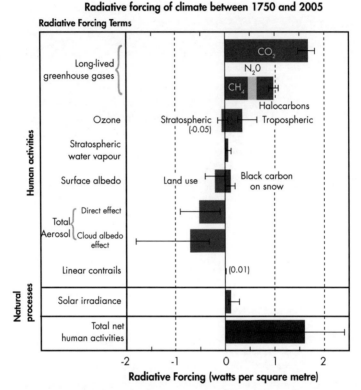

When we look at radiative forcing (the imbalance between the energy flowing into the earth's atmosphere from the sun and the energy being reflected and radiated back out into space) between 1750 and 2005, human-generated CO_2 has had the single largest effect, but various other gases and atmospheric species are significant as well.

KEY GREENHOUSE GASES: INCREASES IN ATMOSPHERIC CONCENTRATIONS 1750–2007 AND RADIATIVE FORCING EFFECTS

This table summarizes the pre-industrial (1750) and current (2007) atmospheric levels in parts per million of four key GHGs, as well as their total increase and their radiative forcing effect in watts per square meter. CO_2 accounts for the largest radiative forcing effect, but the others also make significant contributions. Source: Oak Ridge National Laboratory, Carbon Dioxide Information Analysis Center (2009), http://cdiac.ornl.gov.

Greenhouse Gas	Pre-Industrial Level	Current Level	Increase Since 1750	Radiative Forcing (W/m2)
Carbon dioxide	280 ppm	385 ppm	105 ppm	1.66
Methane	700 ppb	1741 ppb	1045 ppb	0.48
Nitrous oxide	270 ppb	321 ppb	51 ppb	0.16
Ozone	25 ppb	34 ppb	9 ppb	0.35
CFC-12	0 ppt	533 ppt	533 ppt	0.17

Research on GHG emissions from fuel cycles related to electric vehicle use dates back to at least the early 1990s, when the introduction of battery electric vehicles (BEVs) by major automakers and growing concern about climate change spurred interest. At that point, most studies focused on criteria air pollutants (carbon monoxide, lead, nitrogen dioxide, ozone, particulate matter, and sulfur dioxide), but some GHGs were occasionally included. Significant research efforts in the 1990s included those by university and government lab research groups[1] and consulting firms.[2] The next decade saw major efforts by automakers,[3] industry research organizations,[4] and other groups. More recently, there has been a series of more sophisticated efforts based on further developments in electric vehicle technology and the concept of plug-in hybrid electric vehicles (PHEVs).

Among the most useful tools for analyzing and comparing emissions from a wide range of vehicle and fuel combinations are two models developed by academic researchers: the Life-cycle Emissions Model (LEM) from UC Davis and the Greenhouse Gases, Regulated Emissions, and Energy Use in Transportation (GREET) model from Argonne National Lab. Both of these are well developed with long histories and are also relatively well documented. Other studies have examined more specific vehicle and fuel pathways involving EVs with regard to their GHG emissions and have yielded interesting insights. Several of these are also discussed in this chapter.

RECENT TRANSPORTATION FUEL-CYCLE OR LIFE-CYCLE MODELING EFFORTS

Various efforts have examined the emissions of GHGs from electric vehicle fuel cycles or life cycles, but focusing on different types of vehicles and fuel feedstock options, and at varying levels of detail. Here we compare the structure and coverage of several of these modeling efforts. This table gives a good sense of key aspects of emission comparisons and the extent to which each of the models encompasses or addresses them.

Project	GM-ANL U.S.	GM–LBST Europe	MIT 2020/2035	EUCAR
Region **Time frame**	North America Near term (about 2010)	Europe 2010	Based on U.S. data 2020/2035	Europe 2010 and beyond
Transport modes	LDV (light-duty truck)	LDV (European mini-van)	LDV (mid-size family passenger car)	LDV (compact 5-seat European sedan)
Vehicle drive-train type	ICEVs, HEVs, BEVs, FCVs	ICEVs, HEVs, FCVs	ICEVs, HEVs, BEVs, FCVs	ICEVs, HEVs, FCVs
Motor fuels	Gasoline, diesel, naptha, FTD, CNG, methanol, ethanol, CH2, LH2, electricity	Gasoline, diesel, naptha, FTD, CNG, methanol, ethanol, CH2, LH2	Gasoline, diesel, FTD, methanol, CNG, CH2, electricity (2020)/plus ethanol (2035)	Gasoline, diesel, FTD, CNG, ethanol, FAME, DME, aptha, methanol, CH2, LH2
Fuel feedstocks	Crude oil, natural gas, coal, crops, ligno-cellulosic biomass, renewable and nuclear power	Crude oil, natural gas, coal, crops, ligno-cellulosic biomass, waste, renewable and nuclear power	Crude oil, natural gas, renewable and nuclear power (2020) plus corn, cellulose, tar sands (2035)	Crude oil, natural gas, coal, nuclear, wind, sugar beets, wheat, oil seeds, wood
Vehicle energy-use modeling, including drive cycle	GM simulator, U.S. combined city/highway driving	GM simulator, European Drive Cycle (urban and extra-urban driving)	MIT simulator (2020)/Advisor (2035), U.S. combined city and highway driving (2020)/various cycles (2035)	Advisor (NREL simulator), New European Drive Cycle
Fuel life cycle	GREET model	LBST E2 I-O model and database	literature review (2020)/ GREET and other sources (2035)	LBST E2 I-O model and database (review & update of GM et al. [2002])
Vehicle and material life cycle	Addressed in GREET 2.7	Addressed in GREET 2.7	Detailed literature review and analysis (2020)/GREET 2.7 (2035)	Not included
GHGs [CEFs]	CO2, CH4, N2O [IPCC] (other pollutants included as non-GHGs)	CO2, CH4, N2O [IPCC]	CO2, CH4 (2020)/ CO2, CH4, N2O (2035) [IPCC]	CO2, CH4, N2O [IPCC]
Infrastructure	Not included	Not included	Not included	Not included
Price effects	Not included	Not included	Not included	Not included

Project	ADL AFV LCA	EcoTraffic	CMU I-O LCA	Japan AFVs CO2	LEM
Region	United States	Generic, but weighted toward European conditions	United States	Japan	Multi-country (primary data for U.S.; other data for up to 30 countries)
Time frame	1996 baseline, future scenarios	Between 2010 and 2015	Near term	Near term?	Any year from 1970 to 2050
Transport modes	Subcompact cars	LDVs (generic small passenger car)	LDVs (midsize sedan)	LDVs (generic small passenger car)	LDVs, HDVs, buses, light-rail transit, heavy-rail transit, minicars, scooters, offroad vehicles
Vehicle drive-train type	ICEVs, BEVs, FCVs	ICEVs, HEVs, FCVs	ICEVs	ICEVs, HEVs, BEVs	ICEVs, BEVs, FCVs
Motor fuels	Gasoline, diesel, LPG, CNG, LNG, methanol, ethanol, CH2, LH2, electricity	Gasoline, diesel, FTD, CNG, LNG, methanol, DME, ethanol, CH2, LH2	Gasoline, diesel, biodiesel, CNG, methanol, ethanol	Gasoline, diesel, electricity	Gasoline, diesel, LPG, FTD, CNG, LNG, methanol, ethanol, CH2, LH2, electricity
Fuel feedstocks	Crude oil, natural gas, coal, corn, ligno-cellulosic biomass, renewable and nuclear power	Crude oil, natural gas, ligno-cellulosic biomass, waste	Crude oil, natural gas, crops, ligno-cellulosic biomass	Crude oil, natural gas, coal, renewable and nuclear power	Crude oil, natural gas, coal, crops, lignocellulosic biomass, renewable and nuclear power
Vehicle energy-use modeling, including drive cycle	Gasoline fuel economy assumed; AFV efficiency estimated relative to this	Advisor (NREL simulator), New European Drive Cycle	Gasoline fuel economy assumed; AFV efficiency estimated relative to this	None; fuel economy assumed	Simple model based on SIMPLEV-like simulator, U.S. combined city/highway driving
Fuel life cycle	Arthur D. Little emissions model, revised	Literature review	Own calculations based on other models (LEM, GREET)	Values from another study	Detailed internal model
Vehicle and material life cycle	Not included	Not included	Economic Input-Output Life Cycle Analysis software (except end-of-life)	Detailed part-by-part analysis	Internal model based on detailed literature review and analysis
GHGs [CEFs]	CO2, CH4 [partial GWP] (other pollutants included as non-GHGs)	None (energy efficiency study only)	CO2, CH4, N2O? [IPCC] (other pollutants included as non-GHGs)	CO2	CO2, CH4, N2O, NOx, VOC, SOx, PM, CO, H2, HFCs, CFCs [own CEFs, also IPCC CEFs]
Infrastructure	Not included	Not included	Not included	Not included	Crude representation
Price effects	Not included	Not included	Not included (fixed-price I-O model)	Not included	A few simple quasi-elasticities

The terms in the model comparison table are defined as follows:

Region	The countries or regions covered by the analysis.
Time frame	The target year of the analysis.
Transport modes	The types of passenger transport modes included. LDVs = light-duty vehicles, HDVs = heavy-duty vehicles.
Vehicle drivetrain type	ICEVs = internal combustion engine vehicles, HEVs = hybrid electric vehicles (vehicles with an electric and an ICE drivetrain), BEVs = battery electric vehicles, FCVs = fuel cell electric vehicles.
Motor fuels	Fuels carried and used by motor vehicles. FTD = Fischer-Tropsch diesel, CNG = compressed natural gas, LNG = liquefied natural gas, CH2 = compressed hydrogen, LH2 = liquefied hydrogen, DME = dimethyl ether, FAME = fatty acid methyl esters.
Fuel feedstocks	The feedstocks from which the fuels are made.
Vehicle energy-use modeling	The models or assumptions used to estimate vehicular energy use (which is a key part of fuel-cycle CO_2 emissions), and the drive cycle over which fuel usage is estimated (if applicable).
Fuel life cycle	The models, assumptions, and data used to estimate emissions from the life cycle of fuels.
Vehicle and materials life cycle	The life cycle of materials and vehicles, apart from vehicle fuel. The life cycle includes raw material production and transport, manufacture of finished materials, assembly of parts and vehicles, maintenance and repair, and disposal.
GHGs and CEFs	The pollutants (greenhouse gases, or GHGs) that are included in the analysis of CO_2-equivalent emissions, and the CO_2-equivalency factors (CEFs) used to convert non-CO_2 GHGs to equivalent amount of CO_2 (IPCC = factors approved by the Intergovernmental Panel on Climate Change [IPCC]).
Infrastructure	The life cycle of energy and materials used to make and maintain infrastructure, such as roads, buildings, equipment, rail lines, and so on. (In most cases, emissions and energy use associated with the construction of infrastructure are small compared with emissions and energy use from the end use of transportation fuels.)
Price effects	The relationships between prices and equilibrium final consumption of a commodity (for example, crude oil) and an "initial" change in supply of or demand for the commodity or its substitutes, due to the hypothetical introduction of a new technology or fuel.

How Emissions Are Estimated

Although battery electric vehicles (BEVs) and fuel cell vehicles (FCVs) are often called zero-emission vehicles, and although most BEV and FCV fuel options do entail significant reductions in GHG and criteria pollutant emissions compared with conventional gasoline vehicles, this is not always the case—for example, if coal without carbon capture is the sole feedstock for the electricity for BEV charging. Here we take a closer look at the components of electric vehicle emissions and how they are estimated.

Emissions of GHGs from conventional gasoline internal combustion engine vehicles (ICEVs) are a combination of "upstream" emissions from fuel production and distribution, and "downstream" emissions from vehicle operation. By contrast, GHG emissions from the life cycle of fuels for BEVs and FCVs are entirely in the form of upstream emissions related to the production of electricity or hydrogen, with no emission from the vehicles themselves (except for water vapor in the case of FCVs, and any emissions related to heating and cooling sytems). The emissions from these vehicles are thus entirely dependent on the manner in which the elcctricity and/or hydrogen is produced, along with the energy efficiency of the vehicle (typically expressed in watt hours per mile or kilometer for BEVs, and miles or kilometers per kilogram for hydrogen-powered vehicles). For PHEVs emissions are a complex combination of upstream and in-use emissions since these vehicles use a combination of grid electrical power and another fuel that is combusted (or potentially converted with a fuel cell) onboard the vehicle. There can be significant tailpipe emissions depending on travel patterns and the type of plug-in hybrid, along with any strategies to prevent criteria pollutant emissions from the catalyst-based control system when it is periodically starting up from low temperatures.

Emissions of CO_2 from fuel combustion are comparatively easy to estimate since virtually all of the carbon in fuel oxidizes to CO_2. In contrast, combustion emissions of all the other greenhouse gases are a function of many complex aspects of combustion dynamics (such as temperature, pressure, and air-to-fuel ratio, among other factors) and of the type of emission control systems used, and hence cannot be derived from one or two basic characteristics of a fuel. Instead, we need to use published emission factors for each combination of fuel, end-use technology, combustion conditions, and emission control system. Likewise, noncombustion emissions of greenhouse gases as part of the fuel cycle (for example, gas flared at oil fields, or N_2O produced and emitted from fertilized soils) cannot be derived from basic fuel properties and instead must be measured and estimated source by source and gas by gas. We have provided a compendium of many of these emission factors,[14] but note that some of them have since been updated based on more recent data than were available at the time our compendium was published.

Upstream emissions

The emissions associated with fuel production, or upstream emissions, dominate the fuel cycles associated with BEVs and FCVs. For BEVs, upstream emissions consist of emissions from the production and delivery of electricity for vehicle charging. These emissions vary regionally based on the fuels and types of power plants used to generate electricity. For FCVs, emissions are again entirely upstream, from the production, delivery, and dispensing of gaseous or liquid hydrogen, with the exception of small amounts of water vapor emitted directly and any emissions of refrigerants used for air conditioning. For PHEVs, on the other hand, total emissions consist of a mix of upstream emissions from electricity generation (proportional to the extent that the vehicle is recharged with electricity) and both upstream and in-use emissions from fuel combustion from the vehicle engine (or potentially conversion in a fuel cell).

Various studies have examined the upstream emissions from vehicle fuel production, especially from gasoline and diesel fuel and electricity production but also for other fuels such as compressed natural gas, ethanol and methanol, hydrogen, and biodiesel. These have been conducted in various regions (mainly in the United States and Europe) and with various emphases (various vehicle type/technology combinations, CO_2 or a whole suite of gases, sometimes including criteria pollutants as well as GHGs, and so on). Emissions from electricity generation processes are generally well known and well studied; this is less true for hydrogen production, but in most cases these emissions also are well understood. Some novel hydrogen production methods, and those that are based on conversion from biofuels, have somewhat complex and certainly not completely understood and established levels of emissions of GHGs.

Combustion or "in-use" emissions

Emissions of GHGs from engine combustion processes result from a complex combination of combustion dynamics and emission controls, and vary widely by fuel type, engine operation, and emission control system applied (if any). For EVs, combustion emissions from the vehicle are limited to PHEVs that either use a combustion engine and generator as a "range extender" for what is fundamentally an electric vehicle driveline, or where the engine is connected in parallel to the driveline with the electric motor. Either way, the combustion engine operates periodically to supplement the electric motor operation and thereby produces GHG emissions. Additional in-use emissions from EVs include those that can occur from a supplemental fuel-fired heater in the passenger cabin for occasional use in colder climates, and from vehicle air-conditioning systems, where GHGs are often used as refrigerants.

Key GHG emission products from combustion engines include CO_2, CH_4, N_2O, CO, NO_x, soot, and various air toxics and other trace chemicals that can play roles in the formation of secondary particulates and other gases (such as ozone) that are known to have climatic effects.

A CLOSER LOOK AT ESTIMATING KEY GHG EMISSIONS

Carbon dioxide

Carbon dioxide is emitted directly from combustion engine vehicles, and these emissions are closely correlated with the total carbon in the vehicle fuel. The U.S. Environmental Protection Agency (EPA) uses a carbon content estimate of 2,421 grams of carbon per gallon of gasoline and 2,778 grams of carbon per gallon of diesel fuel for purposes of estimating CO_2 emissions from combustion of these fuels.[15] To approximate the CO_2 emissions resulting from combustion of these fuels, we multiply the fuel carbon content by an "oxidization factor" and by the ratio of molecular weights of CO_2 (44) to elemental carbon (12). This results in the following sample calculations, assuming a 99-percent oxidization factor (the value used by the EPA):

CO_2 emissions from a gallon (liter) of gasoline = 2,421 grams x 0.99 x (44/12) = 8,788 grams => 8.8 kg/gallon = 2.3 kg CO_2/liter

CO_2 emissions from a gallon (liter) of diesel = 2,778 grams x 0.99 x (44/12) = 10,084 grams => 10.1 kg/gallon = 2.7 kg CO_2/liter

These factors can be used for reasonable first-order approximations of the direct tailpipe emissions of CO_2 from combustion engine vehicles using gasoline and diesel fuel.[16]

Carbon dioxide is also emitted directly from electricity-generating power plants, particularly those that burn fossil fuels or biomass. In the case of biomass-powered facilities, the CO_2 emitted represents a partial or full closed loop, as biomass removes carbon dioxide from the atmosphere as it is grown. Renewable and nuclear facilities emit little or no CO_2 directly but may have significant emissions through other parts of their full fuel cycle (for example, during construction of nuclear plants, uranium mining, or construction of wind turbine systems). In general, these emissions are much lower than lifetime emissions of coal-fired power plants, which are used for up to 50 years and emit GHGs at a level locked in with each new plant built. For example, an estimated 100 million tons of CO_2 are generated by a 500 MW coal-fired power plant over a 40-year lifetime.[17] For purposes of comparison, a 2004 article reports that coal-fired power plants in the United States emit about 1,200 kg CO_2 per MWh, and natural gas combined-cycle plants emit about 700 kg CO_2 per MWh, while renewable and nuclear sources emit on the order of 25 to 75 kg CO_2 per MWh.[18]

Methane

Methane (CH_4) has a 100-year global warming potential (GWP) value of 25, meaning that each gram has 25 times the radiative-forcing impact of a gram of CO_2 over that time period.[19] It is emitted directly by both combustion engine vehicles and power plants.

Methane emissions from combustion engines are a function of the type of fuel used, the design and tuning of the engine, the type of emission control system, the age of the vehicle, and other factors. Although methane emissions per se are not regulated in the United States, the systems used to control emissions of nonmethane and total hydrocarbons from combustion engines do to some extent control CH_4 emissions. Not much data exists on CH_4 emissions from high-mileage gasoline light-duty vehicles, but these emissions seem to increase somewhat as a function of catalyst age, as do N_2O emissions.[20] There are many CH_4 emissions tests for gasoline vehicles, but comparatively few for diesel and alternative-fuel ones.

Power plants also produce relatively small amounts of methane as unburned hydrocarbons, with emission factors that are available in comprehensive databases from the U.S. EPA and the Intergovernmental Panel on Climate Change (IPCC).[21] Natural gas power plants can also produce fugitive methane emissions from pipelines, purging, and venting procedures.[22]

Nitrous oxide

N_2O is a potent GHG with a 100-year GWP value of 298[23] that is emitted directly from motor vehicles and power plants. Emission factors for both sources are available in comprehensive databases from the U.S. EPA and the Intergovernmental Panel on Climate Change (IPCC).[24] Emissions of N_2O from combustion engines have been estimated by other research centers as well.[25] Generally, N_2O emissions from power plants are a small fraction of total fuel-cycle CO_2-equivalent (CO_2e) GHG emissions from vehicles.

N_2O emissions from catalyst-equipped gasoline light-duty vehicles depend significantly on the type and temperature of catalyst, rather than total oxide of nitrogen (NO_x) levels or fuel nitrogen content. Gasoline contains relatively little nitrogen and therefore fuel NO_x and N_2O emissions from autos are low; as a result, cars without catalytic converters produce essentially no net N_2O. However, cars with catalytic converters can produce significant N_2O when the catalyst starts out cold. Essentially, as a vehicle warms up and the catalyst temperature increases, a "pulse" of nitrous oxide is released. This occurs until the catalyst temperature increases beyond the temperature window for N_2O formation, after which emissions of N_2O are minimal. Older catalysts have a wider window for formation, hence older three-way catalyst equipped vehicles tend to emit more N_2O than newer vehicles.

This temperature dependence of N_2O formation has important implications regarding potential emissions from PHEVs. If the combustion engine in a PHEV is cycling on and off, the catalyst may be cooling off and reheating multiple times during a trip instead of a single time, which could result in increased emissions of N_2O. One way to mitigate this would be to electrically heat the catalyst to keep it from cooling off, but this would come at some (perhaps small) net energy penalty for the vehicle. This issue of potentially increased emissions of N_2O from PHEVs appears to be a significant issue for further study.

Power plants also emit N_2O. Although the power plant combustion chemistry of N_2O is quite complex, several general trends are apparent. Higher N_2O emissions are generally associated with lower combustion temperatures, higher-rank fuels, lower ratios of fuel oxygen to fuel nitrogen, higher levels of excess air, and higher fuel carbon contents.[26]

Other greenhouse gases

Emissions of other GHGs from the production and use of EVs include criteria pollutants, such as CO, NMHCs, NO_x, and SO_x, and automotive refrigerants such as CFC and HFC-134a. Criteria pollutants typically have weak direct-forcing GWP values and are emitted in much lower quantities than CO_2 but can contribute to the formation of compounds that do have a strong radiative forcing effect, such as ozone and sulfate aerosol.

Also potentially important are the refrigerants used in automotive air conditioners, which can be released during accidents or improper maintenance procedures, and which can have very high GWP values. Automotive air conditioners used the refrigerant R-12 throughout the 1970s and 1980s and transitioned to HFC-134a in the 1990s, primarily to help protect the earth's ozone layer. HFC-134a is still a potent GHG, however, with a 100-year GWP value of 1430.[27] Other non-ozone-depleting refrigerants such as HFO-1234yf and CO_2 are being investigated as lower GWP options that can still be effective in automotive applications.

Emissions of CO_2 and other GHGs from the vehicle life cycle

What we call the vehicle life cycle includes producing the materials that compose a vehicle and the life cycle of the vehicle itself. The life cycle of automotive materials, such as steel, aluminum, and plastics, extends from production of raw ore to delivery of finished materials to assembly plants, and includes recycled materials as well as materials made from virgin ore. The life cycle of the vehicle itself includes vehicle assembly, transportation of finished motor vehicles and motor-vehicle parts, and vehicle disposal, but not the operational emissions from the vehicle, which we consider separately.

In the vehicle life cycle there are two broad sources of GHG emissions, similar to the emissions sources in the industrial sector in general: (1) emissions related to the use of process energy (for example, fuels burned in industrial boilers to provide process heat), and (2) noncombustion emissions from process areas (for example, emissions from the chemical reduction of alumina to aluminum, or NMHC emissions from painting auto bodies). Energy use and process areas can produce CO_2, CH_4, N_2O, CO, NMHCs, SO_x, NO_x, particulate matter, and other pollutants relevant to life-cycle analysis of GHG emissions. The most extensive of the vehicle life-cycle assessment models, including the LEM and the GREET model, include characterization of these vehicle manufacturing emissions and their contribution to the overall emissions from various vehicle/fuel life cycles.

In general, manufacturing emissions can be somewhat higher for some types of EVs (such as those that use large nickel-based batteries) than for conventional vehicles. The vehicle manufacturing emissions for EVs are often proportionately larger than for conventional vehicles because of their lower vehicle-operation life-cycle (i.e., fuel cycle) emissions. A key point is that because vehicle operational emissions dominate, EVs are often much cleaner than conventional vehicles in an overall sense even if they have slightly to somewhat higher vehicle manufacturing emissions.

PHEV emissions

PHEVs generate GHG emissions from three distinct sources: the life cycle of fuels used in the ICE, the life cycle of electricity used to power the electric drivetrain, and the life cycle of the vehicle and its materials. A number of studies, reviewed later, have estimated GHG-emission reductions from PHEVs relative to conventional ICEVs considering the life cycle of fuels and the life cycle of electricity generation. Because energy use and emissions for the vehicle life cycle are an order of magnitude smaller than energy use and emissions for the fuel and electricity life cycle,[28] and because there are relatively few studies of emissions from the PHEV vehicle life cycle, we do not consider the vehicle life cycle in much detail here.

Various vehicle design and operational strategies are available for PHEVs, and these can have important emissions implications. For example, PHEVs can be designed to be either charge-depleting (CD) or charge-sustaining (CS), and this affects the relative levels of electricity and gasoline used.[29] PHEVs with true all-electric range (AER) could allow drivers to make some trips without the engine turning on at all (or at least very little), relying almost entirely on the energy stored in the battery. However, some PHEVs are not designed for this and instead employ "blended mode" operation, where the engine turns off and on periodically even at relatively high states of battery charge. And in other cases, even for "series type" PHEVs with extensive AER, some engine operation is to be expected both on longer trips and in other cases where the PHEV battery becomes discharged before it can be charged again.

With regard to GHG emissions from the life cycle of petroleum fuels used in ICEs for PHEVs, these depend mainly on the fuel use of the engine and the energy inputs and emission factors for the production of crude oil and finished petroleum products. A number of studies estimate the fuel use of ICEs in PHEVs; for example, Bradley and Frank[30] found a variety of simulated and tested PHEVs to reduce gasoline consumption by 50 percent to 90 percent. GHG emissions from the use of electricity by PHEVs depend mainly on the energy use of the electric drivetrain, the efficiency of electricity generation, and the mix of fuels used to generate electricity.

The energy use of the electric drivetrain in a PHEV is a function of the size and technical characteristics of the electric components (battery, motor, and controller), the vehicle driving and charging patterns, and the control strategy that determines when the vehicle is powered by the battery and when it is powered by the ICE. The studies reviewed here consider two basic control strategies: all-electric and blended. In a PHEV with a large all-electric range (AER), the electric drivetrain is sized to have enough power to be able to satisfy all power demands over the drive cycle without any power input from the engine. By contrast, in the blended strategy, the electric drivetrain and the engine work together to supply the power over the drive cycle. The blended strategy can be either "engine dominant," in which case the electric motor is used to keep the engine running at its most efficient torque/rpm points, or "electric dominant," in which case the engine turns on only when the power demand exceeds the capacity of the electric drivetrain.[31]

The efficiency of electricity generation can be estimated straightforwardly on the basis of data and projections in national energy information systems, such as those maintained by the Energy Information Administration (for the U.S.) (www.eia.doe.gov/fuelelectric.html) or the International Energy Agency (for the world) (www.iea.org). Life-cycle models, such as the GREET model and the LEM, also have comprehensive estimates of GHG emissions from the life cycle of electricity generation for individual types of fuels.

However, it is not straightforward to estimate the mix of fuels used to generate the electricity that actually will be used to charge batteries in PHEVs. The "marginal" generation fuel mix depends on the interaction of supply-side factors, such as cost, availability, and reliability, with anticipated hourly demand patterns, and can vary widely from region to region.[32] This supply-demand interaction can be represented formally with models that attempt to replicate how utilities actually dispatch electricity to meet demand. A few studies, reviewed below, have used dispatch models to estimate the mix of fuels used to generate electricity for charging PHEVs. However, as dispatch models generally are not readily available, most researchers either have assumed that the actual marginal mix of fuels is the year-round average mix or else have reported results for different fuel-mix scenarios.

OVERVIEW OF THE LEM, THE GREET MODEL, AND OTHER MAJOR EFFORTS

The Life-cycle Emissions Model (LEM) uses life-cycle analysis (LCA) to estimate energy use, criteria air-pollutant emissions, and CO_2-equivalent greenhouse-gas emissions from a wide range of energy and material life cycles. It includes life cycles for passenger transport modes, freight transport modes, electricity, materials, heating and cooling, and more. For transport modes, it represents the life cycle of fuels, vehicles, materials, and infrastructure. It calculates energy use and life-cycle emissions of all regulated air pollutants plus GHGs. It includes input data for up to 30 countries, for the years 1970 to 2050, and is fully specified for the United States.

For motor vehicles, the LEM calculates life-cycle emissions for a variety of combinations of end-use fuel, fuel feedstocks, and vehicle types. The fuel and feedstock combinations included in the LEM for light-duty vehicles are shown in the table below.

Fuel --> Feedstock	Gasoline	Diesel	Methanol	Ethanol	Methane (CNG, LNG)	Propane (LPG)	Hydrogen (CH2) (LH2)	Electric
Petroleum	ICEV, FCV	ICEV				ICEV		BEV
Coal	ICEV	ICEV	ICEV, FCV				FCV	BEV
Natural gas		ICEV	ICEV, FCV		ICEV	ICEV	ICEV, FCV	BEV
Wood or grass			ICEV, FCV	ICEV, FCV	ICEV		FCV	BEV
Soybeans		ICEV						
Corn				ICEV				
Solar power							ICEV, FCV	BEV
Nuclear power							ICEV, FCV	BEV

ICEV = internal combustion engine vehicle; FCV = fuel cell vehicle; BEV = battery electric vehicle. Cells with BEVs and FCVs are highlighted in blue.

The LEM estimates emissions of CO_2, CH_4, N_2O, carbon monoxide (CO), total particulate matter (PM), PM less than 10 microns in diameter (PM10), PM from dust, hydrogen (H_2), oxides of nitrogen (NOx), chlorofluorocarbons (CFC-12), nonmethane organic compounds (NMOCs, weighted by their ozone-forming potential), hydrofluorocarbons (HFC-134a), and sulfur dioxide (SO_2). These species are reported individually and aggregated together weighted by CO_2 equivalency factors (CEFs).

These CEFs are applied in the LEM the same way that global warming potentials (GWPs) are applied in other LCA models but are conceptually and mathematically different from GWPs. Whereas GWPs are based on simple estimates of years of radiative forcing integrated over a time horizon, the CEFs in the LEM are based on sophisticated estimates of the present value of damages due to climate change. Moreover, whereas all other LCA models apply GWPs to only CH_4 and N_2O, the LEM applies CEFs to all of the pollutants listed above. Thus, the LEM is unique for having original CEFs for a wide range of pollutants. The following table compares LEM CEFs with IPCC GWPs.

Pollutant	LEM CEFs (year 2030)	IPCC 100-yr. GWPs
NMOC-C	3.664	3.664
NMOC-03/CH4	3	not estimated
CH4	14	23
CO	10	1.6
N2O	300	296
NO2	-4	not estimated
SO2	-50	not estimated
PM (black carbon)	2,770	not estimated
CFC-12	13,000	8,600
HFC-134a	1,400	1,300
PM (organic matter)	-240	not estimated
PM (dust)	-22	not estimated
H2	42	not estimated
CF4	41,000	5,700
C2F6	92,000	11,900
HF	2000	not estimated

CEF = CO_2-equivalency factor; GWP = global warming potential. Source: LEM CEFs from the year 2005 version of the LEM; IPCC GWPs from Intergovernmental Panel on Climate Change, Climate Change 2001: The Scientific Basis, ed. J. T. Houghton, Y. Ding, et al.

The Greenhouse Gases, Regulated Emissions, and Energy Use in Transportation (GREET) model has been under development at Argonne National Laboratory since about 1995. The model assesses more than 100 fuel production pathways and about 75 different vehicle technology / fuel system types, for hundreds of possible combinations of vehicles and fuels. It has more than 10,000 users worldwide and has been adapted for use in various countries around the world.[33] GREET estimates emissions of CO_2, plus CO_2-equivalent emissions of CH_4 and N_2O (based on the IPCC's GWPs), from the fuel cycle and the vehicle life cycle.

GREET 1.8c, released in 2009, is noteworthy for its much expanded treatment of PHEVs along with updated projections of electricity grid mixes in the United States based on the latest projections by the Energy Information Administration. This latest version of the model analyzes PHEVs running on various fuels—not just gasoline and diesel—along with electricity. The additional fuels it analyzes include corn-based ethanol (E85—85 percent blend with gasoline), biomass-derived ethanol (E85), and hydrogen produced by three different methods: (1) steam methane reforming of natural gas (distributed, small scale); (2) electrolysis of water using grid power (distributed, small scale); and (3) biomass-based hydrogen (larger scale). The analysis also examines different regions of the United States, and the United States on average, for power plant mixes and emission factors for BEV and PHEV charging and other electricity demands.

Various other studies of the relative GHG emission benefits of different types of EVs have been done by other university and national laboratory research groups, consulting firms, government agencies, nongovernmental organizations, and industry research groups. Key organizations that have been involved include the Japanese Ministry of Economy, Trade, and Industry; Japanese research universities including the University of Tokyo; the International Energy Agency; the European Union; Natural Resources Canada; and many other government and research organizations around the world. In the United States, in addition to the national laboratories and the University of California, key efforts have been led by the Massachusetts Institute of Technology, Carnegie Mellon University, Stanford University, the Pacific Northwest Laboratory, the National Renewable Energy Laboratory, General Motors, and the Electric Power Research Institute, among others. The results of several of these efforts are discussed and compared below. In our review we emphasize the most extensive studies that included BEVs and FCVs as well as PHEVs, but we note that there are carefully performed studies that look at a narrower range of vehicle technologies (for example, that only compare BEVs to ICEVs).

Estimates of GHG Emissions from EVs

Now that we have explored how the various GHG emissions from EVs are estimated, we will look at the results of a few of the major modeling efforts. We first examine the results of the well-developed Life-cycle Emissions and GREET models regarding emissions of BEVs and FCVs. Then we look at results for BEVs and FCVs from other major modeling efforts, before considering the results of major studies of potential emission reductions from PHEVs.

LEM emission results for BEVs and FCVs

Using the LEM, we find that in the United States in the year 2010, BEVs reduce fuel-cycle GHG emissions by 20 percent (in the case of coal) to almost 100 percent (in the case of hydro and other renewable sources of power). If the vehicle life cycle is included, the reduction is lower, in the range of 7 percent to 70 percent, because emissions from the BEV life cycle are larger than emissions from the gasoline ICEV life cycle due to the production of materials used in the battery. The emission reduction percentages are generally higher in the year 2050, mainly because of the improved efficiency of vehicles and power plants. The emission reductions in Japan, China, and

Germany are similar to those in the United States, except that in those cases the reduction using coal power is higher, due to the greater efficiency of coal plants in Germany and Japan, and to high SO_2 emissions from coal plants in China (SO_2 has a negative CEF).

In the United States in 2010, FCVs using "reformed" gasoline or methanol made from natural gas offer roughly 50-percent reductions in fuel-cycle GHG emissions. FCVs using methanol or hydrogen made from wood reduce fuel-cycle GHG emissions by about 85 percent; FCVs using hydrogen made from natural gas reduce emissions by about 60 percent, and FCVs using hydrogen made from water (using clean electricity) reduce fuel-cycle GHG emissions by almost 90 percent. Again, the reductions are slightly lower if the vehicle life cycle is included, and slightly higher in the year 2050. The patterns in Japan, China, and Germany are essentially the same, because the vehicle technology and the fuel production processes are assumed to be the same as in the United States.

LEM: ELECTRIC VEHICLE VS. GASOLINE ICEV EMISSIONS FOR FOUR COUNTRIES, 2010 AND 2050

These tables present the final gram-per-km emission results from the LEM by vehicle/fuel/feedstock, and percentage changes relative to conventional gasoline vehicles, for the United States, China, Japan, and Germany, for the years 2010 and 2050.

ICEV = internal combustion engine vehicle; BEV = battery electric vehicle; FCV = fuel cell vehicle; NG = natural gas; Hydro = hydro power; Other = solar, geothermal power; RFG = reformulated gasoline; Ox = oxygenate (ETBE, MTBE, ethanol, methanol) (volume percent in active gasoline); M = methanol (volume percent in fuel for methanol vehicle; remainder is gasoline); CNG = compressed natural gas; LNG = liquefied natural gas; CH2 = compressed hydrogen; E = ethanol (volume percent in fuel for ethanol vehicle; remainder is gasoline).

The vehicle life cycle includes emissions from the life cycle of materials used in vehicles, vehicle assembly and transport, the life cycle of refrigerants, the production and use of lube oil, and brake wear, tire wear, and road dust.

Baseline ICEV

2010	U.S.	Japan	China	Germany
Fuel cycle (g/km)	332.5	329.4	337.8	333.0
Fuel and vehicle life cycle (g/km)	392.9	389.8	408.6	392.3
2050				
Fuel cycle (g/km)	280.0	273.3	281.0	273.4
Fuel and vehicle life cycle (g/km)	316.5	307.8	321.5	306.9

United States—BEVs—By Type of Power Plant Fuel

2010	Coal	Fuel Oil	NG Boiler	NG Turbine	Nuclear	Biomass	Hydro	Other
Fuel cycle (g/km)	266.0	231.9	141.3	143.6	14.6	24.2	10.4	7.7
Fuel cycle (% change)	-20%	-30.2%	-57.5%	-56.8%	-95.6%	-92.7%	-96.9%	-97.7%
Fuel and vehicle life cycle (g/km)	365.9	331.8	241.2	243.5	114.5	124.0	110.3	107.6
Fuel and vehicle life cycle (% change)	-6.9%	-15.5%	-38.6%	-38.0%	-70.9%	-68.4%	-71.9%	-72.6%
2050								
Fuel cycle (g/km)	227.5	197.2	105.9	107.8	7.8	(-3.2)	5.2	3.0
Fuel cycle (% change)	-18.7%	-29.6%	-62.2%	-61.5%	-97.2%	-101.1%	-98.1%	-98.9%
Fuel and vehicle life cycle (g/km)	262.4	232.1	140.8	142.7	42.7	31.7	40.1	37.9
Fuel and vehicle life cycle (% change)	-17.1%	-26.7%	-55.5%	-54.9%	-86.5%	-90.0%	-87.3%	-88.0%

United States—FCVs—By Fuel and Feedstock

	General fuel -->	Gasoline	Methanol	Methanol	Ethanol	H2	H2	H2	H2
	Fuel spec -->	RFG-Ox10	M100	M100	E100	CH2	CH2	CH2	CH2
	Feedstock -->	Crude oil	NG	Wood	Grass	Water	NG	Wood	Coal
2010									
Fuel cycle (g/km)		163.9	164.1	47.9	85.4	35.7	135.1	47.8	83.6
Fuel cycle (% change)		-50.7%	-50.7%	-85.6%	-74.3%	-89.3%	-59.4%	-85.6%	-74.8%
Fuel and vehicle life cycle (g/km)		223.6	224.1	107.9	145.3	96.6	196.0	108.7	144.6
Fuel and vehicle life cycle (% change)		-43.1%	-43.0%	-72.5%	-63.0%	-75.4%	-50.1%	-72.3%	-63.2%
2050									
Fuel cycle (g/km)		134.0	122.6	18.3	13.2	27.5	113.3	24.3	61.6
Fuel cycle (% change)		-52.1%	-56.2%	-93.5%	-95.3%	-90.2%	-59.5%	-91.3%	-78.0%
Fuel and vehicle life cycle (g/km)		163.7	152.5	48.1	43.0	59.7	145.6	56.6	93.9
Fuel and vehicle life cycle (% change)		-48.3%	-51.8%	-84.8%	-86.4%	-81.1%	-54.0%	-82.1%	-70.3%

Japan—BEVs—By Type of Power Plant Fuel

2010	Coal	Fuel Oil	NG Boiler	NG Turbine	Nuclear	Biomass	Hydro	Other
Fuel cycle (g/km)	215.2	185.0	175.8	140.0	11.2	17.7	10.3	7.8
Fuel cycle (% change)	-34.7%	-43.8%	-46.6%	-57.5%	-96.6%	-94.6%	-96.9%	-97.7%
Fuel and vehicle life cycle (g/km)	305.6	275.4	266.2	230.5	101.7	108.1	100.7	98.2
Fuel and vehicle life cycle (% change)	-21.6%	-29.3%	-31.7%	-40.9%	-73.9%	-72.3%	-74.2%	-74.8%
2050								
Fuel cycle (g/km)	175.4	142.0	130.5	111.8	5.4	(-1.0)	5.2	3.0
Fuel cycle (% change)	-35.8%	-48.1%	-52.2%	-59.1%	-98.0%	-100.4%	-98.1%	-98.9%
Fuel and vehicle life cycle (g/km)	207.3	173.8	162.4	143.7	37.3	30.9	37.1	34.9
Fuel and vehicle life cycle (% change)	-32.6%	-43.5%	-47.2%	-53.3%	-87.9%	-90.0%	-88.0%	-88.7%

Japan—FCVs—By Fuel and Feedstock

General fuel -->	Gasoline	Methanol	Methanol	Ethanol	H2	H2	H2	H2
Fuel spec -->	RFG-Ox10	M100	M100	E100	CH2	CH2	CH2	CH2
Feedstock -->	Crude oil	NG	Wood	Grass	Water	NG	Wood	Coal
2010								
Fuel cycle (g/km)	161.6	169.9	45.0	106.4	27.3	138.8	39.2	79.4
Fuel cycle (% change)	-50.9%	-48.4%	-86.3%	-67.7%	-91.7%	-57.9%	-88.1%	-75.9%
Fuel and vehicle life cycle (g/km)	221.0	229.5	104.6	165.9	87.9	199.5	99.9	140.1
Fuel and vehicle life cycle (% change)	-43.3%	-41.1%	-73.2%	-57.4%	-77.4%	-48.8%	-74.4%	-64.1%
2050								
Fuel cycle (g/km)	130.3	126.1	10.6	36.8	17.1	111.9	12.0	51.2
Fuel cycle (% change)	-52.3%	-53.8%	-96.1%	-86.5%	-93.7%	-59.1%	-95.6%	-81.3%
Fuel and vehicle life cycle (g/km)	158.2	154.2	38.6	64.8	47.7	142.5	42.5	81.7
Fuel and vehicle life cycle (% change)	-48.6%	-49.9%	-87.5%	-78.9%	-84.5%	-53.7%	-86.2%	-73.5%

China—BEVs—By Type of Power Plant Fuel

2010	Coal	Fuel Oil	NG Boiler	NG Turbine	Nuclear	Biomass	Hydro	Other
Fuel cycle (g/km)	216.2	217.5	183.0	133.5	15.3	55.5	9.8	7.3
Fuel cycle (% change)	-37.2%	-36.8%	-46.8%	-61.2%	-95.5%	-83.9%	-97.1%	-97.9%
Fuel and vehicle life cycle (g/km)	321.2	322.5	288.1	238.5	120.3	160.5	114.9	112.3
Fuel and vehicle life cycle (% change)	-21.4%	-21.1%	-29.5%	-41.6%	-70.5%	-60.7%	-71.9%	-72.5%
2050								
Fuel cycle (g/km)	201.9	155.4	132.9	97.2	7.3	3.9	4.9	2.8
Fuel cycle (% change)	-28.1%	-44.7%	-52.7%	-65.4%	-97.4%	-98.6%	-98.3%	-99.0%
Fuel and vehicle life cycle (g/km)	240.6	194.1	171.6	135.8	46.0	42.5	43.5	41.4
Fuel and vehicle life cycle (% change)	-25.2%	-39.6%	-46.6%	-57.7%	-85.7%	-86.8%	-86.5%	-87.1%

China—FCVs—By Fuel and Feedstock

	General fuel -->	Gasoline	Methanol	Methanol	Ethanol	H2	H2	H2	H2
	Fuel spec -->	RFG-0x10	M100	M100	E100	CH2	CH2	CH2	CH2
	Feedstock -->	Crude oil	NG	Wood	Grass	Water	NG	Wood	Coal
2010									
Fuel cycle (g/km)		151.3	151.6	65.8	117.9	44.6	124.1	62.0	82.2
Fuel cycle (% change)		-56.0%	-55.9%	-80.9%	-65.7%	-87.0%	-63.9%	-82.0%	-76.1%
Fuel and vehicle life cycle (g/km)		215.7	216.5	130.6	182.5	109.9	189.4	127.4	147.5
Fuel and vehicle life cycle (% change)		-47.2%	-47.0%	-68.0%	-55.3%	-73.1%	-53.6%	-68.8%	-63.9%
2050									
Fuel cycle (g/km)		122.0	114.1	23.6	27.8	31.0	106.2	29.4	60.9
Fuel cycle (% change)		-56.6%	-59.4%	-91.6%	-90.1%	-89.0%	-62.2%	-89.5%	-78.3%
Fuel and vehicle life cycle (g/km)		155.2	147.5	57.1	61.1	66.8	142.0	65.1	96.6
Fuel and vehicle life cycle (% change)		-51.7%	-54.1%	-82.3%	-81.0%	-79.2%	-55.8%	-79.7%	-69.9%

Germany—BEVs—By Type of Power Plant Fuel

2010	Coal	Fuel Oil	NG Boiler	NG Turbine	Nuclear	Biomass	Hydro	Other
Fuel cycle (g/km)	239.5	188.0	164.3	131.1	28.3	23.3	10.4	7.8
Fuel cycle (% change)	-28.1%	-43.5%	-50.7%	-60.6%	-91.5%	-93.0%	-96.9%	-97.7%
Fuel and vehicle life cycle (g/km)	333.8	282.4	258.6	225.4	122.6	117.6	104.7	102.0
Fuel and vehicle life cycle (% change)	-14.9%	-28.0%	-34.1%	-42.6%	-68.8%	-70.0%	-73.3%	-74.0%
2050								
Fuel cycle (g/km)	200.5	144.7	121.5	104.2	15.3	(-0.3)	5.2	3.0
Fuel cycle (% change)	-26.7%	-47.1%	-55.6%	-61.9%	-94.4%	-100.1%	-98.1%	-98.9%
Fuel and vehicle life cycle (g/km)	230.7	174.9	151.7	134.4	45.5	29.9	35.5	33.2
Fuel and vehicle life cycle (% change)	-24.8%	-43.0%	-50.6%	-56.2%	-85.2%	-90.3%	-88.5%	-89.2%

Germany—FCVs—By Fuel and Feedstock

General fuel -->	Gasoline	Methanol	Methanol	Ethanol	H2	H2	H2	H2
Fuel spec -->	RFG-Ox10	M100	M100	E100	CH2	CH2	CH2	CH2
Feedstock -->	Crude oil	NG	Wood	Grass	Water	NG	Wood	Coal
2010								
Fuel cycle (g/km)	163.8	168.7	48.9	100.5	65.5	134.2	49.0	80.9
Fuel cycle (% change)	-50.8%	-49.3%	-85.3%	-69.8%	-80.3%	-59.7%	-85.3%	-75.7%
Fuel and vehicle life cycle (g/km)	222.2	227.5	107.7	159.1	125.1	193.9	108.7	140.6
Fuel and vehicle life cycle (% change)	-43.4%	-42.0%	-72.5%	-59.5%	-68.1%	-50.6%	-72.3%	-64.2%
2050								
Fuel cycle (g/km)	130.4	125.1	9.7	45.7	31.8	96.9	11.8	37.3
Fuel cycle (% change)	-52.3%	-54.3%	-96.4%	-83.3%	-88.4%	-64.6%	-95.7%	-86.4%
Fuel and vehicle life cycle (g/km)	157.2	152.0	36.8	72.6	61.0	126.1	41.1	66.6
Fuel and vehicle life cycle (% change)	-48.8%	-50.5%	-88.0%	-76.3%	-80.1%	-58.9%	-86.6%	-78.3%

GREET emission results for BEVs and FCVs

The GREET model results for BEVs and FCVs are broadly similar to the LEM results discussed above. GREET shows that emission reductions of about 40 percent can be expected from BEVs using the average electricity grid mix in the United States, compared with emissions from conventional vehicles, and that BEVs using a California electricity grid mix would produce reductions of about 60 percent. FCVs using hydrogen derived from natural gas would reduce emissions by just over 50 percent. FCVs using the average grid mix of U.S. electricity to produce hydrogen through the electrolysis process would result in an increase in emissions of about 20 percent. As shown by the LEM as well, BEVs and FCVs using entirely renewable fuels to produce electricity and hydrogen would nearly eliminate GHGs from the fuel cycle.[34]

Emission results for BEVs and FCVs from other major modeling efforts

The various efforts to model electric vehicle fuel-cycle emissions are challenging to compare because of the many different dimensions that they encompass, and because they rarely overlap very well in that regard. Hence there is often the challenge of trying to make "apples to apples" rather than "apples to oranges" comparisons. Later (under "Comparison of GHG emission reductions from various electric vehicle types") we include a figure that does compare the results from a few of the most detailed studies; however we caution that no attempt has been made to correct for key differences in their underlying assumptions (for example, assumed vehicle driveline efficiencies).

Several studies were conducted for California in the 1990s when the introduction of BEVs was being mandated by the state. These studies generally found significant benefits from BEVs in terms of GHG emission reductions, along with more mixed results for the criteria pollutants that were the main focus of the studies.[35] However, the studies were often limited to CO_2 only, as far as the GHGs examined, sometimes along with CH_4 and several air pollutants that were more of concern at the time.

Other studies have been done more recently comparing BEVs and FCVs as alternatives to ICEVs, with results based on more modern assumptions that are better comparisons to the recent work on emissions from PHEVs. One such study by MIT concludes that conventional ICEVs emit about 252 grams of CO_2e per km and that by 2030 this might be reduced to about 156 grams per km. In comparison, 2030 FCVs could emit about 89 grams per km, BEVs could emit 116 grams per km, and a PHEV-30 (with a 30-mile/50-km AER) might emit about 86 grams of CO_2e per km. Thus, the study finds that the PHEV-30s and FCVs have the largest emission reductions relative to the 2030 ICEV (44 percent and 42 percent), followed closely by the BEVs (26 percent). Hence, all three options (as well as a 2030 advanced conventional hybrid in this analysis) are significantly better than the advanced 2030 ICEV.[36]

Another recent comparison of BEVs and FCVs found that GHG emissions from lithium-ion BEVs were much lower than from either nickel-metal hydride or lead-acid battery based vehicles, ranging from about 235 grams per km for a 100-km-range vehicle to about 375 grams per km for a 600-km-range vehicle. This study found that FCV emissions are relatively unchanged by driving range, at about 180 grams per km. This assumes the electricity is from the U.S. marginal grid mix and that hydrogen for the FCVs is made from natural gas. Hence this study suggests that FCVs operating on hydrogen from natural gas can have lower GHG emissions than even relatively low-range BEVs in the United States,[37] a finding that is consistent with most other studies.

A major ongoing European study, the EUCAR study, makes detailed estimates of life-cycle GHG emissions from alternative-fuel ICEVs, hybrid vehicles, and FCVs.[38] The study estimated life-cycle emissions for methanol FCVs, using wood, coal, and natural gas as feedstocks, and for compressed-hydrogen vehicles, using wood and natural gas as feedstocks. FCVs using hydrogen made from natural gas had about 55 percent lower well-to-wheels GHG emissions than a conventional gasoline ICEV, and FCVs using hydrogen made from wood had about 90 percent lower emissions.

Emission results for PHEVs from major modeling efforts

What have major modeling efforts revealed about GHG emission reductions that can be expected from PHEVs? We summarize the results of key studies here.

A 2001 report by the Electric Power Research Institute (EPRI) assumed that the marginal electricity load for PHEVs would be met by combined-cycle natural gas plants. The study estimated a grid GHG intensity of 427 grams of CO_2 per kWh, which is the average of the high and low estimates of marginal emissions made by the consulting firm AD Little Inc. for the California Air Resources Board in 2000. In EPRI's average-driving-schedule case with nightly charging, the PHEV-32 emits 144 grams of CO_2 per km and the PHEV-96 emits 112 grams of CO_2 per km, both of which are much lower than the estimated ICEV CO_2 emissions of 257 grams per km.[39]

Samaras and Meisterling[40] performed a hybrid life-cycle analysis of PHEV GHG emissions using GREET 1.7 along with results from the Economic Input-Output Life Cycle Assessment Model developed at Carnegie Mellon University.[41] They defined the low, average, and high electricity grid GHG intensities as 200, 670, and 950 grams of CO_2-eq/kWh, respectively. They estimated that PHEVs would have only 15 percent lower GHG emissions than a comparable ICEV in the high-grid-emissions case, but 63 percent lower emissions in the low-grid-emissions case.

Kromer and Heywood[42] forecasted that the average GHG intensity of the 2030 U.S. electricity grid will be 769 grams of CO_2e GHGs per kWh, based on projections from the Energy Information Administration and emissions calculations from Groode.[43] Gasoline well-to-tank emissions of 21.2 gCO_2e/MJ were adopted from a GM/ANL study,[44] and tank-to-wheels emissions were modeled in the vehicle simulation program ADVISOR, over standard EPA driving cycles. With these assumptions, PHEVs were estimated to have about 45 percent lower GHG emissions than ICEVs.

Another study of PHEVs by Silva et al.[45] concludes that for the United States, charge-depleting (CD) PHEVs with 15 kWh of battery capacity can have GHG emissions on the order of 70–80 grams per km, or about 40 percent less than a conventional baseline vehicle. The reductions would be greater in Japan and Europe, which have a lower-carbon fuel mix for electricity generation than the United States does. Charge-sustaining (CS) PHEVs were found to have considerably higher emissions than the CD designs—in fact, higher than baseline vehicles in the study for the United States and Europe. The study also found that the proportion of emissions attributable to vehicle fueling versus cradle-to-grave manufacturing and maintenance varies strongly with distance driven. For example, for a CS PHEV driving a total of 300,000 km, 15 percent of the emissions are attributable to the vehicle manufacturing and maintenance and 85 percent to fuel use; for lower total mileage of 150,000 km, the proportion is 25 percent to manufacturing and maintenance and 75 percent to fuel use. Silva et al. assumed NiMH batteries and used ADVISOR to do the simulation modeling and GREET for emissions estimates.

In another study, Jaramillo et al.[46] compare the GHG emissions of PHEVs with those of FCVs and conventional vehicles, assuming that PHEVs are operated either on conventional gasoline or coal-to-liquids (CTL) fuels and electricity and that FCVs use hydrogen made from coal gasification. Under varying assumptions about the level of carbon capture and sequestration from the CTL and gasification processes, they find that PHEVs could reduce emissions by up to 46 percent compared with conventional vehicles and up to 31 percent compared with hybrid vehicles. FCVs could decrease GHG emissions by up to 50 percent compared with conventional vehicles or could increase them considerably, depending on the level of carbon capture and the source of electricity used for hydrogen compression. Meanwhile, CTL fuels used in conventional and hybrid vehicles would significantly increase emissions compared with conventional gasoline and diesel vehicles.

Analysts at Pacific Northwest National Laboratory (PNNL)[47] used a simplified dispatch model to estimate the impacts of PHEV charging on GHG emissions. PNNL estimated the average hourly demand for an average winter day and an average summer day in each of twelve electricity-generating regions of the United States, with no PHEV recharging. The analysts then assumed that the difference between the available hourly electricity-generating capacity and the estimated hourly electricity demand without PHEVs would be used to charge PHEVs. They assumed that only natural gas and coal power would be available to supply this "marginal" electricity demand. They used version 1.6 of the GREET model to estimate fuel-cycle GHG emissions for a gasoline vehicle and for electricity generation. With these assumptions and methods, they estimated that PHEVs operating in all-electric mode would have 0 to 40 percent lower fuel-cycle GHG emissions than gasoline vehicles, with the reduction depending on the share of coal in the regional available capacity mix. (PNNL did not model emissions from operation of the ICE in a PHEV.) For the whole United States, the average reduction was 27 percent.

The approach of Stephan and Sullivan[48] is similar to that of PNNL. They assumed that PHEVs would be supported by "spare utility capacity," which they defined as the difference between 90 percent of peak generating capacity and the actual nighttime demand. However, rather than use a simplified dispatch approach to estimate electricity fuel mix and emissions by region, the authors used what they called "empirical" estimates of CO_2 emission rates in various regions. They estimated that fuel-cycle CO_2 emissions from PHEVs operating in electric mode would be 40 to 75 percent lower than emissions from gasoline vehicles, in the 12 electricity-generating regions of the United States. With the U.S. average electricity generation fuel mix, the reduction would be about 60 percent. They also reported CO_2 emission impacts for current-technology and new-technology coal and natural gas plants.

Parks et al.[49] used the characteristics of Colorado's Xcel energy system in 2004 for their analysis of CO_2 emissions from PHEV charging and use. They used a chronological dispatch model called PROSYM, developed by Global Energy Decisions, to model the operation of the electricity grid. The Xcel region's electricity grid is primarily fossil fuel-based and had an average CO_2 emissions intensity of 884.5 grams of CO_2/kWh (1,950 lb/MWh) in 2004. The study calculated CO_2 emissions under four charging scenarios:
- uncontrolled—no time restrictions, peak around 4 to 6 p.m., 1.4 kW rate
- delayed—charging starts at 10 p.m., 1.4 kW rate
- off-peak—controlled charging starts after 10 p.m. and ends by 7 a.m., 3.2 kW rate
- continuous charging—charging allowed all day, charging stations available, 1.4 kW rate

They found that the CO_2 emissions from PHEV-32 charging were about 454 g/kWh (1,000 lbs/MWh) under all of these scenarios, which results in per-mile emissions of 251 g CO_2/km, about 40 percent lower than the estimated ICEV emissions.

The GREET model has been used to analyze life-cycle GHG emissions from various types of PHEVs. As an example of results from this model, one Argonne National Laboratory (ANL) study[50] focused on three regions (Illinois, New York, and California) that provide a wide range of marginal electricity generation mixes, plus a U.S. average generation case and an all-renewable generation case. To estimate the marginal mix of fuels used to generate electricity in the regions, the study used the results of the region-specific dispatch modeling of Hadley and Tsvetkova.[51] The study examined a scenario in which charging took place in the late evening in the year 2020 at a 2-kW charging rate. It estimated that the GHG emissions of a petroleum-fueled PHEV are 30 to 50 percent lower than those of an ICEV, with the greater reduction corresponding to lower grid emissions. It also estimated the impacts of the grid GHG intensity on the overall emissions of PHEVs powered by other fuels, including biofuels and hydrogen. It found that while the California generation mix reduced CO_2 emissions from all PHEVs relative to the U.S. average mix, PHEVs powered by biomass-based fuels were not affected as greatly. The study also shows that PHEVs charged on a GHG-intensive electricity grid can have greater well-to-wheels GHG emissions than regular HEVs and that this is exacerbated by increasing the battery capacity.[52]

Another set of GREET results for various types of PHEVs—fueled by gasoline, ethanol, or hydrogen fuel cells—shows that use of renewable hydrogen in fuel cells and biomass-derived ethanol result in the largest reductions in both GHG emissions and petroleum use. Fuel cell PHEVs using natural gas-derived hydrogen can also offer significant benefits, along with those using petroleum fuels but with relatively clean electricity—for example, from renewables or the California grid mix.

GREET: GHG EMISSIONS AND PETROLEUM USE OF PHEVS USING VARIOUS FUELS

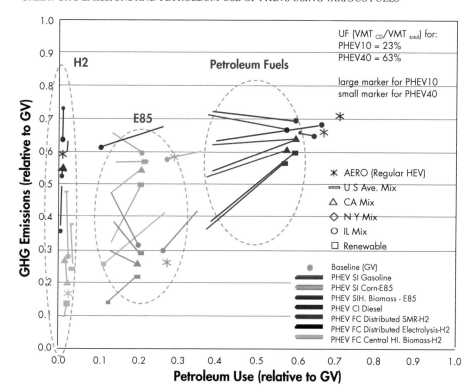

One set of GREET results shows that use of renewable hydrogen in fuel cells and biomass-derived ethanol result in the largest reductions in both GHG emissions and petroleum use. Source: A. Elgowainy, A. Burnham, M. Wang, J. Molburg, and R. Rousseau, Well-to-Wheels Energy Use and Greenhouse Gas Emissions Analysis of Plug-In Hybrid Electric Vehicles, *ANL/ESD/09-2 (Argonne National Laboratory, 2009).*

A 2007 report by EPRI and the Natural Resources Defense Council (NRDC) combines dispatch modeling with scenario analysis to estimate PHEV GHG emissions for the years 2010 and 2050.[53] Grid emissions in this study—97, 199, and 412 grams per kWh in 2050—are much lower than the emissions estimated in the other studies mentioned here because EPRI and NRDC assumed that grid emissions will decrease over time as older plants are retired and are replaced by more efficient ones. Their analysis shows that life-cycle GHG emissions decrease as the range of the PHEV increases, even in the high-grid-emissions case. This is different from the result of (for example) Samaras and Meisterling, who estimate that increased CD range results in higher emissions in their high-grid-emissions case. This difference is due to the large difference in the grid GHG intensities assumed in the two studies.

In sum, PHEVs promise significant reductions in GHG emissions in most regions and under most conditions. This is especially the case in the longer term, when the electricity grid is likely to be cleaner and vehicles are likely to have greater battery storage capacities.

PHEV GHG EMISSION REDUCTIONS PROJECTED BY KEY STUDIES

This table summarizes the results of key PHEV emission studies that can be reasonably compared directly. These studies indicate that PHEVs have 20 to 60 percent lower GHG emissions than their ICEV counterparts, with the lower-end reductions corresponding mainly to relatively low-carbon fuel mixes for electricity generation.

Studies using dispatch modeling of the electricity grid indicate a narrower range of reductions, 30 to 50 percent. By comparison, studies tabulated by Bradley and Frank[54] indicate slightly greater reductions, about 40 to 60 percent. To put the grid GHG emission numbers into perspective, the LEM estimates that in the United States in the year 2020, life-cycle emissions from coal-fired plants are 1,030 grams of CO_2e per kWh generated, and from gas-fired plants are 520 grams of CO_2e per kWh generated, using IPCC GWPs.

CD = charge-depleting; GHG = greenhouse gas; CO_2e = CO_2 equivalent; ICEV = internal combustion engine vehicle; PHEV = plug-in hybrid electric vehicle; AE = all-electric (meaning the vehicle operates solely on the battery until a certain state of charge is reached); blended = vehicle is designed to use both the engine and battery over the drive cycle; n.s. = not specified; n.e. = not estimated; NG = natural gas; C = coal.

Report	Emissions Estimation	CD Range (km)	Control Strategy	Year	Grid GHGs (gCO2e/kWh)	PHEV GHGs (gCO2e/km)	ICEV GHGs (gCO2e/km)	Percent Reduction (vs. ICEV)
EPRI 2001	Average	32.2	AE	2010	427	144	257	44%
		96	AE	2010	427	112	257	57%
Samaras and Meisterling	Scenario	30	AE	NR	200	126	257	51%
					670	183	269	32%
					950	217	276	21%
		90	AE	NR	200	96	257	63%
					670	183	269	32%
					950	235	276	15%
Kromer and Heywood	Average	48	Blended	2030	769	86.2	156	45%
		96	Blended	2030	769	89.8	156	43%
Silva et al.	Average	~57	AE	n.s.	543 (U.S.)	~110	n.s.	n.e.
					387 (Eur.)	~105	n.s.	n.e.
					428 (Japan)	~108	n.s.	n.e.
Jaramillo et al.	Scenario	60	AE	n.s.	883 (coal)	~125–220	~230	~4%–46%
PNNL	Simplified dispatch	53	n.s.	2002	94% NG/6% C	n.s.	n.s.	40%
					1% NG/99% C	n.s.	n.s.	-1%
					U.S. average	n.s.	n.s.	27%
Stephan and Sullivan	Scenario	63	n.s.	current / long term	598 (current NG)	184/119	432	57%/72%
					954 (current coal)	274/192	432	37%/56%
					608 (U.S. average)	177	432	59%
Parks et al.	Dispatch	32	Blended	2004	454	154	251	39%
ANL	Dispatch/ scenario	32	Blended	2020	U.S. average	146	233	37%
					California	140	233	40%
					Illinois	162	233	30%
					Renewable	115	233	51%
EPRI and NRDC	Dispatch/scenario	16	AE	2050	97	140	233	40%
					199	143	233	39%
					412	147	233	37%
		32.2	AE	2050	97	103	233	56%
					199	109	233	53%
					412	119	233	49%

Notes on Table:

In the "Emissions Estimation" column, "Average" = annual average emissions from the entire national electric grid; "Scenario" = the study considered different fuel-mix and hence emission scenarios for the electric grid; "Dispatch" = the study estimated marginal fuel mixes and emissions for PHEV charging based on a dispatch model.

In the "Control Strategy" column, PNNL and Stephan and Sullivan estimate emissions from electric operation only; they do not estimate emissions from the ICE in a PHEV.

In the "Year" column, for Silva et al. and Jaramillo et al. the year of analysis is not specified but appears to be roughly current.

GHG emissions and CO_2 equivalency are estimated as follows:

- *For EPRI 2001, Silva et al., Stephan and Sullivan, and Parks et al.: CO_2 only.*
- *For ANL: 2007 IPCC GWPs for CH_4 and N_2O.*
- *For Samaras and Meisterling, Jaramillo et al., and EPRI-NRDC: 2001 IPCC 100-year GWPs for CH_4 and N_2O.*
- *For Kromer and Heywood and PNNL: 1995 IPCC GWPs for CH_4 and N_2O.*

Samaras and Meisterling and Jaramillo et al. do not explicitly state which GHGs they include in their CO_2e measure; however, they refer to CO_2e estimates from the GREET model, which considers CH_4 and N_2O. Similarly, PNNL does not state which CO_2e measure it uses, but it does state that it uses GREET version 1.6, with year 2001 documentation, so we assume that the 1995 IPCC GWPs apply.

For EPRI 2001, the 32.2 km CD range uses the "unlimited" case, which allows the maximum number of electric miles.

For Silva et al., the numbers preceded by "~" were estimated from Figures 2 and 4 of the study report; for Jarmillo et al., from Figure 4 of the study report.

For Stephan and Sullivan, where there are two numbers given, the number before the slash is the result for "current technology" electricity generation, and the number after the slash is the result for "new technology" electricity generation, in the long term. The new technology is more efficient than the current technology.

Some of the results shown in this table merit further explanation. For example, Kromer and Heywood report a higher grid GHG intensity than several other cases, but lower emissions per km than Samaras and Meisterling and EPRI 2001. The high grid GHG intensity comes from DOE-EIA projections, and the lower emissions per km are likely due to the assumed improvement in efficiency and emissions in the 2030 ICEV. The relatively large reductions estimated by Stephan and Sullivan are due to several factors: (1) they start with a relatively high-emitting gasoline vehicle; (2) they consider electric operation of the PHEV only; (3) they assume relatively efficient power plants in the long term; and (4) they consider only CO_2 emissions.

Comparison of GHG emission reductions from various electric vehicle types

Now that we have reviewed the results of various specific studies, we can make an overall comparison of the emission reductions estimated for the various types of EVs. However, again we note that these emissions vary widely by location and vehicle type/design and are only generally characterized in the following discussion.

BEVs have the potential to reduce well-to-wheels GHG emissions by about 55 to 60 percent using either natural gas power plants or the California grid mix (which is heavily dependent on natural gas). Using coal-based power, BEVs may reduce emissions by about 20 percent or slightly increase them (model results vary somewhat), and using the U.S. grid mix (which is about half coal-based) emission reductions on the order of 25 to 40 percent appear possible. For FCVs using hydrogen produced from natural gas steam reformation, GHG emissions can be reduced by 30 to 55 percent according to the various studies. Once again, when entirely or almost entirely powered by completely renewable fuels such as wind, solar, and hydro, GHG emissions from both BEVs and FCVs can be almost entirely eliminated.

COMPARISON OF GHG EMISSION-REDUCTION ESTIMATES FOR BEVS AND FCVS

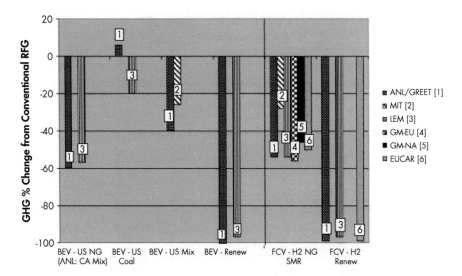

When we compare estimates of the well-to-wheels GHG reductions (from conventional reformulated gasoline) to be expected from BEVs and FCVs, we see that findings vary by study and that emission reductions vary by energy source. When entirely or almost entirely powered by completely renewable fuels such as wind, solar, and hydro, GHG emissions from both BEVs and FCVs can be almost entirely eliminated.

BEV = battery electric vehicle; CA = California; FCV = fuel cell vehicle; H2 = hydrogen; NG = natural gas; Renew = renewable fuel; SMR = steam methane reforming.

Sources:

1. M. Wang, Y. Wu, and A. Elgowainy, GREET1.7 Fuel-Cycle Model for Transportation Fuels and Vehicle Technologies *(Argonne National Laboratory, 2007)*; M. Wang, "Well to Wheels Energy Use Greenhouse Gas Emissions and Criteria Pollutant Emissions—Hybrid Electric and Fuel Cell Vehicles," presented at the SAE Future Transportation Technology Conference, Costa Mesa, CA, June 2003.

2. M. A. Kromer and J. B. Heywood, Electric Powertrains: Opportunities and Challenges in the U.S. Light-Duty Vehicle Fleet, *LEFF 2007-02 RP (Sloan Automotive Laboratory, MIT Laboratory for Energy and the Environment, May 2007)*.

3. LEM.

4. *General Motors et al.,* GM Well-to-Wheel Analysis of Energy Use and Greenhouse Gas Emissions of Advanced Fuel/Vehicle Systems—A European Study, *L-B-Systemtechnik GmbH, Ottobrunn, Germany, September 27, 2002.*

5. *General Motors, Argonne National Lab, et al.,* Well-to-Wheel Energy Use and Greenhouse Gas Emissions of Advanced Fuel/Vehicle Systems, *in three volumes, published by Argonne National Laboratory, June 2001.*

6. *EUCAR (European Council for Automotive Research and Development), CONCAWE, and ECJRC (European Commission Joint Research Centre),* Well-to-Wheels Analysis of Future Automotive Fuels and Powertrains in the European Context, *Well-to-Wheels Report, Version 2c, March 2007.*

Emission reductions possible from PHEVs are somewhat more modest than for some BEV and FCV configurations. For a PHEV type considered in several studies that has a 30-mile/50-km electric range, GHG emission reductions compared with a conventional vehicle are estimated to be in the range of 30 to 60 percent using the U.S. grid mix. For the California electricity mix, a range of 40 to 55 percent has been estimated. Also, one estimate shows a 50-percent reduction potential with PHEV-30s running on renewables-based electricity. We note that for PHEVs in particular, these relative emission reduction results vary by assumed driving patterns and distances as well as underlying emission factors for electricity and gasoline used. This leads to further sources of potential variation amongst the studies, along with other variables such as the assumed driveline efficiencies, upstream emission factors, and the type and size of the vehicle itself.

COMPARISON OF GHG EMISSION-REDUCTION ESTIMATES FOR PHEVS AND HEVS

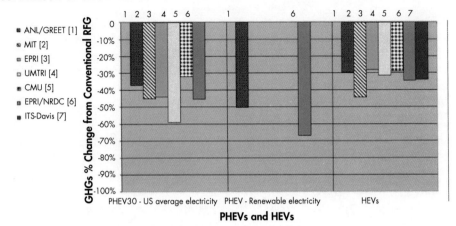

When we compare estimates of the well-to-wheels GHG reductions (from conventional reformulated gasoline) to be expected from PHEVs and HEVs, we see that findings vary by study. For a PHEV that has a 30-mile/50-km electric range, GHG emission reductions compared with a conventional vehicle are estimated to be 30 to 60 percent using the U.S. grid mix. PHEVs running on renewables-based electricity offer greater reductions, in the range of 50 percent to almost 70 percent. For HEVs, most studies typically estimate reductions of about 30 percent, although one study estimates a reduction of about 45 percent.

Sources:

1. M. Wang, Y. Wu, and A. Elgowainy, *GREET1.7 Fuel-Cycle Model for Transportation Fuels and Vehicle Technologies (Argonne National Laboratory, 2007)*; M. Wang, "Well-to-Wheels Analysis of Biofuels and Plug-In Hybrids," presentation at Argonne National Laboratory, June 3, 2009. We calculate GHG reductions for HEVs by weighting their estimated city mpg 55 percent and their estimated highway mpg 45 percent.

2. M. A. Kromer and J. B. Heywood, Electric Powertrains: Opportunities and Challenges in the U.S. Light-Duty Vehicle Fleet, *LEFF 2007-02 RP (Sloan Automotive Laboratory, MIT Laboratory for Energy and the Environment, May 2007)*. Estimates from Table 50, year-2030 U.S. average electricity mix, year-2030 gasoline vehicle, 30-mi PHEV range.

3. R. Graham, Comparing the Benefits and Impacts of Hybrid Electric Vehicle Options, *Report 1000349 (Electric Power Research Institute, 2001)*. Estimates from Table 3-21, U.S. average electricity mix, 20-mi PHEV range.

4. C. H. Stephan and J. Sullivan, "Environmental and Energy Implications of Plug-in Hybrid-Electric Vehicles," Environmental Science and Technology 42 (2008): 1185–90. Estimates from Table 4, U.S. average electricity mix, current technologies, 20- to 40-mi PHEV range.

5. C. Samaras and K. Meisterling, "Life Cycle Assessment of Greenhouse Gas Emissions from Plug-in Hybrid Vehicles: Implications for Policy," Environmental Science and Technology 42 (2008): 3170–76. Estimates from Table 1, U.S. average electricity mix ca. 2007, baseline scenario, 30-mi PHEV range.

6. Electric Power Research Institute and Natural Resources Defense Council, Environmental Assessment of Plug-In Hybrid Electric Vehicles, Volume 1: Nationwide Greenhouse Gas Emissions, *Report No. 1015325 (EPRI and NRDC, 2007)*. Estimates from Figure 5-1, year 2010, 20-mi PHEV range. We estimate the U.S. average electricity

case by scaling the electricity emissions in the "old 2010 CC" results in Figure 5-1 by the ratio of emissions from U.S. average electricity (Table 3-2) to emissions from "old NG CC" (Table 2-1).

7. Based on Burke et al., Chapter 4 in this volume. The reduction shown here is the reduction in fuel use per mile, which we calculate by weighting their estimated city mpg 55 percent and their estimated highway mpg 45 percent, for midsize 2015 vehicles.

How Fast Can the GHG Reductions Promised by EVs Be Achieved?

The emission estimates just discussed demonstrate that EVs can offer significant GHG reductions when compared on a one-to-one basis with conventional vehicles. How fast, then, can the electric vehicle industry scale up? When we pose this question we run into a major issue: the availability of advanced electric vehicle battery packs in the numbers needed for a major commercial launch of vehicles by several automakers at once.

A 2009 analysis examined the potential of various options to scale up to become a "gigaton solution"—that is, to account for reducing CO_2 by a gigaton on a global annual basis—by 2020.[65] The study found that achieving "gigaton scale" with a strategy based largely on a massive introduction of grid-connected EVs would require about 1,000 times as many batteries in the near term as are expected to be available (that is, tens of millions globally rather than tens of thousands), growing to a need for hundreds of millions of battery packs by 2020. This implies a massive investment in battery production capacity at a time when battery designs are still being improved and perfected to the point where commercially acceptable PHEVs and BEVs can be produced—which suggests that achieving gigaton scale with EVs is not possible by 2020. However, much larger gains are possible by 2030 and especially 2050, given the relative slowness of motor vehicle fleet stock turnover.[66]

The need to scale up battery production in the cell sizes and configurations required for different types of EVs is accompanied by several other needs to support the introduction of elecric vehicles into consumer households. These include:
- improving the procedures for installing recharging facilities for EVs at household and other sites,
- better understanding of the utility grid impacts of significant numbers of grid-connected vehicles,
- better understanding of the consumer and utility economics of electric vehicle ownership (and/or leasing of car or battery), and
- better education of consumers and tools to assist them to determine whether their driving habits would be a good fit for the characteristics of the different types of EVs.

These and other related issues are being explored by the University of California and other groups as new EVs are being introduced into the market.[67]

Additional issues related to vehicle scale-up include provision of hydrogen for FCVs, currently an expensive proposition for low volumes of dispensed fuel, development and dissemination of appropriate safety procedures for first responders in dealing with accidents with vehicles with high voltage electrical systems and/or hydrogen fuel storage, and additional education and outreach programs for mechanics and fleet managers.[68] These measures will be needed to help EVs become more established and acceptable to consumers in various market segments.

Still, it is important to note that more generally, PHEV and other electric vehicle technologies can scale fairly rapidly. Typical automotive volumes run to several hundred thousand units per year for individual popular models (for example, the combined U.S. and Japanese sales of the Toyota Prius are around 275,000 to 300,000 per year), and there is the potential to incorporate electric drive technology into many vehicle models. The rate of scaling is mainly limited by the growth of supplier networks and supply chains, and by the dynamics of introducing new vehicles with 15-year lives into regional motor vehicle fleets, along with economic and market response constraints on the demand side.

Given these dynamics of the transportation sector and that a significant percentage of new vehicles sold today will still be on the road in the next 10 years, it is much easier to foresee large reductions in LDV emissions by 2030, 2040, and 2050 than by 2020. For example, the EPRI-NRDC study noted earlier concludes that under the most optimistic U.S. scenario assessed—high PHEV fleet penetration and low electric sector CO_2 intensity—612 million megatons of emissions could be reduced annually by 2050. Extrapolated globally, these emission reductions could be on the order of 2 to 3 gigatons annually.

ANNUAL GREENHOUSE GAS REDUCTIONS POSSIBLE FROM PHEVS IN THE YEAR 2050

The EPRI-NRDC study noted earlier includes scenario estimates of future GHG reductions from vehicle fleets in the United States and finds that reductions of up to about 500 megatons per year are possible by 2050, depending on the level of PHEV fleet penetration and the CO_2 intensity of the electricity sector. This table presents some of the key results of the study. Source: Electric Power Research Institute and Natural Resources Defense Council, Environmental Assessment of Plug-In Hybrid Electric Vehicles, Volume 1: Nationwide Greenhouse Gas Emissions, *Report No. 1015325 (EPRI and NRDC, 2007).*

2050 Annual GHG Reduction (million metric tons)		Electric Sector CO2 Intensity		
		High	Medium	Low
PHEV Fleet Penetration	Low	163	177	193
	Medium	394	468	478
	High	474	517	612

KEY UNCERTAINTIES AND AREAS FOR FURTHER RESEARCH

Because GHGs are produced in myriad ways from electric-vehicle fuel cycles, including both upstream and vehicle-based emissions (in the case of PHEVs and HEVs), and because electric-vehicle technologies are still evolving, there are considerable uncertainties involved in the analysis of life-cycle CO_2e GHG emissions from advanced EVs. Since 1990 many of these uncertainties have been narrowed—for example, the manufacturing cost and performance of electric vehicle motors and motor controllers has become better established—but many still remain. Exploring these uncertainties in much detail is beyond the scope of this chapter but is done to some extent in some of the studies referenced here. The GREET model in particular now has the ability to include estimates of the levels of uncertainty in key input variables, and it incorporates this capability through a graphical user interface version of the model that runs in a PC Windows environment. This can be useful, but of course we can still benefit from additional efforts to characterize and narrow the remaining uncertainties themselves.

Some of the key remaining uncertainties are these:

- Emission rates of high-GWP-value gases (such as N_2O, CH_4, and refrigerants) that are emitted in lower quantities than CO_2 from vehicle fuel cycles but that can still be significant
- Emission impacts of the increased use of power plants to charge BEVs and PHEVs
- Secondary impacts such as the "indirect land use change" impacts of biofuels, where production of biofuels implies cultivation of land that in some cases can displace its use for other purposes, and how emissions from power plants and other combustion sources actually result in exposures and potential harm to humans and the Climate impacts of emissions of typically overlooked but potentially important pollutants such as oxides of sulfur, ozone precursors, and particulate matter
- Rate of future vehicle and fueling-system performance improvements
- Driveline efficiencies of various types of alternative fuel vehicles, and efficiencies
- involved in key upstream fuel production processes
- Potential "wild cards" in future fuel-production processes, such as the successful introduction of carbon capture and sequestration
- Breakthroughs in electricity, advanced biofuel, or hydrogen production

Uncertainty about the exact levels of emissions is compounded by uncertainty about the overall impacts of GHGs, as some aspects of climate dynamics are still not completely understood. But as time goes on, we can expect more to be learned about these key areas, and for the remaining uncertainties to be narrowed. At the same time, new fuel cycles based on evolving technology (for instance, diesel-type fuels from algae, new types of PHEVs running on various fuels, other new types of synthetic Fischer-Tropsch process and bio-based fuels) are likely to become available but with potentially significant uncertainties until more is learned about them in turn. The significant amount of research currently under way is encouraging, but given the pressing nature of the energy and climate challenges facing many nations, one could argue that more attention should be paid to this critical area.

Summary and Conclusions

- Electric-drive vehicles, based on batteries, plug-in hybrid, and fuel cell technology, have been found to significantly reduce emissions of GHGs compared to conventional vehicles in most cases and settings studied. Various types of hybrid-electric and all-electric vehicles can offer significant GHG reductions when compared to conventional vehicles on a full fuel-cycle basis. In fact, most EVs used under most conditions are expected to significantly reduce life-cycle CO_2e GHG emissions. Under certain conditions, EVs can even have very low to zero emissions of GHGs when based on renewable fuels. However, at present this is more expensive than other options that offer significant reductions at lower costs based on the use of more conventional fuels.

- BEVs reduce GHGs by a widely disparate amount depending on the type of power plant used and the particular region involved, among other factors. Reductions typical of the United States for BEVs are on the order of 20 to 50 percent, depending on the relative level of coal versus natural gas and renewables in the regional power plant feedstock mix. However, much deeper reductions of more than 90 percent are possible for vehicles using renewable or nuclear power sources. PHEVs running on gasoline can reduce emissions by 20 to 60 percent, again depending strongly on electricity source. FCVs are found to reduce GHGs by 30 to 50 percent when running on natural gas-derived hydrogen and up to 95 percent or more when the hydrogen is produced using renewable feedstocks.

- Emissions from all of these electric-vehicle types are highly variable depending on the details of how the electric fuel or hydrogen is produced. When coal is heavily used to produce electricity or hydrogen, GHG emissions for EVs tend to increase significantly compared with conventional fuel alternatives. Unless carbon capture and sequestration (CCS) becomes a reality, using electric-drive systems in conjunction with a heavily coal-based fuel supply offers little or no benefit.

CHAPTER 6: COMPARING GREENHOUSE GAS EMISSIONS

- Overall, EVs offer the potential for significant and even dramatic reductions in GHGs from transportation fuel cycles. Pursuing further development of this promising set of more efficient technologies is thus of paramount importance, given the rapidly spiraling growth in motor vehicle ownership and use around the globe.

Notes

1. See M. A. Delucchi, "Emissions of Greenhouse Gases from the Use of Transportation Fuels and Electricity," ANL/ESD/TM-22 (Argonne National Laboratory, 1991); M. Q. Wang, "GREET 1.0 – Transportation Fuel Cycles Model: Methodology and Use," ANL/ESD-33 (Argonne National Laboratory, 1996); M. A. Delucchi, "Emissions of Non-CO_2 Greenhouse Gases from the Production and Use of Transportation Fuels and Electricity," UCD-ITS-RR-97-05 (Institute of Transportation Studies, UC Davis, 1997); and J. M. Ogden and R. H. Williams, *Solar Hydrogen* (World Resources Institute, Washington DC, 1989).
2. See S. Unnasch, L. Browning, M. Montano, S. Huey, and G. Nowell, *Evaluation of Fuel-Cycle Emissions on a Reactivity Basis: Volume 1 - Main Report*, FR-96-114 (Mountain View, CA: Acurex Environmental Corporation, 1996); S. Unnasch, "Greenhouse Gas Analysis for Fuel Cell Vehicles," 2006 Fuel Cell Seminar, Honolulu, HI (2006); and C. E. Thomas, B. D. James, F. Lomax, and I. F. Kuhn, "Fuel Options for the Fuel Cell Vehicle: Hydrogen, Methanol or Gasoline?" *International Journal of Hydrogen Energy* 25 (2000): 551–67.
3. See General Motors and Argonne National Lab, *Well-to-Wheel Energy Use and Greenhouse Gas Emissions of Advanced Fuel/Vehicle Systems* (2001); and General Motors et al., *Well-to-Wheel Analysis of Energy use and Greenhouse Gas Emissions of Advanced Fuel/Vehicle Systems—A European Study* (2002).
4. See R. Graham, *Comparing the Benefits and Impacts of Hybrid Electric Vehicle Options*, Report 1000349 (Electric Power Research Institute, 2001).
5. General Motors, Argonne National Lab, et al., *Well-to-Wheel Energy Use and Greenhouse Gas Emissions of Advanced Fuel/Vehicle Systems*, in three volumes, published by Argonne National Laboratory, June 2001, www.transportation.anl.gov/software/GREET/index.html.
6. General Motors et al., GM *Well-to-Wheel Analysis of Energy Use and Greenhouse Gas Emissions of Advanced Fuel/Vehicle Systems—A European Study*, L-B-Systemtechnik GmbH, Ottobrunn, Germany, September 27, 2002, available from www.lbst.de/gm-wtw; General Motors et al., Annex *"Full Background Report"—Methodology, Assumptions, Descriptions, Calculations, Result— to the GM Well-to-Wheel Analysis of Energy Use and Greenhouse Gas Emissions of Advanced Fuel/Vehicle Systems—A European Study*, L-B-Systemtechnik GmbH, Ottobrunn, Germany, September 27, 2002, available from www.lbst.de/gm-wtw; and General Motors et al., *Annex to Chapter 3 of Annex "Full Background Report" to the GM Well-to-Wheel Analysis of Energy Use and Greenhouse Gas Emissions of Advanced Fuel/Vehicle Systems—A European Study*, L-B-Systemtechnik GmbH, Ottobrunn, Germany, September 27, 2002, available from www.lbst.de/gm-wtw.
7. M. A. Weiss et al., *On the Road in 2020: A Lifecycle Analysis of New Automotive Technologies*, Report EL 00-003 (MIT Energy Laboratory, Massachusetts Institute of Technology, October 2000), http://web.mit.edu/sloan-auto-lab/research/beforeh2/files/weiss_otr2020.pdf; and A. Bandivadekar, K. Bodek, L. Cheah, C. Evans, T. Groode, J. Heywood, E. Kasseris, M. Kromer, and M. Weiss, *On the Road in 2035: Reducing Transportation's Petroleum Consumption and GHG Emissions*, Report LFEE 2008-05 RP (MIT Laboratory for Energy and the Environment, Massachusetts Institute of Technology, July 2008), http://web.mit.edu/sloan-auto-lab/research/beforeh2/otr2035/.
8. EUCAR (European Council for Automotive Research and Development), CONCAWE, and ECJRC (European Commission Joint Research Centre), *Well-to-Wheels Analysis of Future Automotive Fuels and Powertrains in the European Context*, Well-to-Wheels Report, Version 1b, January 2004. Version 2c, March 2007 update, and version 3, November 2008, are available at http://ies.jrc.ec.europa.eu/our-activities/support-for-eu-policies/well-to-wheels-analysis/WTW.html.
9. J. Hackney and R. de Neufville, "Life Cycle Model of Alternative Fuel Vehicles: Emissions, Energy, and Cost Trade-offs," *Transportation Research Part A* 35 (2001): 243–66.
10. P. Ahlvik and A. Brandberg, *Well to Wheels Efficiency for Alternative Fuels from Natural Gas or Biomass*, Publication 2001:85, Swedish National Road Administration, October 2001.
11. H. L. Maclean, L. B. Lave, R. Lankey, and S. Joshi, "A Lifecycle Comparison of Alternative Automobile Fuels," *Journal of the Air and Waste Management Association* 50 (2000): 1769–79.

12. K. Tahara et al., "Comparison of CO_2 Emissions from Alternative and Conventional Vehicles," *World Resources Review* 13 (2001): 52–60.
13. M. A. Delucchi et al., *A Lifecycle Emissions Model (LEM): Lifecycle Emissions from Transportation Fuels, Motor Vehicles, Transportation Modes, Electricity Use, Heating and Cooking Fuels, and Materials*, UCD-ITS-RR-03-17 (Institute of Transportation Studies, UC Davis, December 2003), main report and 13 appendices, http://www.its.ucdavis.edu/people/faculty/delucchi.
14. See T. E. Lipman and M. A. Delucchi, "Emissions of Nitrous Oxide and Methane from Conventional and Alternative Fuel Motor Vehicles," *Climatic Change* 53 (2002): 477–516.
15. U.S. Environmental Protection Agency, EPA420-F-05-001 (2005).
16. For further discussion, see M. A. Delucchi, *A Lifecycle Emissions Model (LEM): Lifecycle Emissions from Transportation Fuels, Motor Vehicles, Transportation Modes, Electricity Use, Heating and Cooking Fuels, and Materials*, UCD-ITS-RR-03-17 (Institute of Transportation Studies, UC Davis, 2003); M. Q. Wang, GREET 1.0—Transportation Fuel Cycles Model: Methodology and Use, ANL/ESD-33 (Argonne National Laboratory, 1996); M. Wang, Y. Wu, and A. Elgowainy, GREET1.7 *Fuel-Cycle Model for Transportation Fuels and Vehicle Technologies* (Argonne National Laboratory, 2007).
17. M. C. Bohm, H. J. Herzog, J. E. Parsons, R. C. Sekar, "Capture-Ready Coal Plants: Options, Technologies and Economics," *International Journal of Greenhouse Gas Control* 1 (2007): 113–20.
18. R. Dones, T. Heck, and S. Hirschberg, "Greenhouse Gas Emissions from Energy Systems: Comparison and Review," in *Encyclopedia of Energy* vol. 3 (Philadelphia: Elsevier Science, 2004), 77–95.
19. Intergovernmental Panel on Climate Change (IPCC), *2006 IPCC Guidelines for National Greenhouse Gas Inventories*, ed. H. S. Eggleston, L. Buendia, et al. (Institute for Global Environmental Strategies, Japan, 2006).
20. See Lipman and Delucchi, "Emissions of Nitrous Oxide and Methane."
21. U.S. Environmental Protection Agency, Compilation of Air Pollutant Emission Factors, AP-42 (2009); IPCC, *2006 IPCC Guidelines for National Greenhouse Gas Inventories*.
22. For estimates, see Delucchi, *A Lifecycle Emissions Model (LEM)*.
23. IPCC, *Climate Change 2007: The Physical Science Basis*, Contribution of Working Group I to the Fourth Assessment Report of the IPCC, ed. by S. Solomon, D. Qin, et al. (Cambridge: Cambridge University Press, 2007).
24. U.S. EPA, Compilation of Air Pollutant Emission Factors; IPCC, *2006 IPCC Guidelines for National Greenhouse Gas Inventories*.
25. E. Behrentz, R. Ling, P. Rieger, and A. M. Winer, "Measurements of Nitrous Oxide Emissions from Light-Duty Motor Vehicles: A Pilot Study," *Atmospheric Environment* 38 (2004): 4291–4303.
26. J. C. Kramlich and W. P. Linak, "Nitrous Oxide Behavior in the Atmosphere, and in Combustion and Industrial Systems," *Progress in Energy and Combustion Science* 20 (1994): 149–202.
27. IPCC, *Climate Change 2007: The Physical Science Basis*.
28. G. A. Keoleian, G. M. Lewis, R. B. Coulon, V. J. Camobreco, and H. P. Teulon, "LCI Modeling Challenges and Solutions for a Complex Product System: A Mid-Sized Automobile," SAE Technical Paper #982169 (Society of Automotive Engineers, 1998); J. L. Sullivan, R. L. Williams, S. Yester, E. Cobas-Flores, S. T. Chubbs, S. G. Hentges, and S. D. Pomper, "Life Cycle Inventory of a Generic U.S. Family Sedan: Overview of Results USCAR AMP Project," SAE Technical Paper #982160 (Society of Automotive Engineers, 1998); M. A. Delucchi, "A Lifecycle Emissions Analysis: Urban Air Pollutants and Greenhouse-Gases from Petroleum, Natural Gas, LPG, and Other Fuels for Highway Vehicles, Forklifts, and Household Heating in The U. S.," *World Resources Review* 13 (2001): 25–51; A. Burnham, M. Wang, and Y. Wu, Development and *Applications of GREET 2.7—The Transportation Vehicle-Cycle Model*, ANL/ESD/06-5 (Argonne National Laboratory, 2006).
29. For further discussion of operating strategies for PHEVs, see J. Gonder and T. Markel, "Energy Management Strategies for Plug-in Hybrid Electric Vehicles," SAE Technical Paper #2007-01-0290 (Society of Automotive Engineers, 2007), and T. Katrasnik, "Analytical Framework for Estimating the Energy Conversion Efficiency of Different Hybrid Electric Vehicle Topologies," *Energy Conversion and Management* 50 (2009): 1924–38.
30. T. H. Bradley and A. A. Frank, "Design, Demonstrations, and Sustainability Impact Assessments for Plug-in Hybrid Electric Vehicles," *Renewable and Sustainable Energy Reviews* 13 (2009): 115–28.
31. Gonder and Markel, "Energy Management Strategies for Plug-in Hybrid Electric Vehicles." See T. Katrasnik, "Analytical Framework for Estimating the Energy Conversion Efficiency of Different Hybrid Electric Vehicle Topologies," for a comprehensive formal framework for analyzing energy flows in hybrid electric vehicle systems.
32. Edison Electric Institute, "Diversity Map" (2009), http://www.eei.org.

33. M. Wang, "Fuel Choices for Fuel-Cell Vehicles: Well-to-Wheels Energy and Emission Impacts," *Journal of Power Sources* 112 (2002): 307–21; Wang, *GREET 1.0—Transportation Fuel Cycles Model: Methodology and Use*; General Motors, Argonne National Lab, et al., *Well-to-Wheel Energy Use and Greenhouse Gas Emissions of Advanced Fuel/Vehicle Systems*; General Motors et al., *Well-to-Wheel Analysis of Energy use and Greenhouse Gas Emissions of Advanced Fuel/Vehicle Systems—A European Study*; M. Wang, Y. Wu, and A. Elgowainy, *GREET 1.7 Fuel-Cycle Model for Transportation Fuels and Vehicle Technologies* (Argonne National Laboratory, 2007).
34. M. Wang, "Well to Wheels Energy Use Greenhouse Gas Emissions and Criteria Pollutant Emissions—Hybrid Electric and Fuel Cell Vehicles," presented at the SAE Future Transportation Technology Conference, Costa Mesa, CA, June 2003.
35. Unnasch et al., *Evaluation of Fuel-Cycle Emissions on a Reactivity Basis: Volume 1*; N. S. Rau, S. T. Adelman, and D. M. Kline, *EVTECA—Utility Analysis Volume 1: Utility Dispatch and Emissions Simulations*, TP-462-7899 (National Renewable Energy Laboratory, 1996).
36. M. A. Kromer and J. B. Heywood, *Electric Powertrains: Opportunities and Challenges in the U.S. Light-Duty Vehicle Fleet*, LEFF 2007-02 RP (Sloan Automotive Laboratory, MIT Laboratory for Energy and the Environment, May 2007), http://web.mit.edu/sloan-auto-lab/research/beforeh2/files/kromer_electric_powertrains.pdf.
37. C. E. Thomas, "Fuel Cell and Battery Electric Vehicles Compared," *International Journal of Hydrogen Energy* 34 (2009): 6005–6020.
38. EUCAR, CONCAWE, and ECJRC, *Well-to-Wheels Analysis of Future Automotive Fuels and Powertrains in the European Context*.
39. Graham, *Comparing the Benefits and Impacts of Hybrid Electric Vehicle Options*.
40. C. Samaras and K. Meisterling, "Life Cycle Assessment of Greenhouse Gas Emissions from Plug-in Hybrid Vehicles: Implications for Policy," *Environmental Science and Technology* 42 (2008): 3170–76.
41. See http://www.eiolca.net.
42. Kromer and Heywood, *Electric Powertrains: Opportunities and Challenges in the U.S. Light-Duty Vehicle Fleet*.
43. T. A. Groode, *A Methodology for Assessing MIT's Energy Use and Greenhouse Gas Emissions*, master's of science thesis (Massachusetts Institute of Technology, 2004).
44. N. Brinkman, M. Wang, T. Weber, T. Darlington, *Well-to-Wheels Analysis of Advanced Fuel/Vehicle Systems—A North American Study of Energy Use, Greenhouse Gas Emissions, and Criteria Pollutant Emissions* (Argonne National Laboratory, 2005).
45. C. Silva, M. Ross, and T. Farias, "Evaluation of Energy Emissions, Consumption, and Cost of Plug-in Hybrid Vehicles," *Energy Conversion and Management* 50 (2009): 1635–43.
46. P. Jaramillo, C. Samaras, H. Wakeley, and K. Meisterling, "Greenhouse Gas Implications of Using Coal for Transportation: Life Cycle Assessment of Coal-to-Liquids, Plug-in Hybrids, and Hydrogen Pathways," *Energy Policy* 37 (2009): 2689–95.
47. M. Kintner-Meyer, K. Schneider, and R. Pratt, *Impact Assessment of Plug-in Hybrid Vehicles on Electric Utilities and Regional U.S. Power Grids* (Pacific Northwest National Laboratory, 2006).
48. C. H. Stephan and J. Sullivan, "Environmental and Energy Implications of Plug-in Hybrid-Electric Vehicles," *Environmental Science and Technology* 42 (2008): 1185–90.
49. K. Parks, P. Denholm, and T. Markal, *Costs and Emissions Associated with Plug-In Hybrid Electric Vehicle Charging in the Xcel Energy Colorado Service Territory*, NREL/TP-640-41410 (National Renewable Energy Laboratory, 2007).
50. A. Elgowainy, A. Burnham, M. Wang, J. Molburg, and R. Rousseau, *Well-to-Wheels Energy Use and Greenhouse Gas Emissions Analysis of Plug-In Hybrid Electric Vehicles*, ANL/ESD/09-2 (Argonne National Laboratory, 2009).
51. S. Hadley and A. Tsvetkova, *Potential Impacts of Plug-in Hybrid Electric Vehicles on Regional Power Generation*, ORNL/TM-2007/150 (Oak Ridge National Laboratory, 2008).
52. M. Wang, "Well-to-Wheels Analysis of Biofuels and Plug-In Hybrids," presentation at Argonne National Laboratory, June 3, 2009.
53. Electric Power Research Institute and Natural Resources Defense Council, *Environmental Assessment of Plug-In Hybrid Electric Vehicles, Volume 1: Nationwide Greenhouse Gas Emissions*, Report No. 1015325 (EPRI and NRDC, 2007).
54. Bradley and Frank, "Design, Demonstrations, and Sustainability Impact Assessments for Plug-in Hybrid Electric Vehicles."
55. Graham, *Comparing the Benefits and Impacts of Hybrid Electric Vehicle Options*.
56. Samaras and Meisterling, "Life Cycle Assessment of Greenhouse Gas Emissions from Plug-in Hybrid Vehicles."
57. Kromer and Heywood, *Electric Powertrains: Opportunities and Challenges in the U.S. Light-Duty Vehicle Fleet*.
58. Silva et al., "Evaluation of Energy Emissions, Consumption, and Cost of Plug-in Hybrid Vehicles."
59. Jaramillo et al., "Greenhouse Gas Implications of Using Coal for Transportation."

60. Kintner-Meyer et al., *Impact Assessment of Plug-in Hybrid Vehicles on Electric Utilities and Regional U. S. Power Grids.*
61. Stephan and Sullivan, "Environmental and Energy Implications of Plug-in Hybrid-Electric Vehicles."
62. Parks et al., *Costs and Emissions Associated with Plug-In Hybrid Electric Vehicle Charging in the Xcel Energy Colorado Service Territory.*
63. Elgowainy et al., *Well-to-Wheels Energy Use and Greenhouse Gas Emissions Analysis of Plug-In Hybrid Electric Vehicles.*
64. EPRI and NRDC, *Environmental Assessment of Plug-In Hybrid Electric Vehicles, Volume 1.*
65. S. Paul and C. Tompkins (eds.), *Gigaton Throwdown: Redefining What's Possible for Clean Energy by 2020* (San Francisco, CA: Gigaton Throwdown Initiative, 2009).
66. T. Lipman, "Plug-In Hybrid Electric Vehicles," in Paul and Tompkins, *Gigaton Throwdown.*
67. E. Martin, S. A. Shaheen, T. E. Lipman, and J. Lidicker, "Behavioral Response to Hydrogen Fuel Cell Vehicles and Refueling: A Comparative Analysis of Short- and Long-Term Exposure," *International Journal of Hydrogen Energy 34* (2009): 8670–80; K. S. Kurani, T. S. Turrentine, and D. Sperling, "Testing Electric Vehicle Demand in 'Hybrid Households' Using a Reflexive Survey," *Transportation Research: Part D* 1 (1996): 131–50.
68. California Environmental Protection Agency, *Hydrogen Blueprint Plan: Volume 1* (2005).

Chapter 7: Comparing Land, Water, and Materials Impacts

Sonia Yeh, Gouri Shankar Mishra, Mark A. Delucchi, and Jacob Teter

The environmental impact of transportation fuels and vehicles doesn't stop at GHG emissions but also includes impacts on land, water, and materials used in their production. Local land-use impacts occur where biofuel feedstocks are grown; these must be acknowledged and weighed against the land-use impacts of oil production. (Note that in addition to its direct local impacts, biofuel production can have important indirect impacts; these are considered in Chapter 12.) Production of fossil fuels, biofuels, electricity, and hydrogen all have water footprints that must be considered in any comprehensive assessment of environmental impacts. In addition, advanced vehicle technologies use materials that might become a barrier to development if they are either scarce or else concentrated in a few countries. This chapter focuses on work that has been done so far comparing the sustainability of different fuel/vehicle pathways along these lines.

Local Land-Use Impacts of Transportation Fuel Production

Government support of major biofuel programs in the United States and other countries has intensified discussion of the land-use implications of biofuels, among other impacts.[1] However, our understanding and measurement of these impacts are at present rather limited. We do know that any land disturbance caused by fuel production, whether of biofuels or oil, not only has an impact on the ecological integrity of the land and its wildlife but also results in GHG emissions. Here we compare the local land-use impacts of biofuel and oil production.

Local land-use impacts of biofuel production

Recent studies point out that if biofuels are produced on carbon-rich lands such as forest or tropical peatlands, this can release large amounts of greenhouse gases that may take decades of biofuel production to sequester back.[2] Land-use impacts from biofuel production can also occur farther afield due to the global reach of commodity markets; this topic—indirect land-use change—is taken up in Chapter 12.

Aside from the question of emissions, the use of monocultural feedstocks (such as corn) to make biofuels can reduce biological diversity and the associated biocontrol services in agricultural landscapes. A simple land-use intensity metric (such as acres per energy unit of fuel produced) is not a good indicator of these impacts, in part because it does not reflect the impact of the land use on habitat integrity, wildlife corridors, and interactions at the edges of the affected area.

By any of these measures, biofuels made from agricultural crops can severely degrade natural habitats. To mitigate these effects, monocultures should be replaced by "natural, diversified and multifunctional vegetation that could meet the broad demand for goods and other resource functions in a sustainable fashion."[3]

Biofuel-crop harvesting practices can affect soil erosion and the nutrient and organic content of the soil, which in turn can affect the use of fertilizer. For example, if crop residues are removed from the field and used as a source of energy in the production of a biofuel, soil erosion might increase and fewer nutrients and less organic matter might be returned to the soil. Additional fertilizer might be required to balance any loss, and the use of additional fertilizer will result in additional environmental impacts.

Land-use impacts of oil development

Many studies examining the land-use impacts of oil and gas production have found significant levels of habitat loss, fragmentation, and other ecological impacts associated with these developments.[4] Yeh et al. were the first to study GHG emissions associated with the land disturbance caused by oil production.[5]

Using oil wells in California and Alberta as examples of conventional oil production, and oil sands production in Alberta as an example of unconventional oil production, Yeh et al. found that the land-use impacts of oil production in Canada can be substantial, as it disturbs large tracts of land in the boreal region. Since a large portion of the disturbed area is on peatlands (a special formation of soil that slowly accumulates carbon over thousands of years and stores ten times more carbon than regular soil found in most places), the carbon emissions can be quite high.

Conventional oil development causes land disturbance when infrastructure such as well pads, pipelines, and access roads are installed, and when seismic surveys are done. Typically, few oil wells are drilled during exploration. During development, well density increases until oil production rates drop below economically recoverable levels. Wells are shut in and abandoned afterward. In Canada, oil wells need to be reclaimed and certified to ensure that abandoned wells have a land capability that is equivalent to predrilling conditions, though the compliance rate has been declining since 2000.

Oil sands projects are generally located in northeast Alberta, with some development extending to the northwest of the province in the Peace River region and east into Saskatchewan, an area classified as boreal forest. Bitumen is extracted from oil sands using in situ recovery or surface mining. In situ recovery involves drilling wells into deposits typically deeper than 100m and injecting steam into the reservoir, reducing the bitumen's viscosity and allowing it to be pumped to the surface. Infrastructure such as central processing facilities and networks of seismic lines, roads, pipelines, and well pads must be built to support in situ recovery.

Surface mining of bitumen, used for shallower deposits, requires the clearing and excavation of a large area; it involves draining peatlands, clearing vegetation, and removing peat, with subsoil and overburden being removed and stored separately. The total land disturbance includes a mine site, overburden storage, and tailing ponds. Disturbed peat is stockpiled and stored until reclamation, when it may be used as a soil amendment. The drained and/or extracted peat will begin to decompose, releasing a combination of CO_2 and CH_4 (methane) depending on peat moisture conditions.[6] When the functional vegetation layer at the surface of a peatland is removed, the disturbed ecosystem loses its ability to sequester CO_2 from the atmosphere, so foregone sequestration must also be factored in. Reclamation of surface mines typically involves reconstructing self-sustaining hydrology and geomorphology on the landscape.[7] A mixture of peat and soil from the original lease and surrounding sites is used to cover the end substrates. The landscape is subsequently seeded and revegetated.

Yeh et al. calculated the amount of land disturbed per unit of fuel produced for both historical and current production of conventional oil in California and Alberta. They used image analysis to determine the land area disturbed per well, dividing the total disturbed area by the number of distinguishable well pads counted in each image. They found that the land area disturbed per well is almost three times larger in Alberta than in California, averaging 1.1 hectares per well (ha/well) in California compared with 3.3 ha/well in Alberta. As a result, the energy yields (PJ per ha of disturbed land) are roughly two times higher for California oil production compared with Alberta conventional oil production. In both places, oil production peaked around 1985 and has been declining ever since. Thus the marginal land-use impact of oil production has increased, with more land disturbance and less energy output.

IMAGES OF LAND DISTURBANCE FROM FOSSIL FUEL PRODUCTION

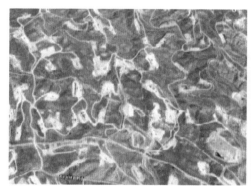 

These images—extracted from Google Earth and attributed to Telemetrics, TeleAtlas and Digital Globe 2009—show the land disturbance resulting from oil production in Elk Hills, California (left), and Alberta (right).

LOCATION OF OIL SANDS DEPOSITS AND PEATLANDS CARBON DEPOSITS IN ALBERTA'S BOREAL REGION

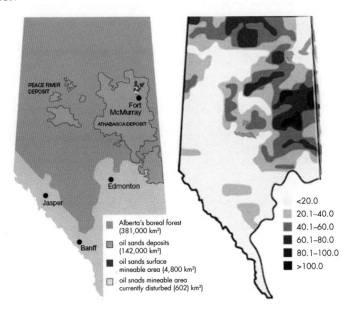

Alberta's conventional and oil sands deposits (left figure) sit right on top of one of the world's largest carbon deposits (right figure, contour interval is 20 kg•m2). Source for the figure on the left: Government of Alberta, Alberta's Oil Sands, Reclamation, http://www.oilsands.alberta.ca/reclamation.html#JM-OilSandsArea. Source for the figure on the right: D. H. Vitt, L. A. Halsey, I. E. Bauer, and C. Campbell, "Spatial and Temporal Trends in Carbon Storage of Peatlands of Continental Western Canada Through the Holocene," Canadian Journal of Earth Sciences (2000): 683–93. Boreal peatlands store 85 percent of global peat and contain about six times more carbon than tropical peatlands. Yeh et al. estimate that 15 percent of conventional oil and 23 percent of oil sands development occurs in peatland.

In addition to having large environmental and ecological impacts, land disturbance also contributes to GHG emissions. Natural carbon stocks increase and decrease as a result of land disturbance through a variety of mechanisms. The mechanisms Yeh et al. examined include clearing of vegetation, loss of soil carbon, foregone sequestration, and resequestration due to reclamation and/or forest regrowth. They also assessed CH_4 emissions from tailings ponds and peat stockpiled during oil sands surface mining operations. Though CH_4 emissions from tailings ponds are different from biological carbon typically included in land-use analysis, these emissions were included because of the large land areas covered by tailings ponds, the high CH_4 emissions, and the extent to which emissions can be affected by mitigation decisions related to land-use management.

Peatland conversion and tailing ponds are the largest sources of GHG emissions of oil production examined in the study. As Canadian oil sands production may reach 1.5 billion barrels per year in 2030[8], this may result in an additional 50,000–96,000 hectares of cumulative land disturbance and 47–580 megatonnes of CO_2e emissions resulting from surface mining between

2010 and 2025; in situ production may add 9,100–21,000 hectares of land disturbance and 0.1–10 megatonnes of CO_2e emissions during the same period (not including upstream disturbance from the use of natural gas). These findings emphasize the importance of restoration activities after oil sands production has been completed, not only to reduce land-related CO_2 emissions but more importantly to recover ecological landscapes and sustain high biodiversity, hydrologic cycles, and forest ecosystems.

NET GHG CHANGES OVER 150 YEARS FROM LAND DISTURBED BY OIL PRODUCTION

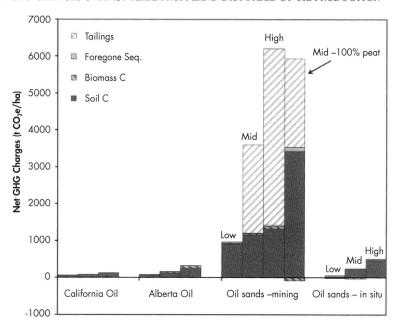

Yeh et al. quantified changes in carbon stock and CH_4 emissions per unit of area disturbed by conventional oil production and oil sands over a modeling period of 150 years, assuming reclamation back to a natural state after project completion. Oil sands surface mining is far and away the largest contributor on this score. Source: S. Yeh, S. M. Jordaan, A. M. Brandt, M. R. Turetsky, S. Spatari, and D. W. Keith, "Land Use Greenhouse Gas Emissions from Conventional Oil Production and Oil Sands," Environmental Science and Technology, (2010): 8766–8772.

Comparing the land-use GHG impact of oil and biofuels

Three important variables determine the direct land-use greenhouse gas (GHG) impact of liquid transportation fuels:
- energy yield—that is, the amount of energy produced per unit of land disturbed
- GHG emissions produced per unit of land disturbed
- GHG emissions produced per unit of energy output

When we compare the land disturbance from fossil fuel and biofuel production, it is the energy yield that greatly distinguishes the two. Due to the significantly lower energy output per unit of land used for crop production versus fossil energy production, biofuels require orders of magnitude more land than do petroleum fuels for the same amount of energy produced. Thus, although GHG emissions per unit of land disturbed by oil production can be comparable to or higher than emissions from biofuel production, land-use GHG emissions per unit of energy output for oil can be significantly lower than for biofuels.

COMPARISON OF DIRECT LAND-USE IMPACTS, BIOFUEL VS FOSSIL FUEL PRODUCTION

Energy Source		Energy Yield (PJ/ha)	GHG Emissions (t CO2e) per Hectare	GHG Emissions (g CO2e) per MJ
Fossil fuel				
California oil	historical impacts	0.79 (0.48–2.6)	73 (59–117)	0.09 (0.02–0.25)
	marginal impacts	0.55 (0.33–1.8)		0.13 (0.03–0.35)
Alberta oil	historical impacts	0.33 (0.16-0.69)	157 (74-313)	0.47 (0.12-1.98)
	marginal impacts	0.20 (0.092-0.40)		0.78 (0.20-3.39)
oil sands—surface mining		0.92 (0.61-1.2)	3596 (953-6201)	3.9 (0.83-10.24)
oil sands - in situ		3.3 (2.2-5.1)	205 (23-495)	0.04 (0.0-0.23)
Biofuel				
palm biodiesel (Indonesia/Malaysia)	tropical rainforest	0.0062	702 +/– 183	113 +/– 30
palm biodiesel (Indonesia/Malaysia)	peatland rainforest	0.0062	3452 +/– 1294	557 +/– 209
soybean biodiesel (Brazil)	tropical rainforest	0.0009	737 +/– 75	819 +/– 83
sugar cane (Brazil)	cerrado wooded	0.0059	165 +/– 58	28 +/– 10
soybean biodiesel (Brazil)	cerrado grassland	0.0009	85 +/– 42	94 +/– 47
corn ethanol (US)	central grassland	0.0038	134 +/– 33	35 +/– 9
corn ethanol (US)	abandoned cropland	0.0038	69 +/– 24	18 +/– 6

Note that values for fossil fuel are single estimates consisting of the mid-range values; the upper-bound and lower-bound estimates are reported in parentheses. Values for biofuels include standard deviations. Source: S. Yeh et al., "Land Use Greenhouse Gas Emissions from Conventional Oil Production and Oil Sands." Biofuel estimates are based on data from J. Fargoine, J. Hill, D. Tilman, S. Polasky, and P. Hawthorne, "Land Clearing and the Biofuel Carbon Debt," Science 319 (2008): 1235–38, which assumes a 50-year biofuel production period.

Water Resource Impacts of Transportation Fuel Production

Another important aspect of the sustainability implications of different fuel/vehicle pathways has to do with their impacts on water resources. Production of fossil fuels, biofuels, electricity, and hydrogen all require consumption and/or withdrawal of freshwater to some extent.[9] However, determining the water footprints of different fuels is a complex topic, complicated by regional and seasonal variations in water availability or scarcity. A direct comparison between fuel pathways cannot be made simply by comparing totals or averages but must be examined at the local and regional level, considering water availability, water quality, and impacts on ecosystem health.

Water impacts of biofuel production

Mishra and Yeh assessed the water requirements of producing ethanol from corn grain and crop residue. They explicitly tracked volumes of water use by different categories throughout the life cycle, including evapotranspiration, application and conveyance losses, biorefinery uses, and water use of energy inputs. They also considered avoided water use due to co-products, which estimates the amount of water that would have been consumed without the production of co-products.

The two categories of water use the researchers examined were (1) consumption of blue water (BW, meaning surface or ground water) and green water (GW, meaning precipitation and soil moisture), and (2) withdrawal of blue water. Consumption is the use of freshwater that is not returned to the watershed but instead is lost as a result of evaporation, evapotranspiration, incorporation into the product, discharge to the sea, or percolation into a salt sink. Withdrawal is the removal of water from a surface water body or aquifer to be used both consumptively and nonconsumptively. BW used nonconsumptively is released back to the environment with or without change in quality, through recycling to water bodies, seepage, and runoff, and is available for alternative uses though these may be in different watersheds or at different times. Unlike BW, use of GW is considered only in a consumptive sense. Water usage is estimated in the form of liters per vehicle kilometer traveled (L/VKT) and hence referred to as water intensity.

LIFE-CYCLE WATER REQUIREMENTS OF BIOFUEL PRODUCTION

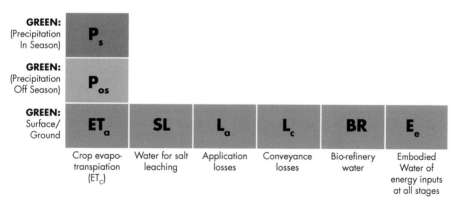

Mishra and Yeh tracked the water required to make ethanol from corn grain and crop residue, including crop evapotranspiration (ETc) or irrigation (ETa), process and cooling water consumed during ethanol conversion (BR, which is included in most water footprint studies), as well as water for uses that haven't been considered by other researchers. These include water for salt leaching (SL), application losses due to irrigation system inefficiencies (La), losses during conveyance of irrigation water (Lc), and water requirements of fuels (Ee)—diesel, electricity, natural gas, and coal—used during corn cultivation, storage, and distribution, and during ethanol production.

Mishra and Yeh focused on ethanol from corn grown in California (CA) and in the U.S. Corn Belt—Illinois (IL), Indiana (IN), Iowa (IA), Kansas (KS), and Nebraska (NE). These states together accounted for more than 50 percent of the corn produced in the United States in 2009 and are likely to witness significant increases in corn cultivation and production of ethanol from both grain and agricultural residue as a result of aggressive targets set forth in the federal renewable fuel standard and the low-carbon fuel standard. For IL, IN, and IA, only rain-fed corn was considered, which accounted for more than 97 percent of the corn produced in those states in 2009. For NE and KS, water requirements of ethanol from rain-fed and irrigation corn were considered separately. All corn grown in California is irrigated.

The researchers found that the GW consumption intensities of rain-fed corn in IL, IN, and IA are similar. The slight differences are entirely due to differences in yields, ET_c requirements, and supply constraints in the form of precipitation and available soil moisture. The team also found that irrigated corn yields are 50 to 60 percent higher than rain-fed yields in KS and NE, resulting in lower GW consumption intensity for irrigated corn in KS and NE, though the total GW and BW consumption intensities are roughly the same. In KS, water was applied at a rate of 40 centimeters (1.6 million liters per acre) for corn irrigation, which is 60 percent higher than in NE.

Though none of the previous studies considered nonconsumptive water withdrawal since the water is released back to the environment through recycling to water bodies, seepage, and runoff, ignoring such use fails to recognize that significant water withdrawals from surface water bodies may exert localized and/or seasonal impacts on the ecosystem. For regions dependent upon groundwater, extraction of groundwater beyond recharge rates could lead to aquifer depletion. Mishra and Yeh found that volumes of water returned (nonconsumptive water) in the form of seepage and deep water percolation account for 8 to 15 percent of total irrigation water withdrawn, which is attributable to the inefficiencies of furrow irrigation in CA and to the conveyance system (unlined irrigation canals) in NE. Most worrisome is that groundwater is the primary source of BW in both KS and NE, where it constitutes 60 to 80 percent of total water withdrawn. Evidence suggests that increased water use for corn is accelerating water-level declines in the Mississippi River Valley alluvial aquifer at an alarming rate.[10]

Since the production of solid, liquid, and gaseous biofuels may in itself generate co-products that displace other products requiring water for their supply, Mishra and Yeh contend that recognizing water requirements displaced by co-products significantly expands the system boundary of water use analysis and considers, albeit partially, the indirect water use associated with bioenergy expansion. In the United States, 88 percent of corn grain conversion to ethanol occurs through biochemical conversion using dry mill technology. A by-product of this process is distillers' grain soluble (DGS), which is used as an animal feed and can substitute for other animal feeds—namely corn grain, soybean meal (SBM), and urea. SBM in turn displaces raw soybeans. Production of DGS thus precludes the need to produce such other animal feed, so corn ethanol should be credited for water saved from not producing them. Similarly, electricity demands during production of cellulosic ethanol from cob are met internally through combustion of the lignin component of the cob, and the surplus—around 220 kWh/dry metric ton of cob—is exported to the grid. Surplus electricity is assumed to displace grid electricity, which has an average water intensity of 2.46 liters/kWh. However, very few cellulosic conversion technologies are currently operating commercially and data on ethanol yield and water consumption are uncertain.

Overall, ethanol from irrigated corn consumes 50–146 L/VKT of BW and 1–60 L/VKT of GW (90–211 L/VKT and 48–124 L/VKT of BW and GW respectively without co-product credits). For ethanol from rain-fed crops, the corresponding numbers are 0.6 L/VKT BW and 70–137 L/VKT GW (0.6 L/VKT and 140–255 L/VKT without co-product credits). Ethanol from cob consumes very little BW: 0.85 L/VKT after co-product credits. Harvesting and converting the cob to ethanol reduces both the BW and GW intensity by 13 percent.

LIFE-CYCLE WATER INTENSITY OF CORN, AND OF ETHANOL FROM CORN GRAIN AND RESIDUES

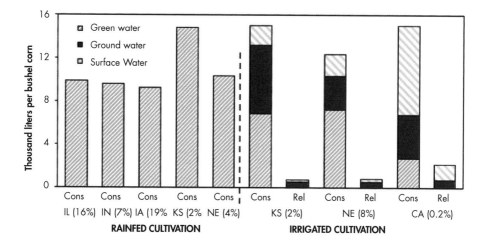

The volume of water required for corn cultivation—consumptive (Cons) and nonconsumptive (released water, Rel) use—is shown here. The values in parentheses are the share of corn produced in 2009. Irrigated corn grown in KS and NE has a 50- to 60-percent higher yield than non-irrigated corn but also requires more use of ground and surface water. The U.S. Geological Survey has found that increased water use for corn is accelerating water-level declines in the Mississippi River Valley alluvial aquifer.

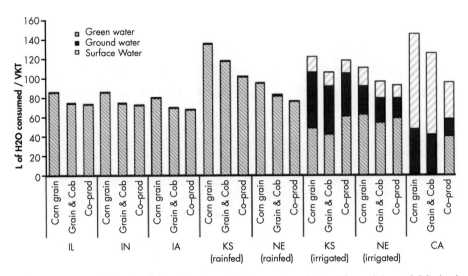

The water consumption intensity of ethanol from corn grain versus grain and crop residue, and the avoided/displaced water use credits assigned to co-products, is shown here. These results suggest that harvesting and converting the cob to ethanol reduces both the BW and GW intensity by 13 percent. Cellulosic ethanol from cob only (not shown in the figure) has a BW consumption intensity of 0.85 L/VKT and zero GW intensity, which is entirely contributed from biorefinery water use. On average, co-product credits are around 5 percent and 45 percent of total BW used to produce ethanol from rain-fed and irrigated corn, respectively; and around 50 percent of GW in both cases. The results reflect the lower yields and hence higher water intensity of soybeans—for example, statewide average applied water for soybean cultivation was around three-quarters that of corn in 2008, but average dry matter yield was less than 40 percent.

Mishra and Yeh also estimated the water consumption of large-scale biofuel production at the state level and found that without accounting for co-product credits, 13 to 15 percent of irrigation water is used to produce the corn required for ethanol in the states of KS and NE, and 7 to 8 percent after credits. In IL, IN, and IA, where corn is largely rain-fed, biorefinery water consumption is less than 0.5 percent of overall BW use.

The researchers argue that the marginal effects of water requirements will be higher given the renewable fuel standards, which have led to higher corn prices as a result of ambitious production mandates. Higher corn prices could lead to expansion of corn production to marginal lands with lower yield potentials. It could also result in intensification of corn cultivation on existing lands, which could lower future yields. Since water intensity is negatively correlated with yield, such expansion and intensification will increase the water intensity of ethanol. Further, corn expansion is occurring disproportionately on land that requires irrigation, which according to these researchers' results has higher average total water due to seepage, application and conveyance losses (GW+BW) and irrigation water consumptive intensity, as well as high nonconsumptive water requirements due to seepage, application and conveyance losses.

Water use associated with other fuel pathways

Water is also used in fossil fuel production. The BW consumption intensity of gasoline from conventional crude oil and Canadian oil sands is 0.41–0.78 L/VKT and 0.29–0.62 L/VKT, respectively.[11] Oil recovery using technology such as water flooding, enhanced oil recovery (EOR) via steam injection, and oil sands in-situ production is the major water consumption step in the petroleum gasoline life cycle. A recent report from the U.S. Government Accountability Office suggests the water intensity of gasoline from shale oil from large deposits found in Colorado, Utah, and Wyoming could be in the range of 0.29–1.01 L/VKT.[12]

Electricity production also withdraws a large amount of water, but the amount of water withdrawn and the impacts on water resources vary by region. The average water withdrawal intensity of thermal electric plants in a region typically correlates with the amount of water resources available within the region. Therefore, an important consideration for assessing the water resource impacts of fuels is the relative water intensity compared to the regional water shortage level. In addition, technology choices, water management, and technological change also explain variation in water use. The national average freshwater withdrawal per unit of electrical energy has decreased more than 35 percent since 1985 despite an increase in the total electricity produced, resulting in the total thermal electric freshwater withdrawal remaining constant over the same period.

Similar to the work on biofuels, Mishra, Glassley and Yeh[13] estimated the fresh and degraded water requirements of geothermal electricity. The research found that geothermal electricity is, in general, less water-efficient than other forms of electricity such as coal- and gas-fired power plants and renewables like solar thermal (i.e., water requirements of electricity from geothermal resources are substantially higher than those of both thermoelectricity and solar thermal electricity for the same amount of electricity generated). Mishra, Glassley and Yeh also conducted a scenario analysis to measure the potential impact of potential scaling up geothermal electricity on water demand in various western states with rich geothermal resources but stressed water resources. Electricity from enhanced geothermal systems (EGS) could displace 8–100% of thermoelectricity generated in most western states.[14] Such displacement would increase stress on water resources if re-circulating evaporative cooling, the dominant cooling system in the thermoelectric sector, is adopted. Adoption of dry cooling, which accounts for 78% of geothermal capacity today, will limit changes in state-wide freshwater abstraction, but increase degraded water requirements.

The research by Mishra, Glassley and Yeh identified the need for R&D to develop advanced geothermal energy conversion and cooling technologies that reduce water use without imposing energy and consequent financial penalties. Further, their results highlighted the need for policies to incentivize the development of higher enthalpy resources, and support identification of non-traditional degraded water sources and optimized siting of geothermal plants.

WATER IMPACTS OF DISPLACING THERMOELECTRIC WITH GEOTHERMAL ENERGY IN FOUR WESTERN STATES

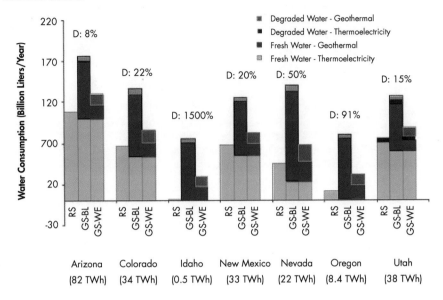

The figure above estimates the impact of displacement of thermoelectricity by EGS electricity on consumptive water requirements. The percentage of thermoelectricity produced in reference scenario (RS) and displaced by electricity from Enhanced Geothermal resources in Geothermal Scenario (GS) is represented by "D". Two geothermal sub-scenarios are envisaged—the baseline (GS-BL) and water efficient (GS-WE) scenarios. In the GS-BL scenario, where evaporative re-circulating cooling dominates, statewide water requirements increase substantially. In the GS-WE scenario, where dry cooling is used in 78% of geothermal electricity, as is the scenario today in the U.S., water requirements increase by a smaller magnitude.

Overall comparisons

Direct comparison of the water demands of biofuels and fossil fuels is much more complicated than simply comparing a commonly used, yet oftentimes erroneous due to its simplicity, water footprint indicator. The BW consumption of biofuels from rain-fed crops and residue is lower than that of gasoline, but it is orders of magnitude higher if the biofuels are from irrigated crops. Ethanol from corn grain has a high groundwater requirement, and groundwater use impacts terrestrial ecosystems and BW availability. Though the water intensity of fossil fuels is on average low compared with biofuels, it has been widely reported that oil sands production and potential shale oil development could result in substantial streamwater withdrawals and significant alteration of water flows during critical low river flow periods; groundwater depletion and contamination; and wastewater discharges.[16] A detailed comparison of biofuel versus fossil fuel water use should carefully examine the impacts of water use on changes in water availability and quality and other ecosystem health effects at the local level and/or accounting for season variability, though such comparison is often missing in the literature and also unfortunately beyond the scope of this analysis.

Mishra and Yeh caution that their assessment necessarily employs spatial and temporal aggregation by summing across types of water consumption (BW and GW consumption and avoided water credits) in locations where the relative importance of water-related aspects may differ; thus, some results may carry no clear indication of potential social and/or environmental harm or trade-offs. Similarly, temporal aggregation of water use estimates ignores the interseasonal variability of water use and water scarcity and can therefore yield erroneous conclusions concerning seasonal water use competition. Recent literature on freshwater life-cycle analysis has developed regionally differentiated characterization factors that measure water scarcity at a watershed level and also account for temporal variability in water availability. For example, in future studies volumetric estimates of green and blue water can be converted to characterization factors, providing a "stress-weighted" or "ecosystem-equivalent" water footprint estimate that can be compared across regions. Such work is still ongoing.[17]

Anticipating Material Use in New Vehicle Technologies

In a sustainable transportation system, the key new technologies will be electric motors and controllers, batteries, and fuel cells. An important question is whether any of these technologies use materials that are either scarce or else concentrated in a few countries and hence subject to price and supply manipulation, in which case the need for such materials might become a barrier to development. Here we focus on rare-earth elements (REEs) for electric motors, lithium for lithium-ion batteries, and platinum for fuel cells.[18]

Neodymium for electric motors

Some permanent-magnet alternating-current motors can use significant amounts of REEs. For example, the motor in the Toyota Prius uses 1 kg of neodymium (Nd) or 16-kg/MW (assuming that the Prius has a 60-kW motor).[19] In a worldwide fleet of EVs with permanent-magnet motors, the total demand for Nd might be large enough to be of concern, especially because permanent-magnet motors with Nd are also used in generators for wind-power turbines. A highly electrified world in which 50 percent of global electricity was provided by wind turbines and two-thirds of light-duty vehicles had electric motors could require up to 200,000 metric tons of Nd oxide per year. This rate of consumption would exhaust known global Nd-oxide reserves in less than one hundred years and would exhaust the more speculative potential resource base in perhaps a few hundred years. Therefore, it seems likely that a rapid global expansion of wind power and electric vehicles eventually will require generators and motors that do not use Nd or other REEs. However, this is not likely to be a serious constraint, because there are a number of alternatives to Nd for use in motors and generators.

Lithium for batteries

Roughly half of the world's identified lithium resources are in Bolivia and Chile. However, Bolivia does not yet have any economically recoverable reserves or lithium production infrastructure, and to date has not produced any lithium. A little more than half of the world's known economically recoverable reserves are in Chile, which is also the world's leading producer. Both Bolivia and Chile recognize the importance of lithium to battery and carmakers, and are hoping to extract as much value from it as possible. This concentration of lithium in a few countries, combined with rapidly growing demand, could cause increases in the price of lithium. In 2010, lithium carbonate (Li_2CO_3) sold for $6–7/kg, and lithium hydroxide (LiOH) sold for about $10/kg, prices which correspond to about $35/kg-Li. Given that lithium is 1–2 percent of the mass of lithium-ion batteries, a battery in an electric vehicle with a relatively long range (about 100 miles) might contain on the order of 10 kg of lithium. At 2010 prices this amount of lithium would contribute $350 to the manufacturing cost of a vehicle battery, but if lithium prices were to double or triple, the lithium raw material cost could approach $1,000. This could have a significant impact on the cost of an electric vehicle.

If one considers an even larger electric vehicle share of a growing future world car market and includes other demands for lithium, it is likely that the current lithium reserve base will be exhausted in less than twenty years in the absence of recycling. As demand grows the price will rise, and this will spur the hunt for other sources of lithium, most likely from recycling. The economics of recycling depend in part on the extent to which batteries are made with recyclability in mind. Ultimately the issue of how the supply of lithium affects the viability of lithium-ion-battery electric vehicles boils down to the price of lithium with sustainable recycling.

Platinum for fuel cells

The production of 20 million 50-kW fuel cell vehicles annually might require on the order of 250,000 kg of platinum (Pt)—more than the total current world annual production. How long this output can be sustained, and at what platinum prices, depends on at least three factors: (1) the technological, economic, and institutional ability of the major supply countries to respond to changes in demand; (2) the ratio of recoverable reserves to total production, and (3) the cost of recycling as a function of quantity recycled.

The effect of recycling on platinum price depends on the extent of recycling. It seems likely that a 90-percent-plus recycling rate will keep platinum prices significantly lower than will a 50-percent recycling rate. We cannot predict when and to what extent a successful recycling system will be developed. Nevertheless, we believe that enough platinum will be recycled to supply a large fuel-cell vehicle (FCV) market and moderate increases in the price of platinum, until new, less costly, more abundant catalysts or fuel-cell technologies are found. Indeed, catalysts based on inexpensive, abundant materials may be available relatively soon; research on iron-based catalysts suggests that a worldwide FCV market will not have to rely on precious-metal catalysts indefinitely.[20]

Preliminary work by Sun et al. (2010) supports this conclusion that platinum recycling will moderate the cost of platinum for FCVs.[21] They developed an integrated model of FCV production, platinum loading per FCV (a function of FCV production), platinum demand (a function of FCV production, platinum loading, and other factors), and platinum prices (a function of platinum demand and recycling). Based on this model, they found that in a scenario in which FCV production was increased to 40 percent of new light-duty vehicle output globally in the year 2050, the average platinum cost per FCV was $500, or about 13 percent of the cost of the fuel-cell system.

Summary and Conclusions

- This chapter has explored sustainability issues associated with land, water, and materials impacts of production along alternative fuel pathways compared with petroleum-based gasoline and diesel. The studies discussed here are representative, but this discussion is by no means comprehensive. It is also important to note that much work remains to be done on understanding and measuring these impacts.

- When biofuels are produced on carbon-rich lands such as forest or tropical peatlands, the resulting GHG emissions may take decades of biofuel production to sequester back. Because biofuels require orders of magnitude more land than do petroleum fuels for the same amount of energy produced, land-use GHG emissions per unit of energy output can be significantly higher than for oil. This is aside from the issue of indirect land-use impacts of biofuels, which are considered in Chapter 12.

- GHG emissions from land-use disturbance caused by fossil fuel exploration and extraction can be significant. In heavily mined areas after oil sands production has been completed, efforts should be focused on post-mining reclamation such as the restoration of habitat to reduce land-related CO_2 emissions, recover ecological landscapes, sustain high biodiversity, and maintain hydrologic cycles and forest ecosystems.

- The sustainability impacts of fuel production on water resources need to be compared at the local and regional levels. Concerns about local impacts on water availability, water quality, and ecosystem health should be carefully evaluated. The relative importance of water aspects compared to other aspects of the shift to a new transportation energy system—such as effects on GHG emissions, soil quality, biodiversity, and economic sustainability—must be weighed.

- The research on life-cycle material use by new vehicle technologies suggests that it is unlikely that material use will impose serious constraints on technology development in the long term. However, short-term price volatility and sustainability impacts due to extraction activities need to be considered and mitigated whenever appropriate.

Notes

1. J. E. Fargione, R. J. Plevin, and J. D. Hill, "The Ecological Impact of Biofuels," *Annual Review of Ecology, Evolution, and Systematics* 41 (2010): 351–77.
2. J. Fargione, J. Hill, D. Tilman, S. Polasky, and P. Hawthorne, "Land Clearing and the Biofuel Carbon Debt," *Science* 319 (2008): 1235–38; H. K. Gibbs, M. Johnston, J. A. Foley, T. Holloway, C. Monfreda, N. Ramankutty, and D. Zaks, "Carbon Payback Times for Crop-Based Biofuel Expansion in the Tropics: The Effects of Changing Yield and Technology," *Environmental Research Letters 3* (2008): 034001.
3. A. Kläy, *The Kyoto Protocol and the Carbon Debate*, Development and Environment Report No. 18 (Switzerland Center for Development and Environment, University of Berne, 2000).
4. R. R. Schneider, J. B. Stelfox, S. Boutin, and S. Wasel, "Managing the Cumulative Impacts of Land-uses in the Western Canadian Sedimentary Basin: A Modeling Approach," *Conservation Ecology* 7 (2003), Article 8; H. I. Jager, E. A. Carr, and R. A. Efroymson, "Simulated Effects of Habitat Loss and Fragmentation on a Solitary Mustelid Predator," *Ecological Modelling* 191 (2006): 416–30; S. M. Jordaan, D. W. Keith, and B. Stelfox, "Quantifying Land Use of Oil Sands

Production: A Life-Cycle Perspective," *Environmental Research Letters* 4 (2009): 024004; P. Lee and S. Boutin, "Persistence and Developmental Transition of Wide Seismic Lines in the Western Boreal Plains of Canada," *Journal of Environmental Management* 78 (2006): 240–50; U.S. Fish and Wildlife Service, *Potential Impacts of Proposed Oil and Gas Development on the Arctic Refuge's Coastal Plain: Historical Overview and Issues of Concern* (Fairbanks, AK: Arctic National Wildlife Refuge, 2001), available from http://arctic.fws.gov/issues1.html.

5. S. Yeh, S. M. Jordaan, A. M. Brandt, M. R. Turetsky, S. Spatari, and D. W. Keith, "Land Use Greenhouse Gas Emissions from Conventional Oil Production and Oil Sands," *Environmental Science and Technology*, (2010): 8766–8772, doi:10.121/es1013278.
6. J. Cleary, N. T. Roulet, and T. R. Moore, "Greenhouse Gas Emissions from Canadian Peat Extraction, 1990–2000: A Life-cycle Analysis," *Ambio* 34 (2005): 456–61.
7. E. A. Johnson and K. Miyanishi, "Creating New Landscapes and Ecosystems: The Alberta Oil Sands," *Annals of the New York Academy of Sciences 2008* 1134: 120–45.
8. Energy Information Administration (EIA), "Natural Gas Consumption in Canadian Oil Sands Production," http://www.eia.doe.gov/oiaf/archive/aeo04/issues_4.html.
9. King and Webber estimated the water needed for different fuel/vehicle pathways. See C. W. King and M. E. Webber, "Water Use and Transportation," *Environmental Science and Technology* 42 (2008): 7866–872.
10. H. L. Welch, C. T. Green, R. A. Rebich, J.R.B. Barlow, and M. B. Hicks, "Unintended Consequences of Biofuels Production: The Effects of Large-Scale Crop Conversion on Water Quality and Quantity," Open-File Report 2010–1229, U.S. Geological Survey, 2010.
11. M. Wu, M. Mintz, M. Wang, and S. Arora, "Water Consumption in the Production of Ethanol and Petroleum Gasoline," *Environmental Management* 44 (2009): 981–97.
12. United States Government Accountability Office (GAO), *Energy-Water Nexus: A Better and Coordinated Understanding of Water Resources Could Help Mitigate the Impacts of Potential Oil Shale Development* (October 29, 2010), GAO-11-35, p. 70.
13. G. Mishra, W. Glassley, and S. Yeh, "Realizing the Geothermal Electricity Potential—Water Use and Consequences." Submitted to *Environment Research Letters*. (2011).
14. J. W. Tester et al., "The Future of Geothermal Energy: Impact of Enhanced Geothermal Systems (EGS) on the United States in the 21st Century," Massachusetts Institute of Technology and Department of Energy Report, Idaho National Laboratory, INL/EXT-06-11746, 2006.
15. S. Yeh, G Berndes, G. S. Mishra, S. P. Wani, A. E. Neto, S. Suh, L. Karlberg, J. Heinke, K. K. Garg, "Evaluation of Water Use for Bioenergy at Different Scales," *Biofuels, Bioproducts and Biorefining* 5 (2011): 361–374.
16. This material is based on M. Z. Jacobson and M. A. Delucchi, "Evaluating the Feasibility of Meeting All Global Energy Needs with Wind, Water, and Solar Power, Part I: Technologies, Energy Resources, Quantities and Areas of Infrastructure, and Materials," *Energy Policy*, (2010): 1154-1169, doi:10.1016/j.enpol.2010.11.040.
17. S. Gorman, "As Hybrid Cars Gobble Rare Metals, Shortage Looms," Reuters, August 31, 2009, http://www.reuters.com/article/newsOne/idUSTRE57U02B20090831.
18. M. Lefèvre, E. Proietti, F. Jaouen, and J.-P. Dodelet, "Iron-Based Catalysts with Improved Oxygen Reduction Activity in Polymer Electrolyte Fuel Cells," *Science* 324 (2009): 71–74.
19. Y. Sun, M. A. Delucchi, and J. M. Ogden, "The Impact of Widespread Deployment of Fuel Cell Vehicles on Platinum Demand and Price," *International Journal of Hydrogen Energy*, (in press), doi:10.1016/j.ijhydene.2011.05.157.

Part 3: Scenarios for a Low-Carbon Transportation Future

Thus far we have explored and compared the alternative fuel and advanced vehicle pathways that might lead us to a low-carbon transportation future. Now it's time to imagine how those different pathways might be combined to reach specific targets. How will our current transportation system need to change if it's to meet ambitious greenhouse gas reduction targets? How will the transition to a low-carbon transportation system occur, and what will it cost? What part will a low-carbon transportation system play in meeting broader economy-wide carbon reduction goals? The three chapters in this section address those questions.

- **Chapter 8** explores how deep GHG reduction targets (50 to 80 percent) could be met in the transportation sector by 2050, with a focus on California and the United States as a whole. It presents a framework for understanding emission reductions in the transportation sector, lays out the major mitigation options for reducing emissions, and presents scenarios to explore how deep reductions could be achieved. It also looks at potential pathways from the present to the deep-reduction scenarios we need to arrive at by 2050.

- **Chapter 9** analyzes and compares alternative scenarios for adoption of new light-duty vehicle and fuel technologies that could enable deep cuts in gasoline consumption and GHG emissions by 2050. It uses simplified technology learning curve models to estimate the transitional costs for making new vehicle and fuel technologies economically competitive with gasoline vehicles.

- **Chapter 10** considers the role the transportation sector might play under economy-wide CO_2 constraints in the United States. If we see emission reductions achieved in different sectors of the economy—including commercial and residential buildings, industry, agriculture, and electric power, as well as transportation—as wedges that add up to an emission reduction target mandated by policy, how might the transportation wedge be optimized at least cost? To address this question, the authors use an integrated energy-economics model called the MARKet ALlocation (MARKAL) model to examine cost-effective deep emission reductions economy-wide and in the transportation sector.

Chapter 8:
Scenarios for Deep Reductions in Greenhouse Gas Emissions

Christopher Yang, David McCollum, and Wayne Leighty

The Intergovernmental Panel on Climate Change has suggested that annual greenhouse gas (GHG) emissions must be cut 50 to 80 percent worldwide by 2050 in order to stabilize the climate and avoid the most destructive impacts of climate change. California governor Arnold Schwarzenegger and U.S. president Obama lined up behind this goal.[1] Yet the strategies for meeting these ambitious economy-wide targets have not been clearly defined, and the technology and policy options are not well understood. This chapter explores how such deep reduction targets (50 to 80 percent) could be met in the transportation sector by 2050, with a focus on California and the United States as a whole. It presents a framework for understanding emission reductions in the transportation sector, lays out the major mitigation options for reducing emissions, and presents scenarios to explore how deep reductions could be achieved. Additionally, this chapter also presents an analysis that looks at the transition scenarios for vehicles in the light-duty sector to investigate how they may evolve from the present fleet to achieve the deep-reduction scenarios by 2050.

GHG Emissions in the Transportation Sector

Transportation is one of the primary sources of GHG emissions in California (where it accounts for 40 percent), the United States (29 percent), and globally (23 percent).[2] These emissions are growing quickly in each of these regions and in all subsectors—from personal light-duty vehicles (LDVs) and heavy-duty vehicles (HDVs, meaning buses and trucks) to rail, aviation, marine, agriculture, and off-road. The main drivers of transportation GHG emissions are population, transport intensity (passenger or freight miles per person), energy intensity (vehicle fuel consumption), and fuel carbon intensity. We can estimate transportation GHG emissions by plugging these four variables into a simple equation. In this equation, total transportation activity in miles is the product of the total human population (P) and transport intensity (T). The amount of carbon emitted per mile of transport is a product of energy intensity (E) and carbon intensity (C). By working out this equation and summing the results for all vehicle types and subsectors, we can arrive at a figure that describes the total CO_2-equivalent GHG emissions from the entire transportation sector on a full fuel-cycle basis in any given year (whether 1990 or 2050 or some point in between). Further, by comparing these figures we can estimate potential reductions in transportation GHG emissions between 1990 and 2050 for a given region.

> **THE EQUATION WE USED IN OUR ANALYSIS**
>
> A decomposition equation is a useful tool for estimating potential reductions in transportation emissions. We developed a transportation variant of the Kaya identity (Equations 1-3) in our analysis. In this decomposition equation, the main drivers for transportation GHG emissions are population (P), transport intensity (T), energy intensity (E), and fuel carbon intensity (C).
>
> $$(1) \quad CO_{2, Transport} \equiv (Population) \left(\frac{Transport}{Person}\right) \left(\frac{Energy}{Transport}\right) \left(\frac{Carbon}{Energy}\right) \quad (1)$$
>
> $$(2) \quad CO_{2, Transport} \equiv P \times T \times E \times C \quad (2)$$
>
> $$(3) \quad CO_{2, Transport} \equiv \sum_i \sum_j CO_{2i, j} \equiv \sum_i \sum_j P \times T_{i,j} \times E_{i,j} \times C_{i,j} \quad (3)$$
>
> where i = subsector and j = vehicle type.

Emissions in a given region can be classified into two categories: emissions generated by trips occurring entirely within the borders of the region and emissions from trips that cross the borders. This affects the jurisdiction of a given policy. For instance, in our California analysis, in-state emissions are linked to trips that occur entirely within the state's borders, while overall emissions also include half of emissions from trips that cross state boundaries. Similarly, for our U.S. analysis, emissions taking place entirely within the United States are called domestic emissions, whereas overall emissions also include half of emissions from international trips that originate or terminate in the United States.

For smaller regions like California or other U.S. states, within-region (in-state) emissions are a smaller proportion of overall emissions than for a larger region like the United States. In 1990, California's in-state emissions (on a full life-cycle basis) accounted for 73 percent of overall emissions (193 vs. 264 MMTCO$_2$e), whereas for the United States, domestic emissions accounted for 91 percent of overall emissions (1,921 vs. 2,104 MMTCO$_2$e). (MMT = million metric tonnes; CO$_2$e includes CO$_2$, CH$_4$, and N$_2$O weighted by their respective global warming potentials.)

In the United States in 1990 (which is the baseline year used for GHG emission reduction targets in this analysis), light-duty cars and trucks (passenger cars, pickup trucks, SUVs, minivans, and motorcycles) were responsible for about 60 percent of domestic life-cycle GHG emissions in the transportation sector. Heavy-duty vehicles (large trucks and buses) accounted for another 17 percent. Domestic aviation (including commercial passenger, freight, and general aviation) comprised 11 percent of emissions, and the remaining 12 percent were from a combination of rail, domestic marine, agriculture, and off-road equipment. The breakdown of energy use by subsector is very similar to that for GHG emissions because of the overwhelming reliance on various forms of petroleum fuels, all of which have roughly similar carbon intensity values.

U.S. TRANSPORTATION ENERGY USE AND GHG EMISSIONS, 1990

To understand the 1990 baseline for GHG emission reduction targets, we broke down transportation energy use and GHG emissions by subsector and vehicle type. These figures are based on our calculations using data from numerous sources. Emissions estimates reported here are higher than those from other published studies because we include the GHGs produced during upstream ("well-to-tank") fuel production processes.

PJ = petajoule, a measure of energy equivalent to a thousand trillion joules or roughly 30 million kilowatt hours. MMT = million metric tonnes; CO_2e includes CO_2, CH_4, and N_2O weighted by their respective global warming potentials.

Subsector	Vehicle Type	Energy Use				GHG Emissions			
		Domestic		Overall		Domestic		Overall	
		(PJ)	%	(PJ)	%	MMT CO_2e	%	MMT CO_2e	%
Light-duty	Cars and trucks	12,603	60.1%	12,603	54.8%	1,159	60.3%	1,159	55.1%
Heavy-duty	Buses	176	0.8%	176	0.8%	16	0.8%	16	0.8%
	Heavy trucks	3,370	16.1%	3,370	14.7%	304	15.8%	304	14.5%
Aviation	Commercial (passenger)	1,779	8.5%	2,335	10.2%	160	8.3%	210	10.0%
	Freight	365	1.7%	555	2.4%	33	1.7%	50	2.4%
	General	139	0.7%	139	0.6%	13	0.7%	13	0.6%
Rail	Passenger	77	0.4%	77	0.3%	14	0.7%	14	0.6%
	Freight	458	2.2%	458	2.0%	41	2.1%	41	2.0%
Marine	Large marine – intl.	-	0.0%	1,278	5.6%	-	0.0%	115	5.5%
	Large marine – domestic	341	1.6%	341	1.5%	31	1.6%	31	1.5%
	Personal boats	197	0.9%	197	0.9%	18	0.9%	18	0.9%
Agriculture	Agriculture	444	2.1%	444	1.9%	40	2.1%	40	1.9%
Off-road	Off-road	1,017	4.9%	1,017	4.4%	92	4.8%	92	4.4%
Total – All subsectors		20,966		22,990		1,921		2,104	

Options for Reducing Transport GHG Emissions

Three of the four drivers of transportation GHG emissions—transport intensity (T), energy intensity (E), and carbon intensity (C)—can also be thought of as levers that technologies and policies can use in order to reduce transport GHG emissions. (Population growth forecasts are taken as given in our work.)

- Travel demand can be reduced—which in turn can reduce transport intensity (T) in many of the subsectors—by integrated land-use planning, high-density development, and improved public transit.

- Energy intensity (E) can be reduced by improving the efficiency of the vehicle drive train, reducing dissipative forces on the vehicle (for example, by improving aerodynamics, reducing vehicle weight, or lowering rolling resistance), changing drivers' acceptance of smaller vehicles and less powerful engines and driving behavior (reducing "lead-foot" acceleration and deceleration).

- Fuel carbon intensity (C) can be reduced by switching to, or blending in, lower-carbon alternative fuels (including biofuels, hydrogen, or electricity). Of course, in order to accurately assess GHG reductions from fuel switching, emissions must be estimated on a full life cycle (that is, well-to-wheels or cradle-to-grave) basis.

These three levers are, to some extent, interdependent, and synergies between them can be realized—for example, shorter travel distances make highly efficient electric vehicles more attractive. These vehicles could, in turn, be powered by low- or zero-carbon electricity. However, because of the multiplicative Kaya identity, using multiple levers simultaneously reduces the impact of any single mitigation option (for example, doubling vehicle efficiency will have a much smaller impact on the absolute quantity of GHG emissions if vehicles are driving half as much or using fuel with lower carbon intensity.)

We worked with these three levers in order to quantify the emission reduction potential of various GHG mitigation strategies in the transportation sector in California and the United States as a whole. The model we developed is called, fittingly, the Long-term Evaluation of Vehicle Emission Reduction Strategies (LEVERS) model.[3]

Our Three Sets of Scenarios

In our LEVERS model, we created three sets of scenarios to illustrate different potential snapshots of the transportation sector in the United States and in California in 2050 and to estimate the extent to which different GHG mitigation options (technologies and policies) can help meet a deep-reduction target of 50 to 80 percent below 1990 levels.

HOW WE BUILT OUR SCENARIOS

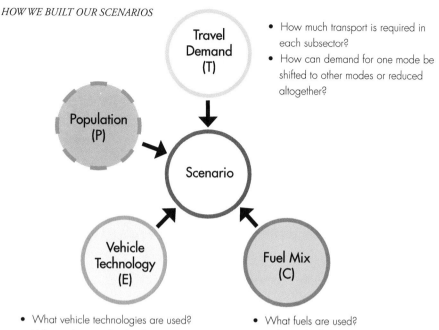

Each of our transportation scenarios is comprised of a unified story line of the future, as well as a variety of individual input assumptions for vehicles, fuels, and travel demand that correspond to the given story line.

- The reference scenario describes a business-as-usual future in 2050.

- The silver-bullet scenarios summarize the extent to which single mitigation strategies alone may reduce emissions.

- The deep-reduction scenarios combine mitigation options to achieve 50-to-80-percent reductions in transportation GHG emissions by 2050.

These scenarios should not be taken as predictions or forecasts of the future, although we have made reasonable judgments—and have included input from external experts—to create snapshots of the future that are technically plausible. It is important to note that political plausibilty is another issue entirely.

While this collection of scenarios is by no means exhaustive, they are nevertheless useful in informing the policy debate because they are clear and transparent. Stakeholders and policy makers can use them to help guide future decision making. These scenarios are meant to highlight the challenges associated with meeting the deep emission reduction targets and promote discussion about the feasibility of the proposed levels of technology and behavioral change and the policies needed to bring these changes about.

The reference scenario

Our reference scenario describes a future in which very little has been done to address climate change, and transportation activity and technology development follow historical trends. The only expected improvement that helps to mitigate growth in GHG emissions in this scenario is a modest reduction (45 percent, roughly 1 percent per year) in energy intensity. In the light-duty sector, this level of improvement is consistent with the entire light-duty fleet achieving 35 mpg on-road new-vehicle fuel economy. However, since both population and transport intensity (travel demand per person) are expected to increase significantly between 1990 and 2050 (a 70-percent increase in the U.S. population, a 100-percent increase in the California population, and an approximate doubling in per-capita transport demand, coming primarily from aviation travel, are forecast), total travel demand increases by a factor of 3.4. The average carbon intensity of transportation fuels is essentially unchanged relative to 1990, as petroleum-based fuels are assumed to remain dominant. Improvements in carbon intensity that result from biofuels being blended into gasoline and diesel in small quantities are balanced by the increased usage of unconventional oil sources, such as oil sands or coal-to-liquids.

In this scenario, U.S. domestic GHG emissions from transportation increase by 82 percent (to 3,496 MMTCO$_2$e) and overall emissions double (to 4,210 MMTCO$_2$e) from 1990 to 2050; California in-state GHG emissions from transportation increase by 61 percent (to 311 MMTCO$_2$e) and overall emissions increase by 86 percent (to 492 MMTCO$_2$e) from 1990 to 2050. Aviation is responsible for the greatest increase in emissions because, in spite of moderately more efficient airplanes, demand for air travel is expected to grow rapidly in the coming decades. Freight transport—by aircraft, heavy trucks, rail, and large marine vessels—is another area where considerable growth is expected. While the exact numbers are slightly different, the same general trends hold true for both the United States and California.

The silver-bullet scenarios

Our silver-bullet scenarios for the United States and California describe futures in which one single mitigation option, such as an advanced vehicle technology or alternative fuel, is scaled up quickly from today and is employed to the maximum extent possible from a technology perspective in 2050. Emissions are calculated in order to understand the GHG reduction potential of particular vehicle and/or fuel technologies or travel demand reduction. The silver-bullet scenarios modify specific individual elements of the reference scenario.

U.S. DOMESTIC GHG EMISSIONS: SILVER-BULLET SCENARIOS

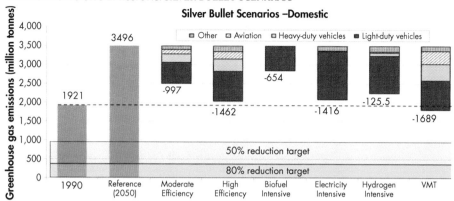

We compared actual GHG emissions from domestic transportation in the United States in 1990, emissions projected for 2050 in the reference scenario, and projected reductions for each subsector in each of six silver-bullet scenarios for 2050. Each silver-bullet scenario describes a future in which one mitigation option is scaled up quickly and employed as fully as technologically possible. Not one of the silver-bullet scenarios by itself achieves the 50-to-80-percent emission reductions goal, implying that a multi-pronged "portfolio" approach is necessary.

In the Biofuel Intensive scenario, the level of biofuels demand is consistent with projected total U.S. supply (~90 billion gge), although this projection may be overly optimistic.[4] Significant uncertainties surrounding indirect land-use change impacts from biofuels production lead to the large variability in potential GHG changes from 1990 levels.

The major take-away message from these silver-bullet scenarios is that none of the individual mitigation options, even ones as encompassing as shifting to the use of biofuels or widespread electrification, can take the transportation sector anywhere close to a 50-to-80-percent reduction. This is in part due to the large projected increase in transportation demand, which counteracts the improvements in efficiency and from fuel switching. However, another factor that prevents a single mitigation option from achieving deep reductions is the diverse nature of the transportation sector. Because of differences in vehicle types, duty cycles, and other application requirements, a given option such as electrification or use of hydrogen and fuel cells cannot be applied universally to all vehicles in each of the subsectors. Aviation and marine are among the most difficult subsectors in which to apply these advanced propulsion systems.

The deep-reduction scenarios

While not one of the silver-bullet scenarios achieves the ambitious 50-to-80-percent reduction goal, several of the options examined in those scenarios are complementary (such as improving efficiency, using low-carbon alternative fuels, and reducing travel demand) and can be combined in a portfolio approach to achieve deep GHG emission reductions.

We developed three different scenarios that represent different potential futures for the United States in which a 50-to-80-percent reduction in domestic GHG emissions might be realized. The scenarios are snapshots of the transportation sector in 2050 and illustrate different mixes of mitigation options in various subsectors. The first scenario relies on moderately high vehicle efficiencies using low-carbon biofuels and the second on higher-efficiency electric-drive vehicles using low-carbon electricity and hydrogen. The third scenario considers a combination of these two strategies. All three assume the same growth in population as in the reference scenario, and each envisions a significant slowing of growth in transport intensity (per-capita VMT) in each subsector to about half of the reference scenario growth, which translates into a 25-percent reduction from the reference scenario in per-capita VMT across all modes. This means that in most cases 2050 transport intensities are still somewhat higher than 2010 levels, but not significantly so.

- US-Efficient Biofuels 50in50 describes a future in which low-carbon biofuels are relatively abundant. In this scenario, a 50-percent reduction in transportation emissions is achieved primarily through the use of low-carbon biofuels, more-efficient internal combustion engine (ICE) vehicles, and travel demand reduction. However, even with relatively optimistic assumptions about the biofuel supply, only 64 percent of total fuel requirements can be met by biofuels, with the remainder coming from petroleum.

- US-Electric-Drive 50in50 describes a future in which significant advances in electric-drive technologies (fuel cells and electric vehicle batteries) reshape the transportation sector, improving vehicle efficiency and advancing low-carbon alternative fuels. Hydrogen and electricity make up 66 percent of total fuel use, with all biofuels (a smaller quantity compared to US-Efficient Biofuels 50in50 due to less optimistic supply estimates) used in the aviation sector.

- US-Multi-Strategy 80in50 is, in essence, a combination of these two 50in50 scenarios, describing a future in which the technology breakthroughs of both are realized, thus leading to an 80-percent reduction in GHG emissions. Extensive biofuels usage and significant penetration of fuel cell vehicles (FCVs) and electric vehicles—plug-in hybrid electric vehicles (PHEVs) and battery electric vehicles (BEVs)—essentially result in the elimination of petroleum consumption.

A similar approach was taken to develop three scenarios representing different potential futures for California. However, underrepresentation of the aviation and marine sectors in in-state emissions allows the state to achieve greater emission reductions than in the corresponding U.S. domestic case for the same level of effort. Consequently, all three California scenarios achieve an 80-percent reduction in in-state transportation GHG emissions from 1990 levels. These are the CA-Efficient Biofuels 80in50, CA-Electric-Drive 80in50, and CA-Multi-Strategy 80in50 scenarios.

ASSUMPTIONS: U.S. DEEP-REDUCTION SCENARIOS

Each of our three deep-reduction scenarios for U.S. domestic emissions in 2050 makes different assumptions about transport intensity (T), energy intensity (E), carbon intensity (C), and the share of transport miles powered by each type of fuel/technology.

		Share of Miles by Fuel Type				T Normalized Transport Intensity (1990=100%)*	E Normalized Energy Intensity (1990=100%)	C Normalized Carbon Intensity (1990=100%)
		Petroleum	Biofuels	Hydrogen	Electricity			
US-Efficient Biofuels 50in50	LDV	0%	100%	0%	0%	137%	33%	13%
	HDV	80%	20%	0%	0%	149%	52%	82%
	Aviation	100%	0%	0%	0%	234%	36%	100%
	Rail	84%	0%	0%	16%	171%	59%	80%
	Marine/Ag/Off-road	100%	0%	0%	0%	117%	40%	101%
	All subsectors combined	35%	64%	0%	1%	152%	37%	53%
	Fuel Demand (billion GGE)	77.2	88.6	0.0	1.3			
	Carbon Intensity (gCO2e/MJ)	90-96	12.3	-	44			
US-Electric-Drive 50in50	LDV	10%	0%	60%	30%	137%	24%	40%
	HDV	72%	0%	22%	5%	149%	60%	100%
	Aviation	20%	75%	5%	0%	234%	37%	32%
	Rail	0%	0%	0%	100%	171%	38%	43%
	Marine/Ag/Off-road	62%	0%	38%	0%	117%	40%	78%
	All subsectors combined	17%	17%	42%	24%	152%	33%	59%
	Fuel Demand (billion GGE)	64.6	21.2	42.2	19.7			
	Carbon Intensity (gCO2e/MJ)	90-96	12.3	24	44			
US-Multi-Strategy 80in50	LDV	0%	10%	60%	30%	137%	22%	30%
	HDV	0%	63%	28%	9%	149%	58%	19%
	Aviation	0%	100%	0%	0%	234%	37%	14%
	Rail	0%	0%	0%	100%	171%	38%	43%
	Marine/Ag/Off-road	2%	79%	20%	0%	117%	40%	28%
	All subsectors combined	0%	36%	40%	24%	152%	32%	24%
	Fuel Demand (billion GGE)	1.9	82.3	39.3	19.1			
	Carbon Intensity (gCO2e/MJ)	90-96	12.3	24	44			

* For example a value of 137% corresponds to a +37% change from 1990, and a value of 34% corresponds to a -66% change.

When we tease out the effects of different mitigation options in the U.S. case, we see that slowing the rapid growth in travel demand makes a major contribution to emission reductions in all three scenarios. US-Multi-Strategy 80in50 is more successful at making deeper emission reductions because it combines the strategies of the two 50in50 scenarios, which are somewhat complementary, and helps to address their key limitations.

U.S. DOMESTIC GHG EMISSIONS: DEEP-REDUCTION SCENARIOS
US-Efficient Biofuels 50in50

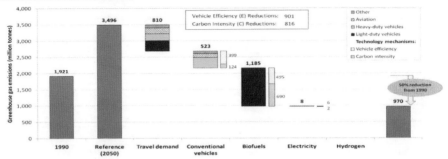

US-Electric-Drive 50in50

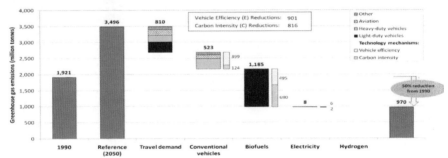

US-Multi-Strategy 80in50

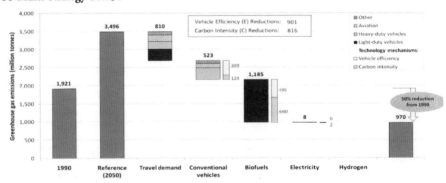

For each of our three US deep-reduction scenarios, we compared actual GHG emissions from domestic transportation in the United States in 1990, emissions projected for 2050 in the reference scenario, and projected reductions for each subsector by 2050 when the various reduction levers—travel demand, vehicle efficiency, and fuel carbon intensity—are used. Slowing the growth of travel demand makes a major contribution to emission reductions in all three scenarios. US-Multi-Strategy 80in50 is more successful at making deeper emission reductions because it combines the strategies of the two 50in50 scenarios (biofuels and electric-drive technologies), which are somewhat complementary, and helps to address their key limitations.

Biofuels are a convenient replacement for liquid fuels that, in theory, can be relatively easily substituted for conventional petroleum fuels in any subsector. However, in US-Efficient Biofuels 50in50, even with relatively optimistic assumptions about the quantity of low-carbon biofuels available, there are limits on biomass resources, which in turn limit how much biofuel substitution can take place. The quantity of low-carbon biofuels is a source of significant uncertainty and one of the most critical parameters in determining the level of GHG reductions possible in the transportation sector. Significant constraints on biofuel availability will require greater contributions from other mitigation options.

Electric-drive vehicles such as FCVs, PHEVs, and BEVs offer the potential for greatly improved vehicle efficiency and the use of low-carbon energy carriers from a variety of primary resources. In US-Electric-Drive 50in50, GHG reductions are not limited by constraints on primary energy resources but rather by the challenges associated with applying electric-drive vehicles to certain subsectors (such as aviation, shipping, and heavy-duty trucks) because of specific technical considerations, most notably energy storage density, as well as temporal limits associated with the market penetration and social acceptance of these vehicles and building their requisite refueling infrastructure.

It should be noted that because the three scenarios rely heavily on very low-carbon-intensive fuels to achieve the GHG target, they are quite sensitive to assumptions about fuels production. The use of higher-carbon-intensive fuels (for example, hydrogen and electricity produced with coal or natural gas without carbon capture and storage, or biofuels associated with significant land-use change impacts) would eliminate many of the emission reductions gained in these scenarios.

The three deep-reduction scenarios can be compared with respect to fuel consumption and primary resource requirements. Increased vehicle efficiencies in Electric-Drive reduce fuel use more than in Efficient Biofuels. Less-efficient biomass-to-biofuels conversion processes and lower internal combustion engine drive train efficiencies lead to increased primary resource requirements in Efficient Biofuels compared to the more-efficient hydrogen and electricity production processes and higher fuel cell vehicle and battery-electric vehicle drive train efficiencies used in Electric-Drive.

The use of hydrogen and electricity in the Electric-Drive scenario leads to a greater diversity of primary energy resources, including contributions from biomass, natural gas, coal, and petroleum, among other resources. The Energy Information Administration's business-as-usual projections suggest that domestic U.S. energy production in 2030 will be sufficient to meet the primary resource demands of the deep-reduction scenarios.[7] For renewable electricity generation, the scenario resource demands are well below the untapped supply potential using domestic resources[8]. Additional analysis should be performed to determine whether there are sufficient energy resources for all energy-consuming sectors (not just transportation) in a given future demand scenario.

U.S. DOMESTIC TRANSPORT FUEL AND PRIMARY RESOURCE USE IN 2050

We compared U.S. domestic transportation fuel use and primary resource consumption in 2050 for each of the three US deep-reduction scenarios. The Total Electricity bar in the Primary Resource Use chart (on the right) refers to the total amount of electricity used for transportation purposes in the given scenario. Because electricity is not a primary resource, the bar is superimposed on top of the primary resource bar.

The deep-reduction scenarios were designed to meet a goal of 50-to-80-percent reduction in U.S. *domestic* and California *in-state* CO_2 emissions by 2050. Reducing U.S. and California *overall* emissions by this amount requires even greater levels of implementation of advanced vehicle technologies, fuels substitution, and/or travel demand reduction. However, since we assume that the aviation and marine sectors will still be powered by liquid fuels in 2050, limitations in biofuel availability appear to preclude these targets from being reached in the overall case.

Limiting the US-Multi-Strategy 80in50 scenario to the same quantity of biofuels and biomass as in the domestic case (82 billion gge, 1.4 billion BDT) would yield overall emission reductions of 68 percent relative to 1990. Achieving an 80-percent reduction in overall emissions in this scenario by increasing the use of biofuels would require an additional 28 billion gge (+34 percent), for a total of 110 billion gge of low-carbon (that is, 12.3 gCO_2e/MJ) biofuels (or 1.8 billion BDT of biomass, including H_2 production). This highlights the fact that achieving these targets for overall emissions will be even more of a challenge than in the domestic case.

How Do We Get There?

The static snapshot scenarios of 2050 just described provide a stark picture of the transformations required in the transportation sector to reduce GHG emissions 50 to 80 percent below 1990 levels. But what does the path to making these changes by 2050 look like, and does it matter which scenario and transition path is followed for the goal of mitigating climate change? We provide some answers to these questions by using the California light-duty vehicle (LDV) subsector as a case study.

We developed the 80in50 PATH model in order to analyze the transition to advanced technologies in the California LDV subsector. The model is a version of the VISION stock-turnover model[9] adapted to California. We applied the model to study of the three deep-reduction scenarios developed for California using the LEVERS model. As mentioned earlier, in each of these scenarios GHG emissions are reduced by 80 percent.

The heart of the VISION model is a stock-turnover module that tracks annually new vehicles entering the fleet, the use and performance of the vehicles in the fleet, and old vehicles exiting the fleet. Inputs to this model include rates of change in new vehicle technology market penetration, vehicle fuel economy, fuel carbon intensity, car and truck market shares, increasing all-electric range for PHEVs, and biofuel blend in gasoline and diesel. These inputs are defined by the current conditions and characteristics defined for 2050 by each scenario from the LEVERS model, are informed by policy requirements and goals between now and 2050 ("waypoints"), and are informed by transition scenarios presented in the literature.

FROM LEVERS TO PATH

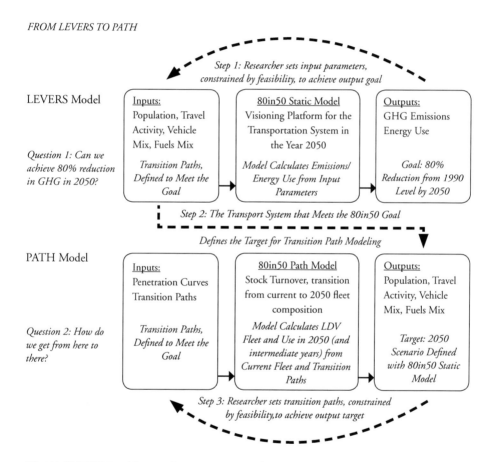

The 80in50 LEVERS model answers the question, Can we achieve an 80-percent reduction in GHG emissions by 2050? The 80in50 PATH model answers the question, How do we get there from here? The inputs required in the LEVERS model to meet emission reduction targets become the desired outputs of the PATH model. Solid arrows indicate the direction of model calculation, and dashed arrows indicate the direction of research inquiry.

We used the 80in50 PATH model to generate transition paths over time for market and fleet share for each vehicle technology, total annual vehicle miles traveled (VMT), vehicle emissions per mile, GHG emissions, fuel carbon intensity, and total energy use. The transition paths spotlighting market shares and annual VMT describe a range of potential answers to the question of how to get from the current transportation system to one in 2050 that meets the 80in50 goal. The transition paths spotlighting GHG emissions reveal that the path taken does matter for cumulative emissions and the potential for continued emission reduction past 2050.

Transition paths: Market and fleet share

To meet our 2050 emission reduction goal and aggressive intermediate waypoints in California, higher-emission vehicles like conventional gasoline ICE vehicles and hybrid electric vehicles (HEVs) must be replaced in the marketplace quickly by lower-emission alternatives like FCVs and BEVs. Of the scenarios we considered, only the CA-Multi-Strategy 80in50 scenario succeeds in reducing light-duty GHG emissions to 1990 levels by 2020 (a policy waypoint). Thus, if binding, intermediate waypoints may begin to constrain the range of acceptable scenarios.[10]

The transitional role of some technologies (such as HEVs and PHEVs) is evident as their market share increases to achieve intermediate waypoints and then decreases. While these vehicles share many components with more advanced electric-drive vehicles (BEVs and FCVs), they do not provide sufficient emission reduction to play a major role in the 2050 transportation system. It is important in any scenario to understand whether the technologies (and resulting infrastructures) used to achieve intermediate emission reduction goals lie along the path to achieving the long-term goals. Further study is needed to determine whether these rapid transitions to multiple technologies, and the investments needed are reasonable.

Although the transitions needed to achieve the 80in50 goal in California are believed to be feasible, they must begin very soon and with rapid rates of market adoption. This takes into account the lag between changes in market share and fleet share due to inertia in the existing fleet of vehicles.

CALIFORNIA LIGHT-DUTY VEHICLES: MARKET AND FLEET SHARE 2000–2050

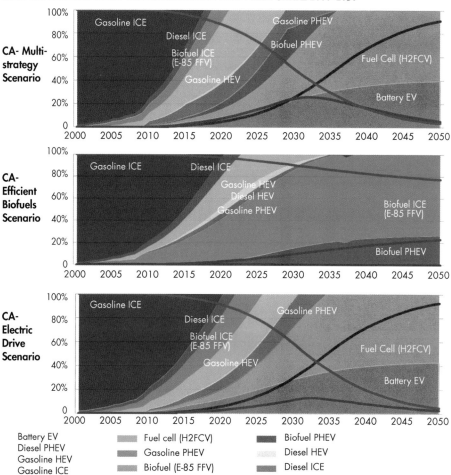

We used the 80in50 PATH model to generate the transition paths for light-duty vehicles (LDVs) in California to reach the market shares (shaded area) and fleet shares (lines) required in each of our three CA deep-reduction scenarios. In all scenarios, higher-emission vehicles like gasoline ICEs and HEVs must be replaced quickly by lower-emission alternatives like FCVs and BEVs.

Transition paths: GHG emissions

Using the 80in50 PATH model, we compared total GHG emissions from LDVs in California along the path to 2050 for each of our three CA deep-reduction scenarios. The annual GHG emission rate from LDVs exceeds the intermediate waypoint for 2010 (that is, emissions at 2000 levels) in all scenarios, and only the CA-Multi-Strategy 80in50 scenario meets the 2020 waypoint.

By 2050, LDVs must reduce their GHG emissions more than 80 percent below 1990 levels in order to compensate for other transportation subsectors (such as aviation) that do not meet the goal.

CALIFORNIA LIGHT-DUTY VEHICLES: GHG EMISSIONS 2000–2050

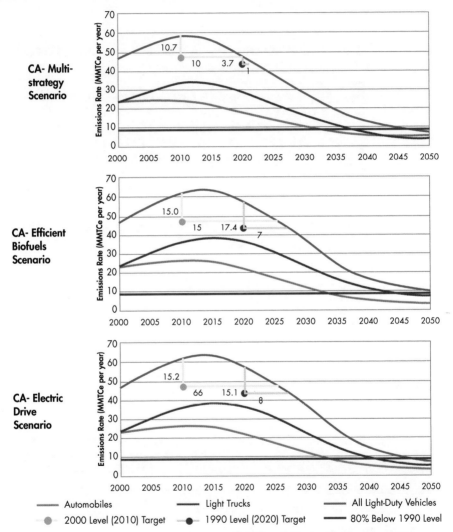

We used the 80in50 PATH model to project GHG emissions for LDVs in California along the path to 2050 in each of our three CA deep-reduction scenarios. The shortfall in GHG emissions reduction for the 2010 and 2020 intermediate waypoints is shown in MMTCe in the target year (vertical line) and in the additional number of years required to meet the target (horizontal line). All scenarios fall short of meeting the 2010 target, and only the CA-Multi-Strategy scenario meets the target for 2020.

Transition paths: Total energy used

The total quantity of energy used by LDVs in California during the transition from 2000 to 2050 decreases most dramatically in the CA-Multi-Strategy 80in50 and CA-Electric-Drive 80in50 scenarios. Electricity and hydrogen become the primary forms of energy used by LDVs in these scenarios, while biofuels dominate in the CA-Efficient Biofuels 80in50 scenario. Biofuels play a transitional role for LDVs in the CA-Multi-Strategy 80in50 and CA-Electric-Drive 80in50 scenarios; over time, the limited supply of low-carbon biofuels shifts to other transportation subsectors (especially aviation and marine) in order to meet the 80in50 goal for the whole transportation sector. Overall use of biofuels in the transportation sector increases steadily over time, consistent with rational expansion of production capacity, while the chemical nature of these biofuels may change over time (from predominantly lighter gasoline-like fuels for LDVs to heavier fuels such as diesel-like and jet-like fuels).

The effect on cumulative GHG emissions of acting early vs. late

Does it matter which path we take to get to the 80in50 goal? Our analysis of cumulative GHG emissions from California LDVs between 2010 and 2050 suggests that it does. The largest difference among scenarios is 439 MMTCe, a 30-percent variation. Furthermore, initiating the transition paths early versus delaying action results in a 22-to-27-percent difference in cumulative GHG emissions from LDVs, depending on the scenario. In other words, delaying action to initiate transitions can increase cumulative emissions by 22 to 27 percent compared to acting early. Thus, even though all scenarios still meet the 80-percent GHG reduction target for the transportation sector in the year 2050, both the scenario path and the transition timing within each scenario matter for effective climate change mitigation.

From a different perspective, acting early to initiate transitions may increase the probability of success in mitigating climate change. If success were defined by a target for cumulative emissions for the period 2010 to 2050 rather than an emission rate in the year 2050, acting early could yield success even if emissions in the year 2050 are higher than the 80in50 goal.

COMPARISON OF CUMULATIVE GHG EMISSIONS

Comparing the cumulative GHG emissions from California LDVs in three different transition path scenarios for the period 2010 to 2050 makes clear that the scenarios differ in climate change mitigation, and acting early can decrease cumulative emissions compared to acting late.

	CA-Multi-Strategy		CA-Efficient Biofuels		CA-Electric-Drive	
Cumulative GHG emissions, 2010–2050 (MMTCe)	1,250		1,518		1,503	
	Act-Early	Act-Late	Act-Early	Act-Late	Act-Early	Act-Late
Cumulative GHG emissions, 2010–2050 (MMTCe)	1,166	1,443	1,375	1,756	1,365	1,777
Change from PATH scenario	-7%	15%	-9%	16%	-9%	18%

Summary and Conclusions

- The major drivers of transportation GHG emissions are population, transport intensity (T), energy intensity (E), and carbon intensity (C); the latter three are the levers that technology and policy can use to reduce these emissions in the future. Low carbon intensity alternative fuels (biofuels, hydrogen, and electricity) appear to be a feasible means of lowering transportation carbon intensity (C), but carbon intensity can vary widely for these fuels based upon the details of their life cycle. There is significant potential for greatly improved vehicle efficiency (reduced E) for use in all of the transportation subsectors.

- Not all vehicle technology and fuel options can be applied to each of the transportation subsectors because of specific requirements for characteristics such as power, weight, or vehicle range. Biofuels appear to be most applicable across all transportation subsectors as a "drop-in" fuel replacement for petroleum-based fuels. However, because they can only be made from biomass, they are likely to be limited by biomass resource availability and may also be limited by land-use change impacts, which may reduce or negate their GHG benefits. Hydrogen and electricity can be made from a wide range of domestic resources, and resource constraints are unlikely to be major impediments to their adoption; however, they may be limited in their applicability to some transportation subsectors (especially aviation, marine, and off-road).

- The scenarios developed in this chapter highlight the level of effort and extent of transformation required to meet an ambitious greenhouse gas emission reduction goal of 50 to 80 percent below 1990 levels by 2050, whether in the United States or California. The scenarios are not meant to show exactly how these reductions should or will be achieved but instead are presented to provide stakeholders with a sense of the enormous challenges ahead. The hope is that these scenarios will provide a useful starting point for stakeholders and policy makers in discussing whether these changes are possible and what steps must be taken in the near term to ensure that we are on a path to meet the long-term goals.

- The silver-bullet scenarios show that while many mitigation options can yield small-to-moderate GHG reductions, no single mitigation option or strategy can meet a 50-to-80-percent reduction goal individually. By contrast, the three deep-reduction scenarios are each able to meet the goal, and each in a different way, requiring very extensive penetration of advanced technologies and large quantities of low-carbon fuels in addition to significant reductions in the growth of per-capita travel demand. Meeting the reduction goals for *overall* emissions is more difficult than meeting the goals for *domestic* and *in-state* emissions because aviation and marine are two of the more challenging subsectors to address from a technology perspective, and demand for these travel modes is growing rapidly, especially in the aviation subsector.

- The transitions in vehicle fleets and energy supply systems necessary to reach the deep-reduction scenarios for 2050 are feasible but must begin soon and progress rapidly, with rates of market penetration and change near feasible limits, because of the lag between market transition and fleet transition. Technologies that play a transitional role (that is, have a relatively short period of high market share) are necessary for meeting intermediate waypoints for GHG emission reduction but may be challenging from an industry perspective, and even then we may not achieve some waypoints.

- Both the scenario and the path taken to 2050 matter for effective climate change mitigation. Based on the 80in50 transition path analysis for California, it appears that although the deep-reduction scenarios are equal in meeting the 80 percent target for the transportation sector in 2050, they differ by as much as 30 percent in cumulative GHG emissions over the period 2010 to 2050. Similarly, initiating transitions early versus delaying action can cause up to a 27-percent difference in cumulative emissions for each scenario.

- From a policy perspective, current vehicle and fuels regulations address only some of the transportation subsectors (mainly light-duty vehicles), and almost none address options for reducing travel demand to a significant extent. These policy gaps may impede the development of options to address transportation GHGs. Furthermore, while this analysis developed and analyzed scenarios that achieve 50-to-80-percent reductions in GHG emissions for the transportation sector as a whole, it is not yet clear what exact role the sector will ultimately play in bringing down total economy-wide emissions from all sources. That said, given the size of the sector and the likely need for even deeper GHG reductions after 2050, transportation is certain to play a major role in the coming decades.

Notes

1. In 2005, California governor Arnold Schwarzenegger signed Executive Order S-3-05, calling for an 80-percent reduction in greenhouse gas (GHG) emissions relative to 1990 levels by 2050 (the "80in50" goal). Later, U.S. president Obama proposed an 80-percent reduction goal for the country as a whole (an 80-percent reduction in annual U.S. GHG emissions below 1990 levels is equivalent to an 83-percent reduction below 2005, since annual GHG emissions in 1990 were 14 percent lower than in 2005), and in fact several climate change bills have been proposed in the U.S. Congress that would set up a domestic cap-and-trade program to help reduce GHG emissions 50 to 80 percent by 2050. See World Resources Institute, "Net Estimates of Emission Reductions Under Pollution Reduction Proposals in the 111th Congress, 2005–2050," http://www.wri.org/publication/usclimatetargets and http://www.wri.org/chart/net-estimates-emission-reductions-under-pollution-reduction-proposals-111th-congress-2005-2050.
2. California Air Resources Board, "California Greenhouse Gas Emission Inventory," 2008; U.S. Environmental Protection Agency, Office of Transportation and Air Quality, "Greenhouse Gas Emissions from the U.S. Transportation Sector, 1990–2003," 2006; International Transport Forum, Organization for Economic Co-operation and Development (OECD), "Greenhouse Gas Reduction Strategies in the Transport Sector: Preliminary Report," 2008.
3. For an expanded description of the LEVERS model and all input assumptions, see the appendix to D. McCollum and C. Yang, "Achieving Deep Reductions in U.S. Transport Greenhouse Gas Emissions: Scenario Analysis and Policy Implications," *Energy Policy* 37 (2009): 5580–96.
4. See N. Parker, P. Tittmann, Q. Hart, R. Nelson, K. Skog, A. Schmidt, E. Gray, and B. Jenkins, "Development of a Biorefinery Optimized Biofuel Supply Curve for the Western United States," *Biomass and Bioenergy* 34 (2010): 1597–607.
5. For an extended discussion of our silver-bullet scenarios and results, including descriptions of the scenarios themselves, see the appendix to McCollum and Yang, "Achieving Deep Reductions in U.S. Transport Greenhouse Gas Emissions."
6. D. McCollum, G. Gould, and D. Greene, "Greenhouse Gas Emissions from Aviation and Marine Transportation: Mitigation Potential and Policies," Pew Center on Global Climate Change, December 2009.

7. EIA, *Annual Energy Outlook 2008 with Projections for 2030* (Washington, DC: Energy Information Administration, U.S. Department of Energy), 2008. EIA's projections for domestic energy production in 2030 include: crude oil (12,699 PJ), natural gas (21,099 PJ), coal (30,202 PJ), biomass (8,570 PJ), total electric generation (17,599 PJ), nuclear power (10,093 PJ), and renewable power (1,991 PJ).
8. National Renewable Energy Laboratory (NREL), "PV FAQs—How Much Land Will PV Need to Supply Our Electricity?" (Washington, DC: Office of Energy Efficiency and Renewable Energy, U.S. Department of Energy), 2004, http://www1.eere.energy.gov/solar/pdfs/35097.pdf.
9. Argonne National Laboratory, "VISION 2008 AEO Base Case Expanded," VISION Model, 2009, http://www.transportation.anl.gov/modeling_simulation/VISION/index.html.
10. It is important to remember, however, that the 2020 and 2050 targets for reducing GHG emissions in California are economy-wide goals that are not specific to the transportation sector; many analysts believe the transportation sector will not play an equal role with other sectors, especially in meeting the 2020 goal.

Chapter 9: Transition Scenarios for the U.S. Light-Duty Sector[1]

Joan Ogden, Christopher Yang, and Nathan Parker

Besides imagining how a combination of alternative fuels and new vehicle technologies can help us meet GHG reduction targets, it is important to consider how transportation—particularly the light-duty sector—might make the transition to a low-carbon future. The light-duty vehicle (LDV) sector accounts for about two-thirds of energy use and greenhouse gas (GHG) emissions from transportation in the United States. Automakers are targeting light-duty markets for advanced electric-drive technologies such as plug-in hybrids and hydrogen fuel cell vehicles. In this chapter, we analyze and compare alternative scenarios for adoption of new LDV and fuel technologies that could enable deep cuts in gasoline consumption and GHG emissions by 2050. We also estimate the transitional costs for making new vehicle and fuel technologies economically competitive with gasoline vehicles. We do this with the caveat that concentrating only on the light-duty fleet may miss important constraints, especially for biofuels—which may be needed to make liquid fuels for air and marine transportation.

Our Scenarios

We analyze and compare these scenarios:

- **Efficiency**—Currently feasible improvements in gasoline internal combustion engine vehicle (ICEV) and hybrid electric vehicle (HEV) technology are introduced.
- **Biofuels**—Large-scale use of low-carbon biofuels is implemented.
- **PHEV success**—Plug-in hybrid electric vehicles (PHEVs) play a major role beyond 2025.
- **FCV success**—Hydrogen fuel cell vehicles (FCVs) play a major role beyond 2025.
- **Portfolio**—More-efficient ICEVs + biofuels + PHEVs + FCVs are implemented in various combinations.

All scenarios assume the same total number of vehicles and vehicle miles traveled, but the vehicle mix over time is different for each scenario. We compare each scenario to a reference scenario where modest improvements in efficiency take place and use of biofuels increases but no electric-drive vehicles are implemented. We estimate future GHG emissions and gasoline use for each scenario.

The reference scenario

Our reference scenario is based on projections to 2030 by the U.S. Department of Energy.[2] We use the Energy Information Administration's high oil price case—where oil prices in the period from 2010 to 2030 are projected to vary from $80 to $120 per barrel—for the number of vehicles and their fuel consumption, oil prices, and other factors. We extend these projections to 2050, assuming that the average growth rate between 2010 and 2030 remains the same for the two decades that follow.

In this scenario, ICEVs continue to dominate the light-duty sector. HEVs gain only about 10 percent fleet share by 2050. The fuel economies of these vehicles (that is, the on-road fuel economies, which are 20 percent lower than EPA sticker fuel economies) follow Energy Information Administration (EIA) projections, meeting 2020 fuel economy standards, with only modest improvements beyond this time. HEVs reach an on-road fuel economy of 44.5 mpg in 2050, while conventional gasoline cars reach 31.7 mpg.

REFERENCE SCENARIO: NUMBERS OF LDVS AND FUEL ECONOMIES TO 2050

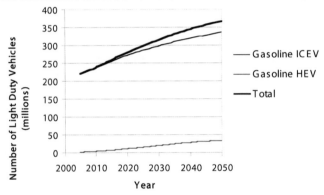

In our reference scenario, gasoline internal combustion engine vehicles (ICEVs) continue to dominate the light-duty sector. Gasoline hybrids (HEVs) gain only about a 10-percent fleet share by 2050.

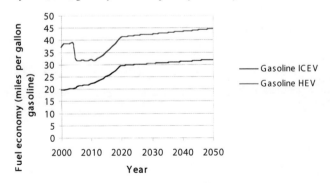

The fuel economy of new LDVs meets 2020 fuel economy standards and improves only modestly beyond that time. HEVs reach an on-road fuel economy of 44.5 mpg in 2050, while conventional gasoline cars reach 31.7 mpg.

In the reference scenario, a significant amount of biofuel is used: 12 billion gallons of corn ethanol are produced by 2015 (and production stays at this level to 2050), and a growing amount of cellulosic ethanol is produced after 2012. In 2050, corn ethanol production is 12 billion gallons and cellulosic ethanol production an additional 12 billion gallons.

The efficiency scenario

In our efficiency scenario, improvements in engines and other vehicle technologies are implemented at a more rapid rate than in the reference scenario. The fuel economy of ICEVs and HEVs is assumed to increase as follows:

- 2.7 percent per year from 2010 to 2025
- 1.5 percent per year from 2026 to 2035
- 0.5 percent per year from 2036 to 2050

In addition, HEVs become the dominant technology, comprising 80 percent of the fleet by 2050. Fuel economy for ICEVs and HEVs approximately doubles by 2050, when HEVs average 60 mpg and ICEVs 42 mpg.

EFFICIENCY SCENARIO: NUMBERS OF LDVS AND FUEL ECONOMIES TO 2050

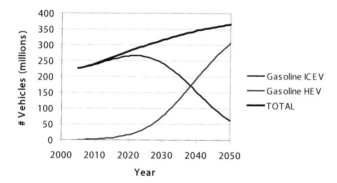

In our efficiency scenario, HEVs become the dominant technology, comprising 80 percent of the fleet by 2050. Numbers of ICEVs on the road drop off sharply after 2025 and are exceeded by numbers of HEVs by 2038.

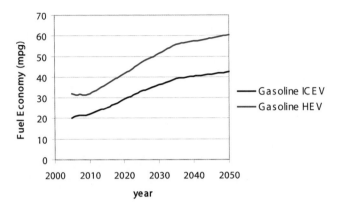

On-road fuel economy for ICEVs and HEVs approximately doubles by 2050, when HEVs average 60 mpg and ICEVs 42 mpg.

The biofuels scenario

In our biofuels scenario, we assume that biofuels are introduced at a rapid rate, reaching an optimistic total of 75 billion gallons per year in 2050. Production of corn ethanol levels off, but cellulosic ethanol grows rapidly, reducing carbon emissions (well-to-wheels GHG emissions for cellulosic ethanol are only 15 percent those of gasoline). Competition with food crops and indirect land-use impacts on GHG emissions are not considered in this analysis. As discussed in Chapter 8, liquid biofuels may be required in heavy-duty aviation and marine applications, where electric battery and hydrogen fuel cell drivetrains are not practical. This could limit the amount of biofuel available for light-duty vehicles. It is important to note that this particular biofuel scenario is not the only feasible path forward: other scenarios are discussed in Chapter 1, using large amounts of "drop-in" biofuels similar to gasoline and diesel in addition to cellulosic ethanol (see Chapter 1). Moreover, the uncertainties in biofuel GHG emissions could influence the amount of carbon reductions that could be achieved with biofuels. These issues are discussed in Chapters 1, 6, and 12.

We assume that ICEVs capable of running on biofuels will have only a small incremental cost compared to gasoline vehicles, and that these vehicles can be mass-produced quickly. Further, we assume that biofuel vehicles will have the same fuel economy as gasoline cars.

BIOFUELS SCENARIO: BIOFUELS PRODUCTION TO 2050

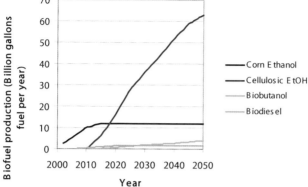

In our biofuels scenario, we assume that biofuels are introduced at a rapid rate, reaching 75 billion gallons per year in 2050. Production of corn ethanol levels off, but cellulosic ethanol grows rapidly.

These are optimistic estimates for implementing large-scale biomass supply systems, as Chapter 1 indicates. Studies by Parker et al.[3] suggest that about 24 billion gallons gasoline equivalent (or 36 billion gallons of ethanol) might be available at less than $3.25 per gallon gasoline equivalent in 2018. Beyond this level of production fuel costs rise rapidly, making biofuels less economically attractive. Advances in energy crop yields, crop yields and conversion efficiencies may increase the production potential by 2050.

The PHEV success scenario

Following the 2009 National Academies report on plug-in hybrids,[4] we analyze an optimistic market penetration scenario for plug-in hybrids, where PHEVs are introduced in 2010 and markets grow rapidly. This case assumes strong policy support for PHEVs so that 1 million vehicles are on the road in 2017 and 10 million by 2023; by 2050, about two-thirds of all light-duty vehicles are PHEVs. The National Academies also analyzed a more pessimistic case where PHEVs account for about 30 percent of the fleet in 2050 and market growth is slower. This case was not economically attractive and for simplicity is not presented here.

Two types of PHEVs are modeled: a PHEV-10, which has a battery large enough to provide a 10-mile all-electric range, and a PHEV-40, with a larger battery that offers a 40-mile all-electric range. We calculated the fuel economies (averaged over the driving patterns of the entire fleet) for gasoline ICEVs and HEVs (based on both reference scenario and efficiency scenario fuel economies), PHEV-10s, and PHEV-40s. The gasoline fuel economy of PHEVs increases over time at the same rate as that of HEVs in the efficiency scenario. We assume that PHEVs will incorporate all the most efficient aspects of evolving HEV technology, as well as lighter-weight materials, streamlining, and so on.

PHEV SUCCESS SCENARIO: NUMBERS OF PHEVS AND FUEL ECONOMIES TO 2050

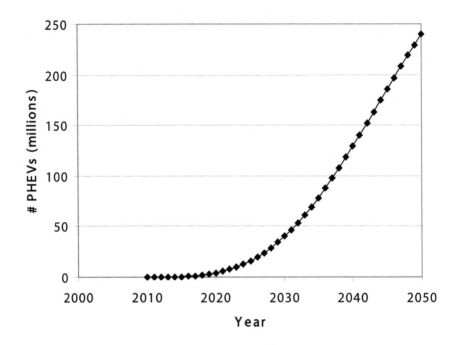

We analyze an optimistic case in our PHEV success scenario. PHEVs are introduced in 2010, and 1 million vehicles are on the road in 2017 and 10 million by 2023; by 2050, about two-thirds of all light-duty vehicles are PHEVs.

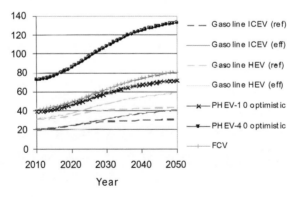

We assume that fleet average fuel economy for all different types of vehicles increases over time. Gasoline fuel economy is highest for PHEVs with a 40-mile electric range (note that this does not include electricity use, which is shown in the next figure). Fuel economy for FCVs is given in gasoline-equivalent energy.

PHEVs also drive some fraction of their total miles on electricity. For a PHEV-10, we assume that about 19 percent of the miles are driven on electricity, and for a PHEV-40, about 55 percent of the miles.[5] The source of electricity has a strong impact on the environmental benefits of PHEVs versus HEVs.[6] We analyze two possibilities for the future electricity system. One is a business-as-usual grid based on projections by the U.S. Department of Energy.[7] The other is a low-carbon grid based on studies by the Electric Power Research Institute (EPRI) and the Natural Resources Defense Council (NRDC).[8] In the low-carbon grid, emissions per kWh are reduced by about two-thirds through a variety of more-efficient and lower-carbon generation technologies, including advanced renewables, carbon capture and sequestration (CCS), and nuclear power.

PHEV SCENARIO: ASSUMED ELECTRICITY USE AND GRID GHG EMISSIONS TO 2050

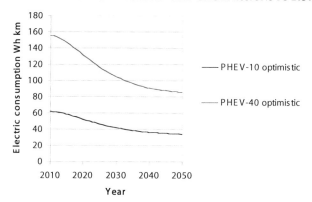

We averaged the assumed electricity use per kilometer over the fleet drive cycle for PHEV-10s and PHEV-40s over time. For a PHEV-10, we assume that about 19 percent of the miles are driven on electricity, and for a PHEV-40, about 55 percent of the miles.

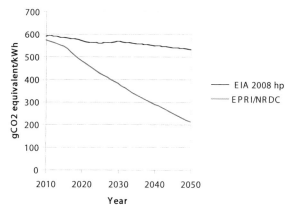

The environmental benefits of PHEVs will hinge on how electricity is generated. We compared GHG emissions per kilowatt hour of electricity for a business-as-usual future grid (EIA) and a low-carbon future grid (EPRI/NRDC). In the low-carbon grid, emissions per kWh are reduced by about two-thirds through a variety of more-efficient and lower-carbon generation technologies, including carbon capture and sequestration (CCS), advanced renewables, and nuclear power.

The FCV success scenario

Finally, we consider a range of cases where hydrogen fuel cell vehicles are successfully developed. We assume that hydrogen fuel cell vehicles (FCVs) are introduced beginning in 2012, reaching 10 million on the road by 2025 (4 percent of the fleet) and 60 percent of the fleet by 2050.[9] Initially, hydrogen is produced from natural gas, but over time energy sources that emit less carbon are used to produce hydrogen: biomass gasification and coal gasification with carbon capture and sequestration (see Chapter 3).

As with electric vehicles, the source of hydrogen makes a difference in the well-to-wheels GHG emissions of FCVs. Following the modeling in "The Hydrogen Fuel Pathway," we assume that hydrogen is made from progressively lower-carbon sources over time. As with electricity in the low-carbon case, we assume that the GHG emissions per megajoule (MJ) of fuel will fall by about two-thirds by 2050 through expanded use of renewables and carbon capture technology in hydrogen production.

FCV SCENARIO: NUMBERS OF VEHICLES AND ASSUMED GHG EMISSIONS FROM H_2 TO 2050

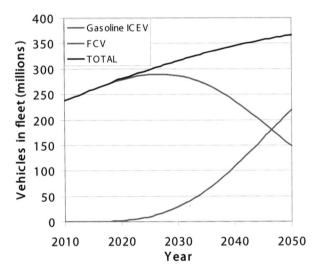

In the FCV success scenario, hydrogen fuel cell vehicles (FCVs) are introduced beginning in 2012, reaching 10 million on the road by 2025 (4 percent of the fleet) and 60 percent of the fleet by 2050.

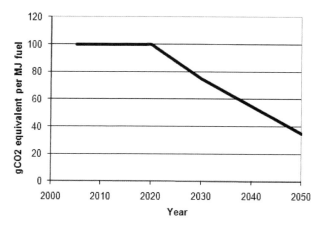

We assume that well-to-wheels GHG emissions per MJ of hydrogen will decrease over time. Before 2025, we assume that H_2 will be made primarily from on-site steam methane reforming. Later, centralized H_2 plants using biomass or coal with CCS will be phased in.

The portfolio scenarios

We have just described single-pathway scenarios based on implementing efficiency, biofuels, PHEVs, and FCVs. But it is more likely that a range of policies will be put into place to incentivize higher-efficiency gasoline vehicles while advanced vehicle technologies (like PHEVs and FCVs) and new fuels (biofuels and hydrogen) are being developed. To model this, we developed a series of portfolio scenarios that combine efficiency and advanced vehicles and fuels in different ways. In one of our portfolio scenarios, we combined the efficiency scenario with the rapid introduction of advanced vehicles. We added the introduction of low-carbon biofuels (similar to the biofuels scenario) to the mix in another portfolio scenario.

PORTFOLIO SCENARIO: NUMBERS OF LDVS TO 2050

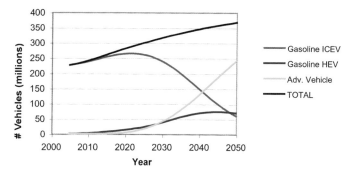

In this portfolio scenario, advanced vehicles (PHEVs or FCVs) are deployed rapidly so that their number surpasses that of ICEVs after 2040, when the number of HEVs peaks.

Comparing Strategies to Reduce Gasoline Use and GHG Emissions

We have outlined our reference scenario, four single-pathway scenarios, and our portfolio approach in terms of numbers of light-duty vehicles and fuel economies. Now let's compare the different scenarios with respect to gasoline use and GHG emissions.

Gasoline use

First we consider gasoline use in our single-pathway scenarios. With rapid deployment of biofuels, it would be possible to displace gasoline use starting before 2020, although the effect of biofuels plateaus because of constraints on production. None of the other options results in noticeable gasoline savings before about 2025, because of the time required to bring new vehicle types into the fleet. After 2025, the efficiency scenario leads to a rapid decrease in gasoline use compared to the reference case. If we replace a certain number of gasoline ICEVs with FCVs or PHEVs without changing the ICEV efficiency, there is a major decrease in gasoline use beyond 2030. In the long term, FCVs yield the greatest reduction in gasoline use of the technologies considered.

What about the portfolio scenarios, where efficiency technologies are implemented in ICEVs and HEVs along with rapid adoption of advanced electric-drive technologies such as PHEVs or FCVs and introduction of low-carbon biofuels? We find in our scenario combining the efficiency case with introduction of advanced vehicles that gasoline use starts to decline rapidly after about 2015. When we combine the efficiency case with introduction of advanced vehicles and low-carbon biofuels, we find that gasoline use starts to decline immediately and reaches 0 before 2050 for the case where FCVs are combined with efficiency and biofuels. Clearly, any portfolio approach is superior to any of the single-pathway approaches in terms of how soon and how much it will reduce gasoline use.

GASOLINE USE FOR OUR SCENARIOS TO 2050

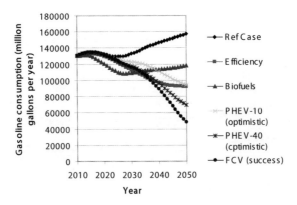

Gasoline use does not decrease noticeably before about 2025 in our single-pathway efficiency, PHEV and FCV scenarios because of the time required to bring these new vehicle types into the fleet. Biofuels could have an impact earlier, but after an initial reduction starting before 2020, the effect of biofuels plateaus because of constraints on production. In the long term, FCVs yield the greatest reduction.

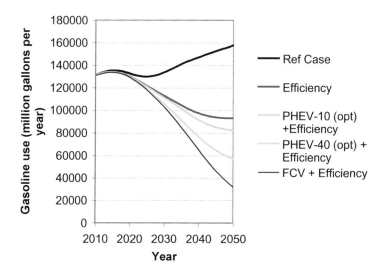

In one portfolio scenario, we combined the efficiency case with introduction of advanced vehicles. Gasoline use starts to decline rapidly after about 2015 in this scenario.

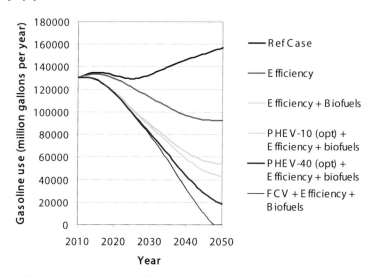

In another portfolio scenario, we combined the efficiency case with introduction of advanced vehicles and low-carbon biofuels. Gasoline use starts to decline immediately and reaches 0 before 2050 for the case where FCVs are combined with efficiency and biofuels.

GHG emissions

Trends similar to those for gasoline use hold for GHG emissions in single-pathway scenarios. PHEV and FCV scenarios don't show a marked decrease in GHG emissions before about 2030. No single pathway can meet societal goals for deep cuts in carbon (such as an 80-percent reduction) by 2050.

The importance of moving to a low-carbon energy supply can be seen when we compare the GHG emissions for the reference case, the efficiency case, and portfolio scenarios combining advanced vehicles with efficiency and with efficiency plus biofuels. Assuming a business-as-usual energy supply (the EIA fossil-intensive electric grid and H_2 made from natural gas), GHG emissions with PHEVs + efficiency are no lower than those from improved ICEV efficiency alone, and FCVs + efficiency show only about a 10-percent reduction by 2050. By contrast, assuming a low-carbon grid and H_2 production from low-carbon sources, GHG emissions trend lower in the three advanced vehicle cases. This highlights the importance of decarbonizing the energy supply (electricity and fuels) as advanced vehicles are introduced.

GHG EMISSIONS FOR OUR SCENARIOS TO 2050

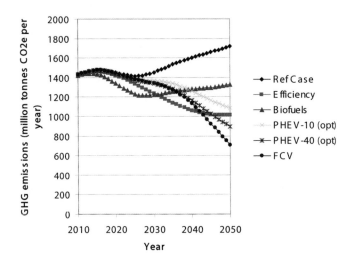

No single pathway can meet societal goals for deep cuts in carbon (such as an 80-percent reduction) by 2050. The PHEV and FCV scenarios don't show a marked decrease in GHG emissions before about 2030.

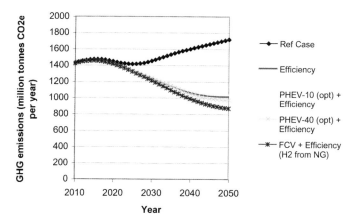

Assuming a business-as-usual energy supply (the EIA fossil-intensive electric grid and H_2 made from natural gas) for a portfolio scenario combining the efficiency case with introduction of advanced vehicles, GHG emissions from PHEVs or FCVs are not much lower than those from improved ICEV efficiency alone.

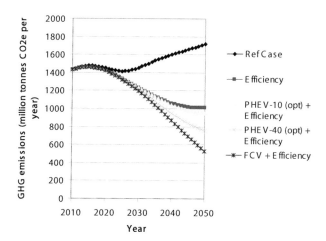

Assuming a low-carbon grid and H_2 production from low-carbon sources for a portfolio scenario combining the efficiency case with introduction of advanced vehicles, GHG emissions trend lower in the three advanced-vehicle cases than for efficiency alone.

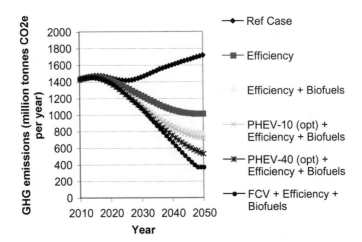

Assuming a low-carbon grid and H_2 production from low-carbon sources for a portfolio scenario combining the efficiency case with introduction of advanced vehicles and biofuels, GHG emissions fall off sharply in the three advanced-vehicle cases.

Comparing Transition Costs

What will it cost to make the transition to biofuels, PHEVs, and FCVs? One of the major challenges facing any new alternative-fuel vehicle is reaching economic competitiveness with gasoline vehicles. Initially, the new vehicles are manufactured in small quantities and the cost is much higher than for a comparable gasoline vehicle, which is a major disincentive to consumers. Getting enough new alternative vehicles on the road to bring down costs is a key issue. For new fuel infrastructure, the analogous problem is putting in enough fueling stations to make it convenient for consumers and to bring down the cost of the fuel. The question is how much money must be invested in the first million or so vehicles and the early infrastructure to reach cost competitiveness.

To study this "buydown" process for alternative vehicles, we developed a transition model to aggregate transition costs over the entire fleet, based on models for evolving vehicle and fuel infrastructure. For PHEVs and FCVs, this includes buying down the vehicle cost and building new infrastructure. For biofuels, the delivered fuel cost is the main concern, since we expect that the cost of vehicles that can run on biofuels will be quite similar to the cost of gasoline vehicles. The issue is scaling up the biofuel supply chain to reach competitive fuel costs.

Vehicle costs

For both PHEVs and FCVs, we assume that vehicle costs will come down with learning and scaled-up manufacturing.

The key enabling technology for PHEVs is the battery. Although current PHEV battery costs are still too high to compete ($1000/kWh with a lifetime of five years; see Chapter 4), battery costs are projected to shrink to $250 to $400/kWh with a lifetime of ten to twelve years assuming technology improvements and economies of scale in mass production. For purposes of plotting the retail price of PHEVs, we assumed that batteries follow a learning curve trajectory from current costs to a "learned-out" mass-produced cost.[10] We analyzed two levels of technical progress for PHEVs, based on the recent National Research Council study: "optimistic" and "DOE goals."[11] In the optimistic case, the learned-out cost of batteries in 2030 is $360/kWh of nameplate capacity. In the DOE goals case, battery cost is roughly half of this, and these goals are achieved by 2020.

KEY PERFORMANCE AND COST ASSUMPTIONS FOR PHEV BATTERIES

	2010	2020	2030
Battery lifetime			
Optimistic	8 years	12 years	15 years
DOE goals	8 years	12 years	15 years
Battery pack cost per kWh nameplate			
Optimistic	$625	$400	$360
DOE goals	$625	$168–280 (DOE 2014 goal)	$168–280

Initially, hydrogen vehicles will be much more costly than gasoline vehicles, but as fuel cell and hydrogen storage technology improve and scale economies of mass production take hold, the price should fall rapidly. We estimated the retail price of a hydrogen fuel cell car based on a learning curve model developed by Greene et al.[12] For "learned-out" technology, the National Research Council H_2 success case finds a retail price difference of $3,600 between a hydrogen and a gasoline car.[13] (For the NRC H_2 partial success case, the price difference is about $6,100.)

KEY COST ASSUMPTIONS FOR FCVS

	2010	2020	2025-2030
Fuel cell system cost per kW			learned-out
H2 success	$1,000	$60	$50
Partial success	$1,000	$100	$75
H2 storage cost per kWh			learned-out
H2 success			$10
Partial success			$15

PROJECTED PRICE PREMIUM FOR PHEVS AND FCVS

This is the premium over the price of a gasoline ICEV that a purchaser of a PHEV or an FCV will pay. For reference, the price of the 2011 Chevy Volt is $41,000, about $24,000 more than a comparable Chevy ICEV car.

	2010	2020	2030
PHEV-10			
Optimistic	$7,700	$5,600	$5,100
DOE goals	$7,700	$4,500	$4,500
PHEV-40			
Optimistic	$20,000	$13,500	$12,200
DOE goals	$20,000	$7,600	$7,600
	2015	**2020**	**learned-out 2025-2030**
FCV			
Partial success	$120,000	$31,000	$6,100
H2 success	$39,000	$7,000	$3,600

PROJECTED RETAIL PRICE OF FCVS AND PHEVS TO 2030

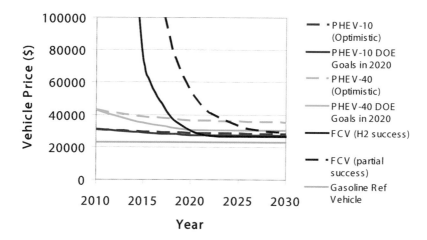

Assuming a maximum practical market penetration scenario, the retail prices of new PHEV-10s and PHEV-40s are projected to drop somewhat in the period from 2010 to 2030 in the NRC optimistic case and DOE goals case but still remain above the price of a conventional ICEV. We assume that the price of FCVs will fall rapidly after their introduction in 2012, to the point where the price premium will be only $3,600 by 2025.

Infrastructure and fuel costs

As described in "The Hydrogen Fuel Pathway," we assume that early hydrogen infrastructure will be built in a phased or regionalized manner where hydrogen vehicles and stations are initially introduced in selected large cities like Los Angeles and New York and move to other cities over time. This so-called lighthouse concept reduces early infrastructure costs by concentrating development in relatively few key areas. We also assume that the delivered hydrogen cost will drop sharply over time and become competitive with gasoline. We used the UC Davis SSCHISM model to design hydrogen infrastructure and find the delivered hydrogen cost over time.[14]

HYDROGEN FUEL SUPPLY IN THE U.S. OVER TIME, HYDROGEN SUCCESS SCENARIO

	2020	2035	2050
Number of cars served (percentage of total fleet)	1.8 million (0.7%)	61 million (18%)	219 million (60%)
Infrastructure capital cost	$2.6 billion	$139 billion	$415 billion
Total number of stations	2,112 (all on-site SMR)	56,000 (40% on-site SMR)	180,000 (44% on-site SMRs)
Number of central plants	0	113 (20 coal, 93 biomass)	210 (79 coal, 131 biomass)
Pipeline length (mi)	0	39,000	80,000
Hydrogen demand (tonnes/day)	1,410 (100% NG)	38,000 (22% NG, 42% biomass, 36% coal w/CCS)	120,000 (31% NG, 25% biomass, 44% coal w/CCS)

Source: National Research Council, Transitions to Alternative Transportation Technologies, A Focus on Hydrogen *(Washington, DC: National Academies Press, 2008).*

For PHEVs, we assume that the electricity cost for vehicle charging is 8 cents per kilowatt-hour. The capital cost for residential charging is $1,000–2,000 per charger.[15] We do not include costs for upgrades in transmission and distribution of electricity or building new power plants.

ASSUMED HYDROGEN COST, GASOLINE PRICE, AND ELECTRICITY PRICE TO 2040

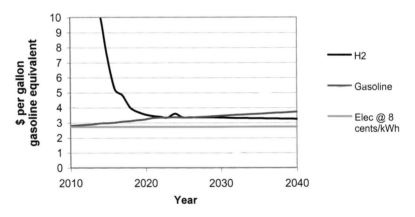

We assume that the cost of delivered hydrogen will decrease rapidly starting in about 2015 and will become competitive with the price of gasoline by about 2025 on a gasoline-equivalent-energy basis. Counting the higher efficiency of fuel cells compared to gasoline cars, hydrogen competes on a fuel-cost-per-mile basis before 2020. Electricity at 8 cents per kWh is already less expensive than gasoline on a gasoline-equivalent-energy basis and we assume electricity prices for charging will stay at this level.

For biofuels, we assume as mentioned earlier that the main issue is scaling up the supply chain to reach cost competitiveness with conventional fuels like gasoline or diesel on an equivalent-energy basis. To estimate investment costs, we assume that the biofuel supply chain is built up over time, at costs determined by a national U.S. model.[16]

Several studies have estimated the costs to meet U.S. policy goals for biofuels. The U.S. Environmental Protection Agency[17] estimated that it would cost about $90 billion to meet the 2022 goal under the Renewable Fuel Standard of producing 36 billion gallons per year of bioethanol (enough to displace about 24 billion gallons per year of gasoline). More than 80 percent of the investment cost was for biorefineries, with the remainder for distributing bioethanol to users. Another study by Sandia National Laboratory[18] found that fuel infrastructure investments of about $390 billion would be required to produce 90 billion gallons of bioethanol per year in 2030 (displacing about 60 billion gallons of gasoline).

Studies by STEPS researchers Nathan Parker and Bryan Jenkins suggest that about $100-360 billion of investments in biorefineries would be needed to meet the U.S. Renewable Fuel Standard goal of 36 billion ethanol equivalent gallons of biofuel (displacing 24 billion gallons of gasoline) by 2022. The cost of biofuels should fall initially, with technology learning and scale economies for biorefineries. But beyond annual production levels of about 34 billion gasoline-equivalent gallons of biofuels, the fuel cost is estimated to climb sharply. The steep climb occurs once low-cost, environmentally desirable biomass resources have been exploited. STEPS analysis suggests an upper limit on the amount of economically competitive domestic biofuels at perhaps 20–30 percent of the fuel demand in the light-duty sector. (This limit is sensitive to assumptions about biomass productivity, biorefinery conversion efficiency, land-use constraints, and interactions with other sectors of economy.)

Because biofueled vehicles could be introduced quickly and at similar cost to gasoline vehicles, the rate of fuel supply build-up will determine the transition time for biofuels. The amount of low-cost biofuels available nationally is limited. Demands for liquid fuels from sectors such as aviation and marine may further limit the amount of biofuels that can be used in the light-duty sector (Chapter 8).

Cash flows

Based on these assumptions about PHEVs and FCVs, we conducted a transition cash-flow analysis to determine the investment costs required for PHEVs and FCVs to reach cost competitiveness with reference scenario gasoline vehicles. We estimated two components of this transition cost over time:

1. The incremental price of buying alternative-fuel vehicles (AFVs) each year, instead of gasoline cars; this is summed over all the AFVs sold in a given year and is the aggregated extra cost paid by consumers each year to buy AFVs instead of gasoline cars.
2. The difference between the annual cost of fuel for these AFVs and the annual cost of gasoline to go the same distance.

The annual cash flow or cost difference between a transition (where the alternative is introduced) and "business as usual" (all gasoline cars) is the sum of the vehicle first cost increment and the fuel cost increment. Cost competitiveness is achieved in the break-even year, when the total incremental cost for all the new AFVs bought that year is balanced by the annual fuel cost savings for all AFVs on the road in comparison to the reference gasoline vehicles.

NET CASH FLOWS FOR PHEVS AND FCVS, TRANSITION YEARS

Positive cash flow values indicate that the cost of advanced vehicles and/or fuel is lower than the cost of gasoline vehicles and/or fuel.

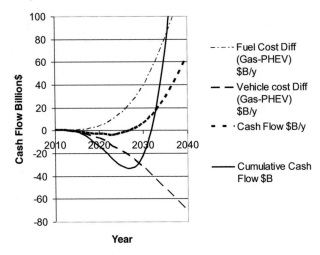

For the PHEV-10, case under optimistic technical assumptions, the break-even year is 2028 and the buydown cost is $33 billion. Source: National Research Council Transitions to Alternative Transportation Technologies, Plug-in Hybrid Electric Vehicles (Washington, DC: National Academies Press, 2010).

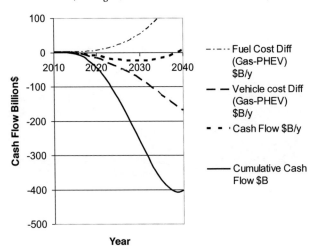

For the PHEV-40, maximum practical case under optimistic technical assumptions, the break-even year is 2039 and the buydown cost is $400 billion. Source: , NRC, Transitions to Alternative Transportation Technologies, Plug-in Hybrid Electric Vehicles (Washington, DC: National Academies Press, 2010).

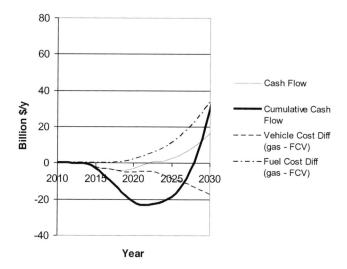

For FCVs, the break-even year is 2023 and the buydown cost is $22 billion. Source: National Research Council, Transitions to Alternative Transportation Technologies, A Focus on Hydrogen (Washington, DC: National Academies Press, 2008).

Transition costs for PHEVs and FCVs compared

In the table below we compare the costs for different scenarios for the introduction of PHEVs and FCVs. The break-even year is the year when annual buydown subsidies equal fuel cost savings for the fleet. The cumulative cash flow difference for PHEVs does not take into account infrastructure costs for home rewiring, distribution system upgrades, and public charging stations, which might average more than $1,000 per vehicle.

The PHEV-10 and H_2 FCV have similar buydown costs and timing, but the NRC PHEV-40 "optimistic" case gives a longer and more costly transition, because of the assumed relatively high battery cost out to 2020 and beyond. To examine what would be required to bring PHEV-40s to competitiveness more rapidly, we carried out two sensitivity analyses: "DOE Goals" assumes the DOE battery cost and technology goals for the PHEV-40 are met by 2020, showing the importance of technology breakthroughs. Reducing costs this rapidly would significantly reduce subsidies and advance the break-even year relative to the NRC "optimistic" technical progress cases. "High Oil" assumes oil costs twice that in the base case, or $160/bbl in 2020, giving results similar to meeting DOE's cost goals.

COMPARISON OF TRANSITION COSTS FOR PHEVS AND FCVS

	PHEV-10 Optimistic	PHEV-40 Optimistic	PHEV-40 Sensitivity Cases		FCV	
			High Oil	DOE Goals	Success	Partial
Break-even year	2024	2039	2025	2024	2023	2033
Cumulative cash flow to break-even	$24 billion	$408 billion	$41 billion	$24 billion	$22 billion	$46 billion
Cumulative vehicle retail price difference to break-even	$82 billion	$1,639 billion	$174 billion	$82 billion	$40 billion	$82 billion
Number of vehicles at break-even	10 million	132 million	13 million	10 million	5.6 million	10 million
Infrastructure cost at break-even	$10 billion (in-home charger @$1,000)	$132 billion (in-home charger @$1,000)	$13 billion (in-home charger @$1,000)	$10 billion (in-home charger @$1,000)	$8 billion (H2 stations for first 5.6 million FCVs)	$19 billion (H2 stations for first 10 million FCVs)

ALTERNATIVES VS. THE COST OF BUSINESS AS USUAL

The last section of this chapter has focused on the costs of making a transition to alternative fuels and vehicles, but we should note that continuing a petroleum-based fuel supply also has significant costs. Oil supply investment costs are growing rapidly, especially for exploration and production, with oil companies drilling deeper wells in more remote areas. Much of the oil capacity that will be needed in 2030 has not been built yet and will require development of new oil fields and investments in refineries that can deal with heavier crudes, oil sands, and gas to liquids. The International Energy Agency's *World Energy Outlook 2008* has estimated that oil supply infrastructure development between 2007 and 2030 will cost about $6.3 trillion globally for a new supply capacity of about 50 million barrels of oil per day (enough to fuel a fleet of about 1.3 billion cars, assuming an average fuel economy of 25 mpg and a vehicle driven 15,000 miles per year). In North America alone, the oil infrastructure costs for that period are estimated to be $1 trillion, or an average of $45 billion per year.

How would the capital outlay compare for alternative fuels versus oil? The IEA estimates that about $1.3 trillion would be for oil refineries and fuel transport, the remaining $5 trillion for exploration and production (drilling oil wells). The investment for oil refineries and transport is then about $1,000 per vehicle served. Counting exploration and production, total oil supply investment costs would be about $5,000 per vehicle.

By contrast, the capital cost for biofuels would be about $90–3600 billion to build biorefineries and biofuel transport capacity to fuel about 30 to 60 million cars, or $3,000–6,000 per vehicle served (assuming a 25-mpg vehicle that travels 15,000 miles per year). The biofuels analogy to oil exploration and production is developing land for biofuel production. However, these costs are likely to be very small compared to drilling oil wells, especially if low-carbon residues are employed (if the land is already developed for another purpose). And even for energy crops, land costs are treated more as rents or operating costs than capital costs.

For hydrogen or electricity, 80 percent of the transition costs are associated with the vehicle, with infrastructure accounting for only 20 percent of the total. The National Research Council has estimated that infrastructure capital costs would be $1,000–2,000 per car for PHEVs (including only the in-garage charger but not electricity transmission or generation or primary resources to make electricity). For hydrogen, infrastructure capital costs are estimated to be $1,400–2,000 per car, including hydrogen production, delivery, and refueling equipment, but not the capital costs for development of primary resources to make hydrogen—for example, natural gas wells, biomass resources, or wind farms.

The average annual transition cost to bring FCVs or PHEVs to cost competitiveness is about $4–8 billion per year over a 10-to-15-year period, and roughly 20 percent of this for infrastructure (or $0.8–1.6 billion per year). The cumulative infrastructure transition cost is roughly $8–12 billion, compared to projected capital expenditures in North America of perhaps $100–150 billion for oil refineries and fuel transport, and an additional $600–800 billion for exploration and production between 2007 and 2030. (This doesn't count oil investments that might be made abroad to serve North American markets). We could launch an alternative fuel infrastructure for much less than we are planning to spend on oil, and at a comparable investment cost per vehicle served.

Summary and Conclusions

- Only a portfolio approach can meet goals for an 80-percent reduction in GHG emissions by 2050. Substantial cuts in gasoline use are possible through improved efficiency of vehicles (about a 40-percent reduction from the reference case in 2050), use of low-carbon biofuels (a 15-percent reduction), and implementation of PHEVs (45 to 55 percent) or FCVs (60 percent). However, no single-pathway approach yields deep enough cuts in carbon emissions to add up to an 80-percent reduction by 2050. On the other hand, portfolio approaches combining improvements in ICEV efficiency with rapid introduction of PHEVs or FCVs and low-carbon biofuels can cut gasoline use to near zero by 2050 (depending on the amount of biofuel available for light-duty vehicles) and meet goals for an 80-percent reduction in GHG emissions.

- To realize the potential GHG benefits of PHEVs and FCVs, it is essential to decarbonize electricity and hydrogen production over time. If we rely on current grid technologies and hydrogen production from natural gas, there is little benefit compared to a strategy that stresses ICEV efficiency without advanced vehicles. For both PHEVs and FCVs, the buydown of vehicles could occur before substantial decarbonization of the fuel supply, but to realize the full benefits of these electric-drive vehicles, a parallel transition to a low-carbon energy supply is needed. We did not cost out this transition to low-carbon energy explicitly—it comes in through the fuel cost.

- The transition timing and costs are similar for PHEVs and FCVs. In each case, it will take fifteen to twenty years and 5 to 10 million vehicles for the new technology to break even with initial purchase and fuel supply costs for a reference gasoline car. Total transition costs are in the range of tens to hundreds of billions of dollars. For radically new types of vehicles like these, there is a need to buy down the cost of the vehicle through improvements in technology and scale-up of manufacturing. (Vehicle buydown costs are typically 80 percent of the total transition cost, and infrastructure costs 20 percent for both PHEVs and FCVs). For hydrogen, the fuel cost is initially high and comes down by focusing scaled-up development in lighthouse regions.

- For biofuels, the main transition cost is for improving second-generation biorefinery technology and scaling up the supply chain to the point where biofuel competes with other liquid fuels. In the United States, the total investment to this point is estimated by various studies to be perhaps $90–360 billion for biorefineries, fuel storage terminals, feedstock, and fuel transport to provide enough fuel for 30 to 60 million cars.

- Bringing new vehicle and fuel technologies to cost competitiveness will require fifteen to twenty years and a total investment of tens to hundreds of billions of dollars. Although this sounds like a lot of money, it is small compared to the investment in the existing gasoline system and the money flows in the current energy system. Maintaining and expanding the existing petroleum infrastructure is projected to cost about $1 trillion in North America alone between 2007 and 2030, an average of $45 billion per year. Perhaps 20 percent of this capital is for building refineries and fuel transport; the remainder is for exploration and production. By contrast, the average infrastructure costs during a transition to hydrogen FCVs or PHEVs would be $0.5–1 billion per year.

Notes

1. This chapter is based on studies done at UC Davis in support of National Academies assessments of hydrogen fuel cell vehicles (NRC, *Transitions to Alternative Transportation Technologies, A Focus on Hydrogen* (Washington, DC: National Academies Press, 2008)) and plug-in hybrid electric vehicles (NRC, *Transitions to Alternative Transportation Technologies, Plug-in Hybrid Electric Vehicles* (Washington, DC: National Academies Press, 2010)), and on N. Parker, Q. Hart, P. Tittmann, C. Murphy, M. Lay, R. Nelson, K. Skog, E. Gray, A. Schmidt, and B. Jenkins, *National Biorefinery Siting Model: Spatial Analysis and Supply Curve Development* (Denver, CO: Western Governors' Association, 2010).
2. U.S. Energy Information Administration (EIA), *Annual Energy Outlook 2008 with Projections to 2030*, DOE/EIA-0383(2008) (Washington, DC: Energy Information Administration, U.S. Department of Energy, 2008).
3. N. Parker et al., *National Biorefinery Siting Model*.
4. National Research Council (NRC), America's NRC, *Transitions to Alternative Transportation Technologies, A Focus on Plug-in Hybrid Electric Vehicles* (Washington, DC: National Academies Press, 2010).
5. M. A. Kromer and J. B. Heywood, Electric Powertrains: Opportunities and Challenges in the U.S. Light-Duty Vehicle Fleet, LEFF 2007-02 RP (Sloan Automotive Laboratory, MIT Laboratory for Energy and the Environment, May 2007), http://web.mit.edu/sloan-auto-lab/research/beforeh2/files/kromer_electric_powertrains.pdf. J. Gonder, A. Brooker, R. Carlson and J. Smart, *Deriving In-Use PHEV Fuel Economy Predictions from Standardized Test Cycle Results*, presented at the 5th IEEE Vehicle Power and Propulsion Conference, Dearborn, Michigan September 7–11, 2009, available at http://www.nrel.gov/docs/fy09osti/46251.pdf.
6. R. W. McCarthy, C. Yang, and J. M. Ogden, "Interactions Between Electric-Drive Vehicles and the Power Sector in California," UCD-ITS-RR-09-11 (Institute of Transportation Studies, University of California, Davis, 2009).
7. EIA, *Annual Energy Outlook* 2008.
8. Electric Power Research Institute / Natural Resources Defense Council, *Environmental Assessment of Plug-In Hybrid Electric Vehicles, Volume 1: Nationwide Greenhouse Gas Emissions*, July 2007. http://my.epri.com/portal/server.pt?open=514&objID=223132&mode=2.
9. NRC, *Transitions to Alternative Transportation Technologies, A Focus on Hydrogen* (Washington, DC: National Academies Press, 2008).
10. This follows NRC, *Transitions to Alternative Transportation Technologies, A Focus on Hydrogen* (Washington, DC: National Academies Press, 2008) and NRC, *Transitions to Alternative Transportation Technologies, Plug-in Hybrid Electric Vehicles* (Washington, DC: National Academies Press, 2010); J. Ogden, "A Comparison of Buydown Costs for Hydrogen Fuel Cell Vehicles and Plug-in Hybrid Vehicles," presented at the 2009 National Hydrogen Association Meeting, Columbia, SC, March 30–April 2, 2009.
11. *This follows NRC, Transitions to Alternative Transportation Technologies, A Focus on Plug-in Hybrid Electric Vehicles* (Washington, DC: National Academies Press, 2010).
12. D. Greene, P. Leiby, and D. Bowman, "Integrated Analysis of Market Transformation Scenarios with HyTrans," ORNL/TM-2007/094 (Oak Ridge National Laboratory, June 2007).
13. NRC, Transitions to *Alternative Transportation Technologies, A Focus on Hydrogen* (Washington, DC: National Academies Press, 2008).
14. C. Yang and J. Ogden, "U.S. Urban Hydrogen Infrastructure Costs Using the Steady State City Hydrogen Infrastructure System Model (SSCHISM)," presented at the 2007 National Hydrogen Association Meeting, San Antonio, TX, March 18–22, 2007. A beta copy of the model is posted on Christopher Yang's website at UC Davis Institute of Transportation Studies, available at www.its.ucdavis.edu/people.
15. K. Morrow, D. Karner, and J. Francfort, "Plug-in Hybrid Electric Vehicle Charging Infrastructure Review," INL/EXT-08-15058 (Idaho National Laboratory, November 2008).
16. The model was developed by N. Parker, Q. Hart, P. Tittmann, C. Murphy, M. Lay, R. Nelson, K. Skog, E. Gray, A. Schmidt, and B. Jenkins in *National Biorefinery Siting Model: Spatial Analysis and Supply Curve Development* (Denver, CO: Western Governors' Association, 2010).
17. *Renewable Fuel Standard Program (RFS2) Regulatory Impact Analysis*, EPA-420-R-10-006 (U.S. Environmental Protection Agency, February 2010).
18. T. West et al., "Feasibility, Economics, and Environmental Impact of Producing 90 Billion Gallons of Ethanol per Year by 2030," preprint version, SAND 2009-3076J (Sandia National Laboratory, August 6, 2009).

Chapter 10:
Optimizing the Transportation Climate Mitigation Wedge[1]

Sonia Yeh and David McCollum

The previous two chapters have looked at scenarios for making deep reductions in GHG emissions in the transportation sector by 2050. We now turn to considering what role the transportation sector might play under economy-wide CO_2 constraints in the United States. If we see emission reductions achieved in different sectors of the economy—including commercial and residential buildings, industry, agriculture, and electric power, as well as transportation—as wedges that add up to an emission reduction target mandated by policy, how might the transportation wedge reduce its emissions to meet the policy goals under optimized least-cost solutions? Will economy-wide carbon taxes and cap-and-trade programs result in emission reductions from the transportation sector commensurate with its contribution to economy-wide emissions? Or are other approaches needed to incentivize the transportation climate mitigation wedge? To address these questions, we used an integrated energy-economics model called the MARKet ALlocation (MARKAL) model to examine least-cost emission reductions scenarios within economy-wide emission cap scenarios.

Background: Other Models and Their Findings

Policymakers rely on integrated climate-energy-economics models to help them identify the most economical way to meet climate mitigation objectives. Few of these models examined in greater detail the role transportation GHG emission reductions will play under economy-wide emission cap policies. These models found that carbon taxes and cap-and-trade programs will have a large effect in the electric power sector but little effect in the transportation sector. In other words, these analyses find that electric power sector responds well to market-based policies such as cap-and-trade, while all the other end-use sectors including residential, commercial, industrial and transportation sectors respond poorly to market-based policies. For example, analyses of proposed U.S. cap-and-trade programs by the U.S. Environmental Protection Agency (EPA) and Energy Information Administration (EIA) suggest that less than 5 percent of total emission reductions would come from the transportation sector by 2030, even though transportation accounts for almost a third of total emissions.[2] If these proposed cap-and-trade policies were implemented, the transport sector would become the single largest emission source by 2050, accounting for more than half of total GHG emissions in the United States.[3]

On the other hand, studies that have used engineering economic analyses to examine the potential of transportation GHG emission reductions suggest that the cost of improving energy efficiencies of light-duty vehicles will be minor. For example, a McKinsey & Company report concluded that a cluster of transportation technologies—including improvement of vehicle efficiency, use of cellulosic biofuels, and hybridization of vehicles—could avoid 340 megatons of GHG emissions at a cost of less than $50 per ton (in 2005 dollars) by 2030.[4]

Why the difference in results between the economy-wide models where transportation emission reductions are estimated to be expensive and unlikely, and studies specifically examining the transport sector that conclude that moderate emission reductions from the transport sector can be achieved with reasonable costs? The contradictions lie in the "energy paradox" that has been widely researched in the literature outside of the energy modelling community, i.e., energy markets are particularly inefficient and ineffective in addressing end-use technology efficiency and demand reduction. Thus, while market-based policies are more effective in reducing GHG emissions on the supply-side, separate policies are needed to reduce GHG emissions from end-uses, including transportation. Policies such as vehicle, building, and appliance efficiency standards, R&D programs targeting advanced technologies, and subsidies for infrastructure development are a few examples of policies needed to overcome market barriers and imperfect decision making in the real world.

In this chapter, however, we use a model that assumes perfect decision making and a perfect market to estimate GHG reductions needed to achieve deeper climate reduction goals. The purpose of the modelling exercise is not to predict the future, but to understand *the least-cost technology mix (given our assumptions) that will be required to meet the policy targets*. The results provide a useful roadmap for policymakers to decide policy solutions and incentive structures needed to overcome market barriers in order to achieve emission reduction goals.

Our Modeling Framework and Scenarios

We used the MARKet ALlocation (MARKAL) model developed by the Energy Technology Systems Analysis Programme (ETSAP) of the International Energy Agency to help us identify the most cost-effective technological pathway to meet GHG emission reduction targets economy-wide while also satisfying future end-use demands and other policy constraints. MARKAL is a bottom-up model that characterizes current and future energy technologies in detail, including variables such as capital cost, operational and maintenance costs, fuel efficiency, emissions, and useful life. The model also accounts for fuel supply, resource potentials, and other user constraints. It assumes rational decision making, with perfect information and perfect foresight, and computes a supply-demand equilibrium where energy demand is responsive to changes in price. The model finds the least expensive combination of technologies to meet future energy demands, subject to resource availability and user constraints such as economy-wide GHG emission reduction targets or technology-specific appliance efficiency/emission standards that become increasingly stringent over time.

We used the model to examine cumulative emission-reduction targets from 10 to 50 percent economy-wide (E scenarios) and from 10 to 30 percent economy-wide and equal percent reduction from the transportation sector (E&T scenarios) for the period 2010 to 2050. These reductions (also referred to as CO_2 avoided) are in comparison to a reference case where no significant GHG policies have been adopted. We assume that under GHG reduction scenarios, complementary policies (such as policies to improve access to transit, incentivize fuel infrastructure development to lower consumers' risk aversion (represented as high hurdle/discount rate) to new technologies, etc.) will be adopted to address market barriers, and consumers will be more likely to respond to price changes by reducing vehicle travel demands (by driving less, taking more transit trips, or using other modes of transportation) and more willing to purchase new technologies that have higher up-front costs and a longer payback time. For example, we assume that when gasoline prices increase by 10 percent, consumers in the reference case (no climate policy) will reduce their travel demand by 1 percent (a demand elasticity of –0.1), while consumers in the climate policy scenarios will reduce their travel demand by 3 percent (a demand elasticity of –0.3). Similarly, we assume that in the policy case consumers will be willing to wait longer (indicated by a lower discount rate) to recover their investment in more advanced and efficient vehicles than they normally would have. Later, we will demonstrate that the first assumption (increased elasticity) has very little effect on the results, while the second (longer payback period) is necessary to broadly adopt advanced, low-carbon vehicles within the policy timeframe.

SCENARIOS EXAMINED

Scenario	Description	Note
Reference case	Projections of the reference case	Travel demand elasticity = -0.1, vehicle technology discount rate = 0.33
10%E, 20%E, 30%E, 40%E, 50%E	10–50 percent economy-wide cap	Travel demand elasticity = -0.3, vehicle technology discount rate = 0.15
10%E&T, 20%E&T, 30%E&T	10–30 percent economy-wide + 10–30 percent transportation cap	

The scenarios examined in this chapter are not intended to project the future with and without climate policies. Instead, our aim is to identify least cost mitigation technology mix based on our assumptions about technology costs and resource availability within an integrated energy system, if society were to act in the least-cost manner with perfect foresight.

There are three things to note about our use of the MARKAL model. First, the MARKAL type of bottom-up model is not suited to analyze nontechnology policies—such as policies encouraging behavior changes or those regarding land use, smart growth, mass transit, carpooling, or telecommuting—even though these mitigation options also play important roles in reducing transportation emissions. Second, most analyses of alternative fuels (except for hydrogen fuel, where transport, delivery, and refueling-station costs are examined in detail) assume a flat rate for transportation and distribution cost and ignore infrastructure hurdles such as refueling stations and transport distance, the classic chicken-and-egg problem. Mitigation strategies involving alternative fuels must take into consideration not only cost but also other social factors and policies that encourage technology adoption. Third, we do not take into account the social and environmental benefits and co-benefits of reducing CO_2 emissions, such as reducing air pollution, improving energy security risk, and reducing the costs of climate change.

Our Modeling Results: Where Emission Reductions Will Come From

Our modeling results suggest that more stringent economy-wide emission caps than currently proposed, or transportation-sector emission caps, will be needed in order to effectively reduce long-term transportation sector CO_2 emissions. We also found that the market penetrations of low-carbon fuels and advanced vehicle technology depend on policy drivers. As the GHG reduction target becomes more stringent, faster penetration of low-carbon fuels and advanced vehicles becomes necessary to achieve the policy target. And finally, our model projects that emission reductions beyond current policy requirements will be contributed almost entirely by the interactions of three mitigation wedges: vehicle efficiency improvement, advanced vehicle technologies, and low-carbon fuels including electricity and biofuels. The role of price-induced VMT reduction in reducing GHG emissions is small in this economic modeling, primarily due to the low elasticity (albeit higher in the policy case), the rebound effect, and the resulting longer payback period with reduced VMT.

Emission caps and emission reductions by sector

Consistent with previous research findings, our analysis shows that when economy-wide emission caps are low to moderate (our 10%E to 30%E scenarios), the transportation sector contributes just a small portion of the overall reductions and the electric power sector contributes the majority. Our 30%E scenario (2,879 million metric tons CO_2 reduction in 2030) is roughly consistent with the EIA analysis of S. 2191 (America's Climate Security Act of 2007), which projects the total CO_2 emission reduction by 2030 with no international offsets at 3,030 million metric tons CO_2-equivalent.[5] The transportation sector starts to make more substantial reduction contributions at the 40-percent reduction target and above (7 percent in the 40%E scenario and 13 percent in the 50%E scenario between 2010 and 2050). If the same percentage emission caps (10 to 30 percent) apply equally to the full economy and to transportation (the E&T scenarios), the transportation sector contributes roughly 30 percent of the overall reductions between 2010 and 2050, while the electric power sector contributes 51 to 66 percent.

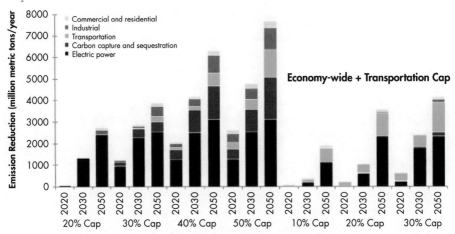

CO$_2$ EMISSION REDUCTIONS BY SECTOR FOR SEVEN SCENARIOS IN 2020, 2030, AND 2050

We compared energy-related CO$_2$ emission reductions in 2020, 2030, and 2050 by sector for seven of our scenarios (the 20%E to 50%E scenarios and the 10%E&T to 30%E&T scenarios). Electric power and carbon capture and sequestration (CCS) account for most of the reduction in the economy-wide scenarios. The transportation sector starts to make more substantial reduction contributions at the 40-percent reduction target and above.

Holding emissions constant to 2050 (constituting an emission stabilization trajectory) roughly corresponds to our 10%E scenario, and the shape of our 50%E scenario roughly corresponds to the 450 ppm early-action mitigation wedge proposed by Stephen Pacala and Robert H. Socolow in their 2004 article in *Science*.[6] Our model, which chooses the least-cost solution with perfect foresight, suggests that most of the emission reduction will come from the electric power sector by fuel switching (increasing use of natural gas, nuclear after 2040, and renewables), adopting more efficient electricity-generating technologies, and employing carbon capture and sequestration (CCS) for the 30 percent and above economy-wide cap scenarios.

MITIGATION WEDGES BY SECTOR FOR SIX SCENARIOS, 2010–2050

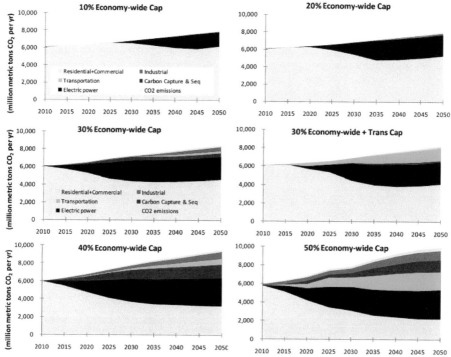

Another way of looking at emission reductions is by picturing each sector as a wedge representing emissions avoided from the reference case. The gray areas here show overall CO_2 emissions. Again we can see that the majority of emission reductions in all scenarios come from the electric power sector. The transportation sector is a significant contributor only in the 30%E&T and the 50%E scenarios.

Fuel and vehicle mix and emission reduction

In all the policy cases that require significant reductions from the transportation sector, gasoline hybrid electric vehicles (HEVs) quickly start replacing conventional gasoline vehicles. In 30%E&T, the scenario that requires the most GHG emission reduction from the transportation sector, plug-in hybrid electric vehicles (PHEVs) are quickly adopted and comprise roughly 68 percent of the total passenger vehicle fleet in 2050. Overall, fleet-average vehicle efficiency increases as the stringency of the CO_2 emission caps increases (the 30%E&T scenario gains up to 92.4 percent in efficiency in 2050 over the reference case), and fuel usage also decreases significantly (up to 48 percent in 2050 in 30%E&T).

In our scenarios, ethanol usage increases from 3.5 billion gasoline-equivalent gallons per year in 2005 to 36 billion gallons per year in 2050 under the reference case, and to the highest level of 88.4 billion gallons per year under 30%E&T. These assumptions about ethanol do not take into account the possibility that there will be policies either to limit the use of biofuel produced from arable land or to phase out food-based ethanol, since biofuels that induce land-use conversion may result in overall greater GHG emissions than gasoline on a life-cycle basis while causing other adverse sustainability impacts. Neither does the scenario take into account the possibility that cellulosic ethanol, which would avoid these pitfalls, will not be commercially successful on a large scale by 2050. In both cases, the mix of fuels and vehicles to meet emission reduction targets will be different from what we have projected above. In a 30%E&T scenario where there is no biofuel mandate, there will also be no ethanol flex-fuel vehicles and a slightly higher PHEV penetration, and a smaller amount of the biofuels will be used in blended gasoline. In a 30%E&T scenario where there is no biofuel mandate and no cellulosic ethanol industry, we see the highest and fastest penetration of PHEVs.

CHAPTER 10: OPTIMIZING THE TRANSPORTATION CLIMATE MITIGATION WEDGE   241

PROJECTED VEHICLE MIX FOR SIX SCENARIOS, 2010–2050

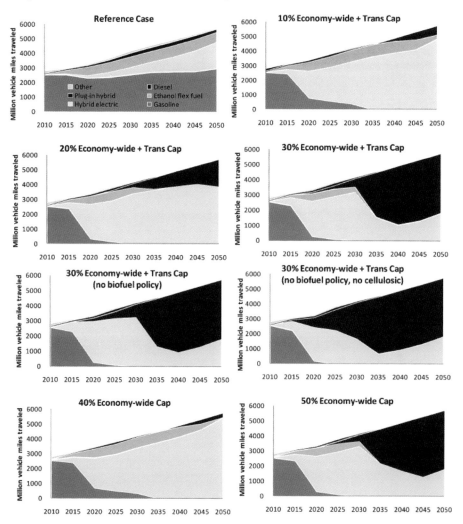

The market penetration of various vehicle types depends on policy drivers. As the GHG reduction target becomes more stringent, faster penetration of advanced vehicles becomes necessary to achieve the policy target. The penetration of ethanol flex-fuel vehicles is entirely driven by biofuel policy that requires 36 billion gallons of biofuels by 2022. To meet deeper reduction goals without biofuels, earlier penetration of PHEVs at higher volumes will be necessary. In almost all of the climate-policy cases, conventional gasoline vehicles need to be replaced by advanced vehicle technology by 2020-2030, depending on the stringency of the targets.

Mitigation strategies and emission reduction

Mitigation strategies for the transportation sector include reducing vehicle miles traveled (VMT), increasing vehicle efficiency, and adopting low-GHG fuels and advanced vehicle technologies. Overall, our model projects that CO_2 emission reductions in all our scenarios are contributed almost entirely by vehicle efficiency improvement, with a growing proportion contributed by switching to electricity and biofuels after about 2030. The travel demand levels are similar in all cases we examined (and contribute nearly nothing to reducing CO_2 emissions in any of the scenarios), reflecting two facts: (1) although we have made consumers *more* willing to change their demand level compared with a no-policy scenario (by increasing the elasticity from -0.1 to -0.3), elasticity of travel demand remains low; (2) improvements in vehicle efficiency and the transition to electricity fuels reduce the cost of driving (in dollars per mile driven), which further decreases consumers' response to the underlying trend of fuel price increases. Though we did not explicitly calculate the rebound effect (as vehicles become more fuel-efficient, it costs less to drive and so VMT increases), it likely explains the lack of response in price-induced VMT reduction as an effective way of contributing to GHG mitigation. It should also be noted that our model cannot simulate the effects of policies that encourage behavior change—such as policies regarding land use, smart growth, mass transit, carpooling, or telecommuting—although such policies are likely to be adopted when climate policies become reality.

TRAVEL DEMAND, FUEL EFFICIENCY, AND TOTAL FUEL USE BY SCENARIO, 2010–2050

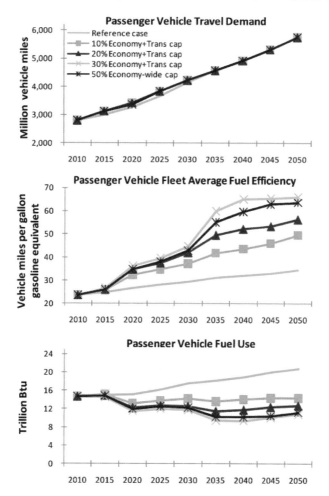

When we compare passenger-vehicle travel demand, fuel efficiency, and total fuel use for our three E&T scenarios and the 50%E scenario, it is clear that fuel efficiency and total fuel use need to improve significantly over the reference case in order to meet the reduction targets. The lack of response in price-induced VMT reduction as an effective way of contributing to GHG mitigation may be explained by low elasticity to travel demand and the decreasing cost of driving per mile due to improvement in vehicle efficiency and the transition to lower cost of alternative fuel, electricity.

We found that even without a specific mandate for biofuel production, cellulosic ethanol can still be a favorable mitigation strategy to achieve significant transportation emission reductions albeit at lower initial volume and slowly ramping up to a higher level by 2050 compared with the reference case. However, if there is neither a biofuel mandate nor commercially successful cellulosic technology on a large scale, more gasoline and electricity, and overall less fuel will be necessary to achieve the required reduction in transportation CO_2 emissions.

FUEL USE BY TYPE OF FUEL FOR FOUR SCENARIOS, 2010–2050

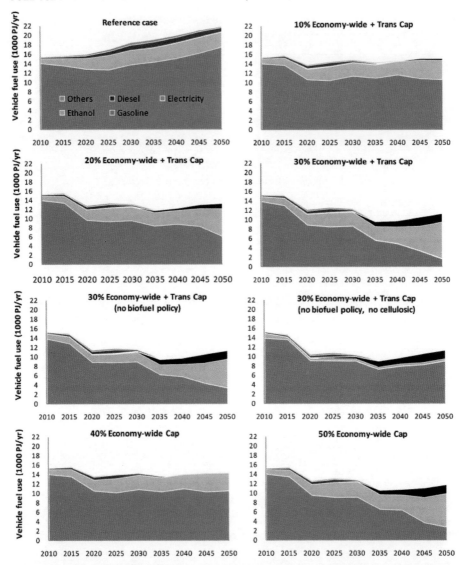

We compared total passenger-vehicle fuel use by type of fuel for the reference scenario, the 30% E&T scenario, and scenarios where (1) there is no biofuel mandate, and (2) there is no successful cellulosic technology to make low-carbon biofuels at the estimated costs. We found that the success of biofuels may not be entirely dependent on a biofuel mandate and can occur without a mandate, although the availability of truly low-carbon biofuels can be a major uncertainty.

RELATIVE CONTRIBUTIONS OF MITIGATION STRATEGIES TO CO_2 EMISSION REDUCTIONS

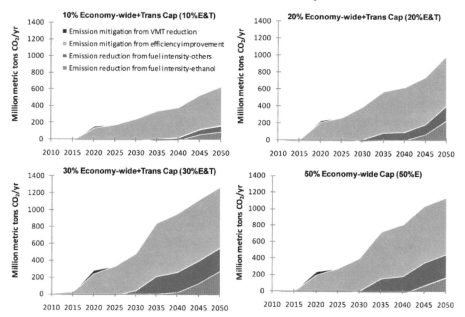

For passenger vehicles, CO_2 emissions can be reduced by reducing fuel CO_2 intensity, improving vehicle efficiency, and reducing vehicle miles traveled (VMT). With the exception of biofuels, fuel CO_2 intensity is based on combustion emissions and is not life-cycle based. Our model projects that vehicle efficiency improvement is the low-hanging fruit in all of the scenarios. Fuel CO_2 intensity reduction can be achieved by increasing the use of electricity (blue wedges) and of biofuel (green wedges) above and beyond the existing mandate. The use of electricity further increases efficiency improvement in the transport sector due to the superior efficiency of electric-drive vehicles compared with conventional internal combustion engines. Because our model cannot simulate behavior changes not related to economic factors, it does not predict significant reductions in travel demand.

Summary and Conclusions

- Mitigation strategies in a number of different sectors might be combined to achieve policy goals in reducing CO_2 emissions. The results illustrated here are by no means predictive of the future outcome of any particular policies. But what we can say with certainty is that much more stringent system-wide CO_2 reduction targets than those that have been discussed in Congress will be required to achieve significant CO_2 reductions in the transportation sector.

- Our study confirms the conclusions of analyses by government agencies, that economy-wide cap-and-trade programs acceptable to politicians are unlikely to reduce transportation emissions, despite the fact that this sector makes a significant contribution to total emissions and is the second largest emission source after the electric power sector. Achieving significant transportation emission reductions over the long term will require much more aggressive economy-wide policies than are currently proposed—or a cap on transportation-specific emissions.

- Realistically, a transportation cap is neither likely nor the most economical approach to reduce economy-wide GHG emissions. The well-known "energy paradox" implies that energy markets are particularly inefficient and ineffective in addressing end-use technology efficiency and demand reduction. Thus, while market-based policies are more effective in reducing GHG emissions on the supply-side, separate policies are needed to reduce GHG emissions from end-uses, including transportation. Policies such as vehicle efficiency standards, R&D programs targeting advanced technologies, and subsidies for infrastructure development are a few examples of policies needed to overcome market barriers and imperfect decision making in the real world.

- Comparable policies could be adopted to achieve the same goal. Examples of these policies include policies to improve vehicle efficiencies (such as the CAFE fuel economy standard or the vehicle GHG emission standard), policies that encourage the reduction of fuel carbon intensities (such as the Low Carbon Fuel Standard discussed in Chapter 11), and polices that encourage the production and adoption of advanced vehicle technologies that both reduce fuel GHG intensity and increase vehicle efficiency (such as the zero-emission vehicle or ZEV program).

- Our model does not predict significant price-induced travel demand reduction. The lack of response in reducing VMT as an effective way of contributing to GHG mitigation may be explained by low elasticity to travel demand and the decreasing cost of driving per mile due to improvement in vehicle efficiency and the transition to lower cost of alternative fuel, particularly electricity. But such reductions should be pursued in parallel in order to reduce congestion, improve air quality, and reduce oil dependence as well as to reduce emissions.

- Though our modelling results do not project the penetration of hydrogen fuel cell vehicles given our scenarios, a portfolio approach is needed to invest in advanced fuel and vehicle technologies including plug-in hybrid electric vehicles, battery electric vehicles and hydrogen fuel cell vehicles to deal with uncertainties in technology costs, market barriers and consumers' preferences.

- More research is needed to help identify robust policies that will achieve the best outcome in the face of uncertainties that we have not addressed here.

FUTURE WORK

In California, the Global Warming Solutions Act of 2006 (AB 32), which enacts former Governor Schwarzenegger's Executive Order #S-3-05, requires the state to reduce its greenhouse gas (GHG) emissions to the 1990 level by 2020 and 80 percent below the 1990 level by 2050. The scoping plan adopted both technology-forcing measures that mandate specific emission reduction strategies/technologies that are estimated to reduce 174 million tonnes GHG emissions by 2020, and a market-based approach (cap-and-trade program) that limits emissions from all major point sources, allowing market mechanisms to determine the most cost-effective strategies to reduce GHG emissions. Though the total emissions reduction needed by 2020 has been adjusted downward due to the economic downturn, the State still faces significant challenges and currently lacks a roadmap to meet its 2050 reduction target. We've developed a California-specific energy-economic-environment model that will help us understand the cost-effective mitigation options needed to achieve the long-term GHG reduction target, and potential impacts of various climate and energy policies adopted or being considered in California. The CA-TIMES (the Integrated MARKAL EFOM System) is a bottom-up, technology-rich model that encompasses all sectors of the economy, including electric, transportation, industrial, commercial, residential, agricultural, and non-energy sectors. Our goal is to understand the impacts of policies on economic costs, energy consumption, and technology portfolios; and to identify market barriers and policies needed to encourage the adoption of advanced technologies. The basic modeling structure is described in McCollum, David L. (2011) *Achieving Long-term Energy, Transport and Climate Objectives: Multi-dimensional Scenario Analysis and Modeling within a Systems Level Framework*, Institute of Transportation Studies, University of California, Davis, Research Report UCD-ITS-RR-11-02. More work is ongoing to improve the modeling as well as scenario analysis in order to learn insights for guiding policy design.

Notes

1. This chapter is a condensed version of S. Yeh, A. Farrell, et al., "Optimizing U.S. Mitigation Strategies for the Light-Duty Transportation Sector: What We Learn from a Bottom-Up Model," *Environmental Science and Technology* 42 (2008): 8202–10.
2. U.S. Energy Information Administration, "Energy Market and Economic Impacts of S. 2191, the Lieberman-Warner Climate Security Act of 2007" (Washington, DC: Energy Information Administration, Office of Integrated Analysis and Forecasting, U.S. Department of Energy, 2008); U.S. Environmental Protection Agency, "United States Environmental Protection Agency's Analysis of Senate Bill S.280 in the 110th Congress, The Climate Stewardship and Innovation Act of 2007" (U.S. Environmental Protection Agency, 2007).
3. Ibid.
4. J. Creyts, A. Derkach, S. Nyquist, K. Ostrowski, and J. Stephenson, *Reducing U.S. Greenhouse Gas Emissions: How Much at What Cost?* (McKinsey & Company, 2007). http://www.mckinsey.com/clientservice/sustainability/pdf/US_ghg_final_report.pdf
5. U.S. EIA, "Energy Market and Economic Impacts of S. 2191, the Lieberman-Warner Climate Security Act of 2007" (U.S. Energy Information Administration, 2008), http://www.eia.doe.gov/oiaf/servicerpt/s2191/.
6. S. Pacala and R. Socolow, "Stabilization Wedges: Solving the Climate Problem for the Next 50 Years with Current Technologies," *Science* 305 (2004): 968–72. See also OECD/IEA, *Energy Technology Perspectives* (Paris, France: International Energy Agency, 2008).

Part 4: Policy and Sustainable Transportation

We have explored and compared advanced vehicle and fuel pathways and imagined scenarios that might get us to specific targets. Now we must ask, What policy measures and tools are needed to encourage progress toward a sustainable transportation system? What are the measurement challenges that must be addressed in order for analysts to be able to predict the full impact of potential policies? The three chapters in this section address those questions.

- **Chapter 11** argues that in the face of petroleum's stubborn dominance as a transportation fuel, a new policy instrument known as a low-carbon fuel standard (LCFS) is the most promising approach to getting the carbon out of fuels. This chapter summarizes failed and ineffective approaches of the past and points out the trouble with current mandates and proposals. It then traces the emergence in Europe and the United States of a GHG performance standard for fuels, looks at the details of California's LCFS, and suggests that these initiatives might lead the way toward a harmonized international effort.

- **Chapter 12** explores four key measurement uncertainties that create challenges in accounting for the climate impacts of biofuels—uncertainties that transportation policies designed to encourage low-carbon fuels must address. First, an accounting of the climate impacts of biofuels should consider the effect over time of GHG emissions from direct land-use change. Second, there is a need to account for non-GHG global warming factors such as albedo, and the effect of non-Kyoto gases and pollutants such as aerosols and black carbon. Third, more work needs to be done on the question of how to account for GHG emissions due to indirect land-use conversions. And fourth, when forest wastes are used as biofuel feedstock, changes to the GHG dynamics within integrated forest systems need to be considered.

- **Chapter 13** takes a critical look at life-cycle analysis, for more than twenty years the conventional method used to estimate emissions of greenhouse gases (GHGs) from the use of a wide range of transportation fuels and one basis of our comparisons of vehicle/fuel pathways in this book. As commonly employed, LCA cannot accurately represent the impacts of complex systems, such as those involved in making and using biofuels for transportation. In order to better represent the impacts of complex systems, this chapter proposes a different tool, one that has the central features of LCA but not the limitations—a tool that starts with the specification of a policy or action and ends with the impacts on environmental systems.

Chapter 11: Toward a Universal Low-Carbon Fuel Standard[1]

Daniel Sperling and Sonia Yeh

Petroleum's dominance as a transportation fuel has never been seriously threatened anywhere—except Brazil, with its sugarcane ethanol—since taking root nearly a century ago. Efforts to replace petroleum, usually for energy security reasons but also to reduce local air pollution, have continued episodically for years—and largely failed. Vehicles, planes, and ships are still almost entirely dependent on petroleum and account for nearly one-third of all greenhouse gas (GHG) emissions in the United States and almost one-fourth of all GHG emissions globally. In the face of this stubborn petroleum lock-in, what is the most effective type of policy to spur technological innovation and investment in alternative fuels?

In this chapter we argue that a new policy instrument known as a low-carbon fuel standard (LCFS) is the most promising approach to getting the carbon out of fuels. We have learned from past failures that to be successful, a policy approach must inspire industry to pursue innovation aggressively; it must be flexible, performance-based, and inclusive so that industry, not government, picks the winners. It should also take account of all greenhouse gas emissions associated with the production, distribution, and use of the fuel, from the source to the vehicle, so that petroleum and alternative fuels such as hydrogen and electricity are compared on a level playing field. (While upstream emissions account for about 20 percent of total GHG emissions from petroleum, they represent almost the total life-cycle emissions for fuels such as electricity and hydrogen; upstream emissions from extraction, production, and refining also comprise a large percentage of total emissions for the very heavy oils and tar sands that oil companies are increasingly embracing to supplement limited supplies of conventional crude oil.) LCFS policies already adopted in California and the European Union fit these requirements and can lead the way toward a harmonized international effort.

Failed and Ineffective Approaches of the Past

No country other than Brazil has been successful at replacing petroleum fuels in the transport sector. Many countries, especially the United States, have provided policy support for one alternative fuel after another, some gaining more attention than others but each one eventually faltering. The fuels du jour in the 1980s and 1990s were coal liquids, methanol, compressed and liquefied natural gas, and electricity for battery vehicles. Early in the 21st century it was hydrogen, followed by corn ethanol, and now electricity for plug-in hybrid electric vehicles. But worldwide, the only nonpetroleum fuels that have gained significant market share are corn ethanol in the United States and sugarcane ethanol in Brazil.

The fuel du jour phenomenon is fed by oil market failures, overblown promises, the inertia of oil industry investments, and the short attention spans of government, the mass media, and the public.[2] Alternatives emerge when oil prices are high but wither when prices fall. They rise when public attention is focused on the environmental shortcomings of petroleum fuels but dissipate when oil and auto companies marshal their considerable resources to improve their environmental performance. When George H. Bush advocated methanol fuel in 1989 as a way of reducing vehicular pollution, oil companies responded by offering cleaner-burning reformulated gasoline (and later, cleaner diesel). And when air regulators in California and the United States adopted aggressive emission standards for engines, vehicle manufacturers diverted resources to improve combustion and emission control technologies.

One key problem is the ad hoc approach of governments to petroleum substitution. The U.S. government provided loan and purchase guarantees for coal and oil shale "synfuels" in the early 1980s when oil prices were high, passed a law in 1988 offering fuel economy credits for flexible-fuel cars, launched the Advanced Battery Consortium and Partnership for a New Generation of Vehicles in the early 1990s to accelerate development of advanced vehicles, promoted hydrogen cars in the early years of this decade, provided tens of billions of dollars in federal and state subsidies for corn ethanol, and is now providing incentives for plug-in hybrids. State initiatives included California's purchases of methanol cars in the 1980s and its zero-emission vehicle requirement of 1990.

But these various alternative fuel initiatives have failed to move us away from petroleum-based transportation, in part because the government did not adopt supporting incentives and plans. More durable policies are needed—ones that are based on performance, that stimulate innovation, and that reduce consumer and industry risk and uncertainty.

FUNDAMENTALS OF EFFECTIVE PROGRAMS

Policies and programs that aim to motivate industry to pursue innovations are more likely to be successful if they are flexible, performance-based, and inclusive. Federal fuel economy standards for cars and light trucks, for example, allow industry to determine the best way to achieve the targets, which stimulates innovation. Experiences with fuel economy standards and other programs suggest several principles for policies that promote low-carbon transportation fuels.

Don't try to pick winners. Programs are more successful if they focus on the goal and not on the specific means to achieve it. If the goal is to lower GHG emissions from fuels, setting GHG performance standards for transportation fuels motivates companies to find the best approach. Although mandating the use of specific fuels such as natural gas or ethanol may reduce GHG emissions, the market generally will achieve that goal at lower cost if allowed the flexibility to choose from the mix of possible fuels. The market can quickly adapt to changes in technology, allowing the introduction of new fuel pathways with greater emissions reduction or lower cost, or both.

Assess the full GHG life cycle. To reduce GHG emissions, all emissions associated with the production, distribution, and use of the fuel must be considered. This well-to-wheels or source-to-wheels life-cycle assessment should include all direct emissions, such as those associated with acquiring, growing, and harvesting the feedstock for biofuels; transporting the feedstock to the fuel-processing facility; turning the feedstock into an acceptable fuel; delivering the fuel to the point of retail sale; and burning the fuel.

Life-cycle analyses should also consider the indirect impacts, which can be large. For biomass-based fuels, for example, indirect emissions are associated with diverting land from food and other uses to energy production; in the case of corn ethanol, additional land is drawn into production to replace the corn diverted to energy use. These effects are controversial because they have never been included in policies or regulations and because the underlying science is still evolving. (See Chapter 12 for more on this.) The indirect land-use effects can be large for food-based feedstocks, which are land-intensive, but small for cellulosic materials, and zero for waste materials.

Be aware of positive and negative side effects. Policies and programs promoting fuels with lower GHG emissions may have other consequences, beneficial or harmful. For example, how is the price of food affected by the diversion of food and animal feed, such as corn and soybeans, to biofuel production? And, more positively, how much does greater reliance on biofuels from feedstock grown in the United States reduce expenditures on imported oil and increase farm incomes and jobs? Perhaps more important, some so-called side effects, such as energy security benefits of reducing dependence on petroleum, may be chief reasons for implementing the policies.

Don't be naïve about real-world responses. Responses may occur outside the jurisdiction of the entity that establishes a low-carbon fuel program. One response, termed "leakage," occurs when fuel suppliers shift their fuels to avoid compliance with California or federal regulations. For instance, a high-carbon transportation fuel made from oil sands or liquefied coal can be shipped to states or countries with no regulations mandating reduced carbon content. Because GHG buildup is a global problem, the benefits of reduction will be lost if the leakage response becomes rampant. The leakage problem diminishes as more states and nations adopt low-carbon fuel policies.

Another potential response is increased consumption of gasoline and diesel fuels in places without low-carbon fuel policies and biofuel mandates, if reduced consumption in California or the United States reduces world oil prices. This rebound effect would probably be small but would nonetheless offset some of the GHG emissions reductions that the program achieves.

Recognize infrastructure and economic barriers. Infrastructure can be slow to change and thus act as a barrier to the widespread introduction of new fuels. For example, ethanol is now used as a blend stock with gasoline. With ethanol use increasing, gasoline in the United States is likely to reach the 10-percent blending limit allowed in vehicles by 2015. Two options exist to expand the use of ethanol. One is to increase

> the blending limit, but this is opposed by manufacturers of off-road equipment, such as lawnmowers, and cars and light trucks, who are concerned about damage to the engines. The second option is to expand the use of flexible-fueled vehicles, which can use ethanol in concentrations of up to 85 percent in gasoline (E85). Yet the number of filling stations now offering E85 is limited, and adding a pump and storage tank for E85 can cost $100,000 and more.

The Trouble with Current Mandates and Proposals

What about using a volumetric standard like the Renewable Fuel Standard (RFS) adopted by the U.S. Congress? What about carbon taxes and the cap-and-trade approach? Although these are steps in the right direction, we do not believe they are the most effective policy instruments to move the transport sector away from petroleum dominance.

Volumetric mandates

Since the start of the 21st century, volumetric mandates have been the preferred policy approach to reduce the use of petroleum fuels. The United States adopted a volumetric mandate for biofuels (the Renewable Fuel Standard or RFS) in 2005 and strengthened it in December 2007 as part of the Energy Independence and Security Act (EISA). RFS2 requires that 36 billion gallons of biofuels be sold annually by 2022; 21 billion gallons of these must be "advanced" biofuels and the other 15 billion gallons can be corn ethanol. To achieve these volumes, the U.S. Environmental Protection Agency (EPA) calculates a percentage-based standard every year. Based on the standard, each refiner, importer, and non-oxygenate blender of gasoline determines the minimum volume of renewable fuel it must use in its transportation fuel mix. The advanced biofuels are required to achieve at least a 50-percent reduction from baseline life-cycle GHG emissions, with a subcategory of cellulosic biofuels required to meet a 60-percent reduction target. These reduction targets are based on life-cycle emissions, including emissions from indirect land-use changes.

Similarly, the United Kingdom's Renewable Transport Fuel Obligation (RTFO) aims to have 3.25 percent of all transport fuel sold in the United Kingdom come from a renewable source by 2009–10 and to reach 5 percent in 2013–14. The European Union's Biofuel Directive (BD) initially set a target of 5.75 percent biofuels by 2010 and 10 percent biofuels by 2020 but has since broadened the target to include all renewable fuels and renamed it the Renewable Energy Directive (RED).

Volumetric biofuel mandates have a number of shortcomings. First, they target only biofuels and not other alternatives. Second, setting GHG reduction targets within the volumetric mandates, as the United States does with its RFS2 program, is a clumsy way to reduce GHGs. It forces biofuels into a small number of fixed categories and thereby stifles innovation. Once the regulatory agency concludes that certain biofuel pathways meet the specified GHG reduction target, there is little incentive for further improvement. As a result, there is less incentive to use very-low-carbon materials, such as waste biomass, or adopt sustainable farming and management practices that reduce direct and indirect land-use emissions.[3] Third, RFS2 exempts existing and planned corn ethanol production plants from the greenhouse gas requirements, essentially mandating a massive unfettered expansion of corn ethanol. Rapid expansion of corn ethanol not only stresses food markets and requires vast amounts of water but also pulls large quantities of prairie lands, pastures, rain forests, and other lands into intensive agricultural production (to replace corn acreage that has been diverted to ethanol production), which means some corn ethanol will likely have higher overall GHG emissions than gasoline or diesel fuels. And fourth, RFS2 could run up against infrastructure barriers. The U.S. EPA estimates that the number of E85 retail facilities may need to expand from approximately 2,000 to between 12,000 and 24,000 nationwide by 2022 if most of the required 36 billion gallons of biofuels are sold as ethanol and the blend limit is not raised. The number of flexible-fueled vehicles on the road capable of using E85 would also need to expand dramatically.

A broader concern is the environmental and social sustainability of biofuels. Unlike the biofuel program in the United States, the European renewable energy mandates are met in large part through imports. In the United Kingdom as of December 2008, 97 percent of the renewable fuels were imports—biodiesel made from American soy, rapeseed from Germany, and palm oil from Malaysia and Indonesia; and ethanol made from Brazilian sugarcane. In the European Union, most of the biofuel imports are ethanol from Brazil and palm oil from Malaysia and Indonesia.[4] Scientists and environmental groups have raised concerns about the local environmental and social impacts of these imported fuels. As a result, the Netherlands, the United Kingdom, and the European Union are adopting sustainability standards for biofuels. These sustainability standards typically address issues of biodiversity, and soil, air, and water quality, as well as social and economic conditions of local communities and workers. They require reporting and documentation but lack real enforcement. The effectiveness of these standards remains uncertain. More science-based research and technical analysis are needed to better quantify the direct effects that the sustainability standards intend to address, as well as the indirect effects and cumulative environmental damages at large scales and over long periods of time that these sustainability standards and certification schemes are ill-equipped to tackle.

Carbon taxes or cap-and-trade

Many argue that a carbon tax or cap-and-trade program would improve the RFS. Economists argue that carbon taxes—taxes on energy sources that emit carbon dioxide—would be a more economically efficient way to introduce low-carbon alternative fuels. Former Federal Reserve chairman Alan Greenspan, car companies, and economists on the left and the right all have supported carbon and fuel taxes as the principal cure for both oil insecurity and climate change. But carbon taxes have shortcomings. Not only do they attract political opposition and public ire, but they are also of limited effectiveness and work better in some situations than others.

For example, even a modest carbon tax works well to reduce carbon from electricity generation. Electricity suppliers can choose among a wide variety of commercially available low-carbon energy sources, including nuclear power, wind, natural gas, and even coal with carbon capture and sequestration. A tax of as little as $25 per ton of carbon dioxide would increase the retail price of electricity made from coal by about 17 percent (in the United States), which would be enough to motivate electricity producers to seek lower-carbon alternatives. The result would be innovation, change, and decarbonization. Politically plausible carbon taxes promise to be effective in transforming the electricity industry.[5]

But transportation is a different story. A $50-a-ton tax, which would raise gasoline prices about 45 cents per gallon (well above what U.S. politicians have been considering), would motivate very little response from consumers or producers, judging by European experience. (Many European countries have had transport fuel taxes equivalent to $4 per gallon for many years, with virtually no effect in decarbonizing fuels—although the taxes are not based on carbon content.) Oil producers wouldn't respond because they've become almost completely dependent on petroleum to supply transportation fuels and can't easily find or develop low-carbon alternatives within a short time frame. Equally important, a transition away from oil depends on automakers and drivers changing their behavior—and they also would be unmotivated by a carbon tax. A tax of $50 a ton (45 cents per gallon) would barely reduce gas consumption, let alone induce drivers to switch to low-carbon alternative fuels when virtually none are available. As a result, oil industries would simply pay taxes and pass the costs to consumers instead of adopting low-carbon fuels.

Carbon cap-and-trade programs suffer the same shortcomings as carbon taxes. This type of policy as usually conceived involves placing a cap on the carbon dioxide emissions of large industrial sources and granting or selling emission allowances to individual companies for use in meeting their caps. Emission allowances, once awarded, could be bought and sold. In the transportation sector, a cap would be placed on oil refineries' emissions, requiring them to reduce carbon dioxide emissions associated with the fuels they produced. The refineries would be able to trade credits among themselves and with others. As the cap was tightened over time, pressure would build to improve the efficiency of refineries and introduce low-carbon fuels. Refiners would likely increase the prices of gasoline and diesel to subsidize low-carbon fuels—creating a market signal for consumers to drive less and for producers of cars to make them more energy efficient. But if the cap were not very stringent, this signal would likely be relatively weak for the transportation sector.

Carbon taxes and/or cap-and-trade should be central to any regional or national initiative to reduce GHG emissions. It is conceivable that in the long run when advanced biofuels and electric and hydrogen vehicles are commercially viable and overcome the infrastructure hurdle, cap-and-trade and carbon taxes will become effective policies within the transportation sector. But until then, more direct forcing mechanisms, such as a low-carbon fuel standard for refiners, will likely be far more effective at stimulating innovation and overcoming the many barriers to change.

Emergence of a GHG Performance Standard for Fuels

The ad hoc approach of the past and current limited mandates and proposals needs to be replaced by durable policies that do not depend on the government's picking winners. A new approach is needed that would ideally be fuel-neutral and performance-based and that would harness market forces. Such an approach has emerged in Europe and the United States. It is farthest along in California, where the Low-Carbon Fuel Standard (LCFS) is a performance-based standard that measures CO_2-equivalent grams per unit of fuel energy. An important feature of the LCFS is that the performance standard applies to *all* fuels, including not just biofuels but also petroleum-based gasoline and diesel, electricity, hydrogen, and other potential fuels that are likely to play a role in the transportation sector in the future.

The LCFS is the first major public initiative to codify life-cycle concepts into law, an innovation that will become more widespread as climate policies are pursued more aggressively.

The point of regulation can occur anywhere along the energy chain, from the individual user all the way upstream to the fuel producers. To ease administration, it is best placed as far upstream as practical—meaning on oil refiners and importers, and fuel producers. An important feature of the LCFS is the ability to buy and sell credits, which will help reduce the cost of achieving the reductions. A tradable credit market will give companies a strong incentive to invest in new and better ways to produce lower-carbon fuels. An oil refiner could, for instance, buy credits (or the fuels themselves) from biofuel producers or from an electric utility that sells power to electric vehicles. Those companies that are most innovative and best able to produce low-cost, low-carbon alternative fuels will thrive, and overall emissions will be lowered at less cost for everyone.

The concepts underlying the LCFS are not unique, but the intellectual and programmatic antecedents of the LCFS are remarkably sparse. The intellectual origin of the LCFS might be Jonathan Rubin's 1993 PhD dissertation at the University of California, Davis, evaluating the use of tradable credits and emission performance standards in transitioning to alternative transportation fuels. Surprisingly, the scholarly literature is otherwise largely quiet on the concept of carbon standards for fuels. John DeCicco and Jason Mark suggested it in various publications in the 1990s, but not until Bob Epstein, a former Silicon Valley entrepreneur, began promoting the concept in 2005 did it gain prominent attention. He and others, especially Roland Hwang of the Natural Resources Defense Council, an advocacy group, pitched the concept to California governor Arnold Schwarzenegger in the autumn of 2006. In January 2007, Governor Schwarzenegger directed the California Air Resources Board (CARB) to develop and implement a low-carbon fuel standard to spur technological innovation and investment in alternative fuels. CARB adopted the LCFS in concept in June 2007 and began a rulemaking process, with the final rule adopted in April 2009; this rule took effect in January 2010.

The European Union unveiled a similar proposal just two weeks after Governor Schwarzenegger did, and in December 2008 its Parliament adopted an amended Fuel Quality Directive (FQD)[6] that is very similar to the California LCFS—with E.U. leaders publicly indicating it was their intent to closely imitate the California standard. In January 2009, 11 northeastern and mid-Atlantic states signed a letter committing to cooperate in developing a regional LCFS.

Compared to biofuel mandates, an LCFS has three key advantages: it inspires industry to pursue innovation aggressively, it is flexible and performance-based so that industry (not government) picks the winners, and it directly targets actual life-cycle GHG emissions associated with the production, distribution, and use of the fuel from the source to the vehicle. An LCFS is a more robust and ultimately more efficient approach than volumetric mandates. Unlike the RFS and other biofuel programs, an LCFS will encourage oil companies to pursue a fuller set of low-carbon fuel options. It will encourage companies to integrate their R&D portfolios across all energy options, including wind, solar, hydrogen, and natural gas, along with carbon capture and sequestration technologies.

On the other hand, an LCFS faces the same concerns about infrastructure barriers and biofuels sustainability that RFS2 and other biofuels mandates face. And some economists characterize the LCFS approach as second best because it is not as efficient as a carbon tax or cap-and-trade,[7] but given the huge barriers to alternative fuels and the limited impact of increased taxes and prices on transportation fuel demand, an LCFS appears to be the most practical way to begin the transition to alternative fuels. Those more concerned with energy security than with climate change might also be skeptical of the LCFS approach, fearing that it might disadvantage high-carbon alternatives such as tar sands and coal liquids. That concern is valid, but disadvantaging does not mean banning. Tar sands and coal liquids could still be introduced on a large scale with an LCFS. The LCFS would require producers of high-carbon alternatives to be more energy efficient and to reduce carbon emissions associated with production and refining. Producers could do so by using low-carbon energy sources for processing energy and could capture and sequester carbon emissions. They could also opt for ways of converting tar sands and coal resources into fuels that facilitate carbon capture and sequestration. For instance, gasifying coal to produce hydrogen allows for the capture of almost all the carbon, since none remains in the fuel itself. In this way, coal could be a nearly zero-carbon option.

HOW TO HANDLE UNCERTAINTY AROUND INDIRECT LAND-USE EFFECTS?

One of the key features of the LCFS approach is that its GHG reduction target takes into account all emissions generated during a fuel's life cycle. This means it takes into account even the emissions generated by indirect land-use changes. But it turns out that this is perhaps the most controversial and challenging issue facing the life-cycle accounting approach adopted by the LCFS and RFS2. The problem is that scientific studies have not yet adequately quantified the indirect land-use effects of increased biofuel production. (You can read more about this in Chapter 12.) So how do regulators add in the emissions from indirect land-use effects when they are measuring the life-cycle GHG emissions of a biofuel? It is a classic challenge: how to handle scientific uncertainty in a policy context.

The prudent approach for regulators is to use available science to assign a conservative value to indirect land-use effects and then to provide a mechanism to update these assigned values as the science improves. Meanwhile, producers should focus on biofuels with low GHG emissions and minimal indirect land-use effects—fuels created from wastes and residues and from biomass grown on degraded or marginal land or with very high yields per unit of land (for example, grasses, some tree species, and algae). Those feedstock materials, instead of intensively farmed food crops like corn, should be the heart of a future biofuel industry—and they will be if producers have to meet a low-carbon fuel standard.

A Closer Look at the LCFS

California's LCFS requires a 10-percent reduction in the greenhouse gas intensity of transport fuels by 2020. The LCFS metric is total carbon and other greenhouse gases emitted per unit of fuel energy. The standard captures all GHGs emitted in the life cycle, from extraction, cultivation, land-use conversion, processing, transport and distribution, and fuel use. The LCFS is imposed on all transport fuel providers, including refiners, blenders, producers, and importers. Aviation and certain maritime fuels are excluded, either because the state does not have authority over them or because including them presents logistical challenges.

To implement the LCFS, each fuel supplier must meet a GHG intensity standard that declines each year, reaching a 10-percent reduction from the baseline year of 2010 by 2020. To maximize flexibility and innovation throughout the energy sector, the LCFS allows for the trading and banking of emission credits. The combination of regulatory and market mechanisms makes the LCFS more robust and durable than a purely regulatory approach and more acceptable and effective than a pure market approach. Companies failing to meet the standard could face monetary penalties and/or legal action via CARB.

There are several ways that regulated parties can comply with the LCFS. Refiners can blend low-GHG fuels, such as biofuels made from cellulose or wastes, into gasoline and diesel. Or they can buy low-GHG fuels such as natural gas, biofuels, electricity, and hydrogen. They can also buy credits from other refiners or use banked credits from previous years. In the EU, producers can also earn credit by improving energy efficiency at oil refineries or by reducing upstream CO_2 emissions from petroleum and natural gas production.

The European Union's FQD requires fuel suppliers to reduce life-cycle GHG emissions by up to 10 percent from the 2010 baseline by 2020. The 10-percent reduction is broader than that mandated by the California LCFS in that it allows credit for upstream reductions in gas flaring and venting and for the use of carbon capture and storage (CCS) technologies. It also allows the purchase of credits under the Clean Development Mechanism (CDM) of the Kyoto Protocol. Upstream emission reductions, CCS, and the CDM can be used to meet up to 4 percent of the 10-percent requirement.

Recent studies suggest that California's LCFS can be met at costs lower than or comparable to oil priced at $60–100 per barrel[8] and that "alternative liquid fuel technology can be deployable and supply a substantial volume of clean fuels for U.S. transportation at a reasonable cost."[9] However, because of market failures, uncertain oil prices, and risk aversion,[10] companies are unlikely to invest in new fuel technologies and infrastructure for alternative fuels. More direct, performance-based policy instruments are needed to overcome carbon lock-in.[11]

A major challenge for the LCFS is avoiding "shuffling," which is similar to leakage but refers specifically to the actions of producers to shift production elsewhere outside of the regulated market. Companies will seek the easiest way of responding to the new requirements, which might involve shuffling production and sales in ways that meet requirements without actually creating a net change in emissions. For instance, a producer in Iowa could divert its low-GHG cellulosic biofuels to California markets and send its high-carbon corn ethanol elsewhere. The same could happen with gasoline made from tar sands and conventional oil. Environmental regulators will need to account for shuffling in their rules. This problem will eventually disappear as more states and nations adopt the same regulatory standards and requirements.

Going National and International with the LCFS Approach

The principle of performance-based standards lends itself to adoption nationally and even internationally. The California program is designed to be compatible with a broader program and in fact will be much more effective if the entire United States as well as other countries also adopt it. Existing volumetric biofuel requirements could be readily converted into an LCFS by converting them to greenhouse gas requirements. In the United States that would not be difficult, since GHG requirements are already imposed on required biofuels. The E.U. biofuel programs could also be converted similarly. Indeed, the evolving carbon and sustainability reporting and certification schemes of the European Union and the U.K. Renewable Transport Fuel Obligation (RTFO) are already gravitating away from a pure volumetric requirement and toward an LCFS.

An important innovation of the California LCFS is its embrace of all transportation fuels. The U.S. RFS2 and E.U. programs, in contrast, include only biofuels, not gaseous fuels or electricity (although biogas is eligible for credits in the European Union, and the December 2008 revisions of the E.U. Fuel Quality Directive envision a future role for electric vehicles). While it is desirable to cast the net as wide as possible, there is no reason why all states and nations must target identical fuels.

Broader LCFS programs are attractive for three reasons. First, it would be easier to include fuels used in international transport modes, especially fuels used in jets and ships. California is excluding these fuels initially because it has only limited jurisdiction over international modes of travel. Second, a broader LCFS would facilitate standardization of measurement protocols. California is currently working with fuel-exporting nations to develop common GHG emissions specifications for their fuels. And third, the broader the pool, the more options are available to regulated entities. More choice means lower overall cost, since there will be a greater chance of finding low-cost options to meet targets.

KEY ELEMENTS OF SUSTAINABILITY STANDARDS FOR FUTURE TRANSPORTATION FUELS

To ensure sustainable development of future transportation fuels, governments—including the Netherlands, the United Kingdom, the European Union, and to some extent the United States with its Renewable Fuel Standard (so called RFS2) program—have begun to impose a variety of sustainability goals and requirements for biofuel production. These sustainability initiatives often include requirements for sustainable management of agricultural production, reduced environmental damage and degradation, and considerations of local community welfare, land rights, and labor welfare. Procedures for certification and verification of sustainability reports, plus requirements to monitor or report progress, are also key elements of sustainability schemes.

California will develop sustainability standards for its low-carbon fuel standard (LCFS). UC researchers have developed a list of recommendations for the implementation of sustainability standards for the LCFS.[12] They assert in their report that a sustainability scheme can be effective only if the proposed framework
- is a multi-stakeholder process,
- is robust but not excessively complicated and acknowledges the limitations of resources, politics, and California's legal jurisdiction,
- sets measurable and verifiable criteria and standards,
- defines methods of enforcement, and
- is consistent with international efforts in sustainability criteria.

Further, the report suggests that government assistance in facilitating information sharing, certification, and capacity building will be crucial for the development of the sustainability criteria. Governments should design incentive mechanisms to encourage the practice of sustainable management and reward practices exceeding minimum standards.

Summary and Conclusions

- The ad hoc policy approach to alternative fuels has largely failed. A more durable and comprehensive approach is needed that encourages innovation and lets industry and consumers pick winners. The LCFS approach does that. It provides a single GHG performance standard for all transport fuel providers and all transport fuels, and it uses credit trading to ensure that the transition is accomplished in a more economically efficient manner.

- Although one might prefer more pure market instruments, such as carbon taxes and cap-and-trade, those instruments are not likely to be effective in the foreseeable future with transport fuels. The envisioned (and politically plausible) price and cap levels would not motivate large investments in electric vehicles, plug-in hybrids, hydrogen fuel cell vehicles, and advanced biofuels. More direct policies, such as an LCFS, are needed to stimulate innovations in low-GHG alternative fuels.

- While an LCFS would be highly effective on its own, to be most effective it must be coupled with other policies—those that address the amount of fuel consumed (since the LCFS is an intensity standard), accelerate the initial provision of infrastructure to supply low-carbon fuels, and assure vehicles are available to use the low-carbon fuels. The LCFS and RFS2 programs are important steps forward. Continued progress will require the concerted efforts of scientists, investors, producers, and elected officials to ensure that wise choices are made in the transition to a different transportation energy future.

Notes

1. This chapter is adapted from and similar to Daniel Sperling and Sonia Yeh, "Low Carbon Fuel Standard," *Issues in Science and Technology* (Winter 2009): 57–66.
2. D. Sperling and D. Gordon, *Two Billion Cars: Driving Toward Sustainability* (New York: Oxford University Press, 2009).
3. D. Tilman, R. Socolow, J. A. Foley, J. Hill, E. Larson, L. Lynd, S. Pacala, J. Reilly, T. Searchinger, C. Somerville, R. Williams, "Beneficial Biofuels—The Food, Energy, and Environment Trilemma," *Science* 325 (2009): 270–71.
4. Organisation for Economic Co-operation and Development (OECD), *Biofuel Support Policies: An Economic Assessment* (Paris, France: OECD, 2008).
5. D. Burtraw (ed.), *Cap and Trade Policy to Achieve Greenhouse Gas Emission Targets* (Newton, MA: Civil Society Institute, 2007); R. Stavins, "Addressing Climate Change with a Comprehensive U.S. Cap-and-Trade System," *Oxford Review of Economic Policy* 24 (2008): 298–321.
6. European Parliament legislative resolution of 17 December 2008 on the proposal for a directive of the European Parliament and of the Council amending Directive 98/70/EC as regards the specification of petrol, diesel and gas-oil and introducing a mechanism to monitor and reduce greenhouse gas emissions from the use of road transport fuels and amending Council Directive 1999/32/EC, as regards the specification of fuel used by inland waterway vessels and repealing Directive 93/12/EEC (COM(2007)0018 – C6-0061/2007 – 2007/0019(COD)).
7. S. Holland, J. Hughes, and C. Knittel, "Greenhouse Gas Reductions under Low Carbon Fuel Standards?" *American Economic Journal: Economic Policy* 1 (2009): 106–46.
8. CARB, *Proposed Regulation to Implement the Low Carbon Fuel Standard*; S. Yeh, N. P. Lutsey, and N. C. Parker, "Assessment of Technologies to Meet a Low Carbon Fuel Standard," *Environmental Science and Technology* 43 (2009): 6907–14.
9. National Academy of Sciences, National Academy of Engineering, and National Research Council, *Liquid Transportation Fuels from Coal and Biomass: Technological Status, Costs, and Environmental Impacts* (Washington, DC: National Academies Press, 2009).
10. D. L. Greene, J. German, and M. A. Delucchi, "Fuel Economy: The Case for Market Failure," in D. Sperling and J. S. Cannon (eds.), *Reducing Climate Impacts in the Transportation Sector* (New York: Springer Science+Business Media, 2009).
11. G. C. Unruh, "Escaping Carbon Lock-in," *Energy Policy* 30 (2002): 317–25.
12. S. Yeh, D. A. Sumner, S. R. Kaffka, J. M. Ogden, and B. M. Jenkins, *Implementing Performance-Based Sustainability Requirements for the Low Carbon Fuel Standard—Key Design Elements and Policy Considerations*, UCD-ITS-RR-09-42 (Institute of Transportation Studies, University of California, Davis, 2009).

Chapter 12:
Key Measurement Uncertainties for Biofuel Policy

Sonia Yeh, Mark A. Delucchi, Alissa Kendall, Julie Witcover, Peter W. Tittmann, and Eric Winford

The previous chapter argued that a policy approach to reducing GHG emissions associated with transportation fuel use should, among other things, take account of all greenhouse gas emissions associated with the production, distribution, and use of a fuel. But as mentioned in that chapter, some areas of scientific uncertainty exist when it comes to quantifying the climate impacts of biofuels. This chapter explores four key measurement uncertainties that create challenges in accounting for such impacts—uncertainties that transportation policies designed to encourage low-carbon fuels should consider addressing. First, instead of treating emissions that occur at different times equally, an accounting of the climate impacts of GHG emissions should consider the effect of emissions over time. Second, there is a need to account for non-GHG global warming factors such as albedo, and the effect of non-Kyoto gases and pollutants such as aerosols and black carbon. Third, more work needs to be done on the question of how to account for indirect land-use effects, which can be large for crop-based feedstocks. And fourth, when forest wastes are used as feedstock for biofuel production, the impacts on forest systems, especially changes in the fire behaviors, forest sinks, soil emissions, and other forest carbon pools should be considered. The last two uncertainties relate to what are often called the leakage and indirect effects that occur when there are dynamic linkages between different carbon pools.

Accounting for GHG Emissions Over Time[1]

When land is cleared in order to grow biofuel crops, carbon that is sequestered in the roots and vegetation below and above ground is released. Although these emissions occur primarily at the outset of land-use change (LUC), current accounting methods typically allocate these emissions evenly over an assumed time horizon (e.g. 20 years).[2] This method underestimates the impact of early emissions and leads to a miscalculation of climate-change effects from LUC emissions. This is due to the fact that the cumulative radiative forcing of GHG emissions, a direct measure of climate warming potential, grows with the time it remains in the atmosphere. The earlier an emission occurs in a product life cycle, the larger its effect at a specific time in the future, unless that time is in the very distant future.

The difference between an earlier and later emission can, and should, be modeled based on the actual climate-change effects of gases. Two methods for doing this are the net present value (NPV) method presented by Delucchi in his lifecycle emissions model (LEM)[3] and the time correction factor (TCF) method proposed by Kendall et al.[4] Both methods aim to address one central question: How do we count the effects/costs of GHG emissions over time and how long do we count them? The two methods offer two distinct approaches, the NPV making an economic valuation of damages and the TCF making an approximation of physical damages over time.

The net present value (NPV) method

Delucchi's net present value (NPV) method for estimating CO_2 emissions from land-use change consists of two steps:
1. Estimate the net present value of the impacts of the actual changes in soil and plant CO_2 emissions, using a time-varying discount rate and accounting for the reversal of the LUC impacts and emissions at the end of the biofuel crop's life cycle.
2. Annualize the NPV—that is, convert it to an annuity—over the assumed life of the crop-to-energy program.

This economic approach is de rigueur in cost-benefit or cost-effectiveness analyses.

Delucchi applied the NPV method in a case study that describes bioenergy crops replacing an originally undisturbed native ecosystem such as a forest or grassland.[5] He laid out four general approaches to estimating LUC emission impacts in grams CO_2 equivalent per KJ of bioenergy (or ton of biomass) produced, depending on how we account for the value of emission impacts as a function of when they occur (with a continuous discounting function or with a discontinuous, threshold time horizon implying zero discounting in the near term and high discounting in the long term), and whether we include emissions that occur after cultivation ends.

CO_2 EMISSIONS FROM PLANTS AND SOILS DUE TO LAND-USE CHANGE

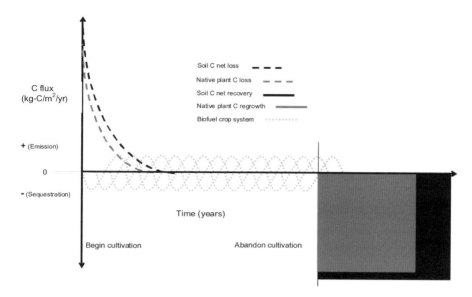

This shows how CO_2 stock and flux change when bioenergy crops replace an originally undisturbed native ecosystem such as a forest or grassland. The start of cultivation of a bioenergy crop creates three streams of CO_2 emission or sequestration: (1) the decay of the original, native ecosystem plant biomass (represented by the dark green dashed line), (2) a change in the CO_2 content of the soil (represented by the brown dashed line), and (3) the growth/harvest cycles of the bioenergy (represented by light green dotted line). Source: M. A. Delucchi, "A Conceptual Framework for Estimating the Climate Impacts of Land-Use Change Due to Energy-Crop Programs," Biomass and Bioenergy 35 (2011): 2337–60.

The total emission impact at the end of cultivation is calculated based on a continuous discounting function to represent the valuation of emissions and impacts over time. The changes in CO_2 fluxes (as shown in the figure above) were converted to changes in CO_2 stocks assuming exponential decay of emission fluxes and atmospheric CO_2 stocks. The change in temperature follows the change in the atmospheric CO_2 stock, but with a time lag of about 50 years (following the FUND model as reported in Warren et al.[6]) that represents the thermal inertia of the system.

With this NPV method, CO_2 emissions impacts from the initial land-use change are at least partially offset by the CO_2 sequestration impacts that occur at the end of the bioenergy program when the land reverts to its original condition. As a result, the method arrives at significantly lower estimates of CO_2-equivalent emissions from land-use change than other models arrive at. Despite offering improvement, the NPV approach also has many uncertainties concerning the treatment of the discount rate (for example, whether the discount rate should be constant or change over time), emission profiles over time (for example, do CO_2 emissions from soil follow an exponential decay pattern, as assumed above), and the lag between changes in concentration and changes in temperature. It nevertheless offers an option to deal with social valuation of CO_2 stock changes as a function of time.

The time correction factor (TCF) method

Another method of accounting for GHG emissions timing, proposed by Kendall et al., is to apply a time correction factor (TCF) that scales the value of an amortized emission to equal the cumulative radiative forcing of the emission at the end of the amortization time horizon. As mentioned earlier, cumulative radiative forcing is a direct measurement of global warming potentials, whereas total cumulative GHG emission is a poor proxy. The cumulative radiative forcing of GHGs is the basis for both global warming potentials (GWPs) and TCF values: the Intergovernmental Panel on Climate Change (IPCC)[7] calculates the relative effects of different gases compared to CO_2 and calls them GWPs; the relative effect of CO_2 emitted at different points in time is captured via TCF.

Applying the TCF increases the relative importance of LUC-derived GHG emissions, which occur predominantly at the outset of the biofuel cultivation life cycle. For example, Searchinger et al. amortized their estimate of LUC emissions for U.S. corn ethanol over a 30-year time horizon[8] and estimated a total life-cycle GHG intensity of 177 g CO_2e/MJ. Applying the TCF to LUC emissions estimates in this case increases the life-cycle GHG estimate by 46 percent (from 177 to 258.6 g CO_2e/MJ) for the 30-year time horizon.

THE TCF APPLIED TO U.S. CORN ETHANOL LIFE-CYCLE EMISSIONS

This table shows how much the life-cycle GHG emissions profile for corn ethanol arrived at by Searchinger et al. changes when LUC emissions estimates are adjusted by the TCF. For the 30-year time horizon, the TCF increases the life-cycle GHG estimates by 46 percent. Source: Adapted from Table 2 in A. Kendall, B. Chang, and B. Sharpe, "Accounting for Time-Dependent Effects in Biofuel Life Cycle Greenhouse Gas Emissions Calculations," Environmental Science and Technology *43 (2009): 7142–47.*

Time Horizon (years)	Time Correction Factor (TCF)	% Increase over Non-TCF Calculation
10	1.730	246%
20	1.778	98%
30	1.785	46%
40	1.775	19%
50	1.769	4%

The TCF has other important applications in life-cycle GHG emissions intensity estimates for biofuels. Many life-cycle analyses of biofuel production omit capital investments required for production, such as factory construction and equipment manufacture, but when included they are straight-line amortized over a time horizon, just like LUC emissions. (For an example of this, see the Energy and Resources Group Biofuel Analysis Meta-Model, also known as EBAMM.[9]) While in many cases the emissions associated with capital investments are small compared to production-related emissions, their importance increases when GHG intensity calculations account for their timing. In addition, as lower-GHG-intensity fuels such as cellulosic ethanol are developed and commercialized, the influence of capital investments on life-cycle GHG intensity is more pronounced.

When EBAMM 1.1 and an average TCF factor of 1.77 are employed,[10] life-cycle GHG intensity estimates increase by slightly more than 1 percent for conventional ethanol (referred to as "ethanol today" in Farrell et al.[11]) and nearly 10 percent for cellulosic ethanol compared to straight-line amortization calculations. This finding suggests that for advanced, lower-carbon fuels, GHG intensity accounting for capital investments and their timing will affect the calculation of the climate change effects of a fuel.

Accounting for Other Non-GHG Climate-Forcing Attributes

Besides ignoring emissions timing, the conventional method of life-cycle analysis of GHG emissions has led to ignoring important climate-forcing effects of other gases and pollutants that are emitted in significant quantities during biofuel life cycles. These include ozone precursors, carbon monoxide (CO), nitrogen oxides (NO_x), sulfur oxides (SO_x), and black carbon (BC). Estimating climate impacts including these non-GHG gases and aerosols will produce comparative assessments that are appreciably different from those that use only the traditional GHGs.[12]

Conventional LCA also does not quantify the climate impact of biofuel-induced changes in biogeophysical characteristics. For example, changes in land use and vegetation as a result of biomass cultivation can change albedo (reflectivity) and evapotranspiration, and these directly affect the absorption and disposition of energy at the surface of the earth and thereby affect local and regional temperatures. Changes in temperature and evapotranspiration can affect the hydrologic cycle, which in turn can affect ecosystems and climate in several ways—for example, via the direct radiative forcing of water vapor, via evapotranspirative cooling, via cloud formation, or via rainfall, affecting the growth of and hence carbon sequestration by plants.

Because of the higher albedo and higher evapotranspiration of many crops, the conversion of mid-latitude (for example, North American) forests and grasslands to agriculture will generally reduce regional temperatures. On the other hand, the biogeophysical effects of a conversion of broadleaf tropical forests to agriculture will lead to a significant warming. In some cases, the climate impacts of changes in albedo and evapotranspiration due to LUC appear to bear an inverse relationship to the climate impacts that result from the associated changes in carbon stocks in soil and biomass due to LUC. For example, Bala et al. find that "the climate effects of CO_2 storage in forests are offset by albedo changes at high latitudes, so that from a climate change mitigation perspective, projects promoting large-scale afforestation projects are likely to be counterproductive in these regions."[13] This suggests that incorporation of these biogeophysical impacts into biofuel LCAs could significantly change estimates of the climate impact of biofuel policies.

Accounting for Indirect Land-Use Effects[14]

Accounting of the GHG effects of biofuels must also consider the indirect, or market-mediated, impacts of biofuel production, which can be large. To understand indirect impacts, we need to realize that when biofuel production from land-using feedstocks increases, prices change in feedstock, energy, and related markets to the extent that land is diverted from growing food crops. Indirect impacts of concern include food security for the poor due to higher food prices and a rebound of increased use of fossil fuels outside the area where the biofuel policy is being implemented due to lower fossil fuel prices caused by decreased fossil fuel consumption inside that policy area. For example, increased biofuel demand in the past 2 years due to U.S. biofuel policies led to lower gasoline consumption in the U.S. and some moderate effect on slowing down global oil price increases. Lower global oil prices can lead to higher demands in other countries, especially fast growing countries such as China.

Indirect land-use change (iLUC) and associated emissions have received the most policy attention thus far. The accounting of emissions associated with iLUC is controversial, since carbon emissions occur outside of the direct biofuel supply chain (including from other domestic agricultural sectors or elsewhere in the world) are counted within the lifecycle emissions of increased biofuel production. This reflects a policy approach counting all direct and indirect emission changes as a result of policy, the so-called consequential LCA. Biofuel-induced price changes cause some food production to be displaced elsewhere, bringing new land that might previously have been pastureland, wetlands, or perhaps even rain forest into agricultural production. When the new land is cleared for production, carbon sequestered in the roots and vegetation below and above ground is released. If rain forests are destroyed or peat is burned, the carbon releases are huge.[15] In the more extreme cases, these land-use shifts can result in each new gallon of biofuel releasing several times as much carbon on a life-cycle basis as the petroleum fuel it is replacing. In the case of corn ethanol, some analyses suggest that under some conditions, indirect land-use changes may increase GHG emissions by 40 percent or more per unit of energy in ethanol compared to the petroleum fuel it is replacing.[16] Cellulosic fuels are expected to have a much smaller effect (mostly because of less direct competition with food-based agricultural production if planted on degraded or marginal land, or derived from waste and residue without disrupting existing production).

Estimates of iLUC emissions associated with specific biofuel feedstocks—"iLUC factors"—have entered the regulatory arena as iLUC regulations have emerged as a way to address the urgent issue of land-use change due to biofuel policy. This consequential LCA GHG emission accounting for biofuels has been adopted in policies such as the California Low Carbon Fuel Standard (LCFS) and the U.S. Renewable Fuel Standard (RFS). Such a policy differs from the more familiar "polluter pays" principle for conventional environmental regulation because the land use emissions can happen far away from where the feedstock is produced. In the absence of systems in effect worldwide to control carbon accounting from land use change of *any* kind (which would eliminate iLUC), a policy that targets iLUC (and other significant indirect emissions) is necessary to address unintended policy consequences.[17]

Modeling systems used to derive regulatory figures have been subjected to scrutiny over their assumptions and readiness—in terms of accuracy and transparency—for a policy role. There are considerable differences in feedstock-specific results from the iLUC models being used by different regulatory bodies, due to the use of different modeling approaches and assumptions, different time frames for policy evaluation, and different methods for allocating effects to specific feedstocks. Results have also changed as the modeling systems themselves have evolved in response to critiques and cross-fertilization.

The piecemeal nature of sensitivity and uncertainty analysis conducted so far by existing iLUC studies means a plausible range of iLUC results has yet to be established. But even with the substantial variation in and uncertainty about results, both short-run and long-run studies find a potentially large impact from iLUC emissions, indicating a need for policy options to mitigate these impacts.

Models used for iLUC analysis

Three main types of models are used for iLUC analysis: economic equilibrium models, causal-descriptive models, and deterministic models. U.S. and California regulations have thus far been based on economic equilibrium models, with each regulatory agency relying on a single modeling system to generate results. Strengths of these types of models include history of policy analysis and theoretical underpinning, but there are drawbacks. Among these are uncertainty about certain model parameters, model transparency, and ease of use (the complicated representation of multi-market adjustments can make it difficult to glean pathways of causation, and the models themselves must be run by those trained in them).

Causal-descriptive and deterministic models stress transparency (making them more amenable to stakeholder input), fewer data requirements, and ease of implementation. By simplifying the characterization of market links, however, they risk missing some market feedbacks that drive iLUC.

TYPES OF MODELS FOR ANALYZING iLUC

Three main types of models are used for iLUC analysis: economic equilibrium models, causal-descriptive models, and deterministic models.

Model Type	Description	Who Uses?	Pros and Cons
Economic equilibrium models (general or partial)	Focus on regional supply and demand for biofuel feedstocks and related agricultural commodities; trade; links to energy market	California LCFS (GTAP model); U.S. Renewable Fuel Standard (RFS2) (FASOM and FAPRI models); European Commission (MIRAGE model)	Pros: History in policy analysis, captures actual economic behavior and linkages. Cons: Many data gaps and uncertainties, false sense of precision, lack of transparency.
Causal-descriptive models	Trace specific market pathways to iLUC change	Under development for UK's Renewable Transport Fuels Obligation (RTFO)	Pros: Transparency. Cons: Can miss complex market feedbacks; relies on historical trends and expert and stakeholder opinion to identify pathways.
Deterministic models	Use externally specified average land-use, trade patterns, land cover	Research institute (Öko-Institut)	Pros: Transparency, ease of implementation. Cons: Can miss complex market linkages and feedbacks; use of averages may not reflect most likely effects; some unsubstantiated assumptions regarding iLUC pathway potential.

Source: Adapted from Table 1 in S. Yeh and J. Witcover, "Indirect Land-Use Change from Biofuels: Recent Developments in Modeling and Policy Landscapes," International Food and Agricultural Trade Policy Council policy brief, 2010.

Some of the critical factors driving model results include the following:
- Yield trends for both crops and livestock—due to technological progress, productivity response to higher prices, and productivity of new areas
- Land competition, or type of land cover displaced by new cropping areas
- Co-products that can substitute for agricultural commodities, easing the need for additional land
- Trading relationships, or whether production flows, via trade, to lowest-cost regions
- Time frame for LUC emissions after clearing, and how to account for the time profile of emissions by methods such as the TCF and NPV, described earlier

Sources of uncertainty in iLUC model analyses

There are many sources of uncertainty in iLUC model analyses. These range from choice of model type, what to include in the model, and level of aggregation, to projections about future developments that provide the without-policy baseline against which the policy effects are measured. Key uncertainties across models of iLUC are as follows:[21]
- Feedstock demand—Fuel yield; co-product markets; price elasticity of demand
- Trade balance—Tariffs and other trade barriers (for example, subsidies); trade impacts of increased biofuel demand (altered trading patterns)
- Area and location of lands converted—Increases in crop yields; productivity of new land; bioenergy-induced additional productivity increase; land-use elasticities; supply of land across different uses; availability of idle, marginal, degraded, abandoned, and underutilized land and unmanaged forest; methodology of allocating converted land (for example, grassland vs. forests)
- GHG emissions from land use and land use change—Biofuel cultivation period; soil and biomass carbon stock data (especially peatlands); soil nitrogen emissions; time accounting of carbon emissions
- Other non-iLUC emissions and climate effects—GHG emissions from agriculture production changes such as cattle, methane emissions from rice cultivation and fertilizer inputs; albedo changes (for example, snow on former boreal or temperate forest land)

Behind the uncertainty and variation lies, in some cases, knowledge gaps due to the difficulty of modeling relationships with no historical track record—the case with various aspects of biofuel markets (market penetration and its dependence on new infrastructure, trade in biofuels or their feedstocks, substitutability between biofuels and petroleum fuels). In other cases, variation results from disagreements over or lack of clear empirical evidence about current patterns (for example, about how yields respond to output price changes, or what determines how and where agricultural land expands). The studies deal with the uncertainty by presenting alternative scenarios and/or undertaking systematic sensitivity analysis across many parameters to create a range of likely results.

POLICY OPTIONS FOR MITIGATING iLUC EMISSIONS

Even though GHG emissions from iLUC cannot be quantified exactly due to the nature of uncertainties in future projections, options for mitigating these emissions are being explored. In addition to the 'iLUC factor' already described, other complementary policy approaches that could be considered include the following:

- **Promoting biofuel feedstocks that avoid or minimize land and resource competition** (for example, agricultural wastes and residues, cellulosic energy crops on marginal land, and forest wastes). Biofuels produced from cellulosic energy crops grown on degraded lands can have lower iLUC effects due to less direct competition for land for food and other agricultural production. This pathway also tends to have better sustainability performance than food crops due to lower intensity of agricultural inputs (fertilizer, irrigation, and pesticides).
- **Improving the overall pool of agricultural resources for food and fuel by investing in higher yields or reducing losses throughout the supply chain.**
- **Linking into existing mechanisms designed to reduce or offset carbon emissions,** such as the Kyoto Protocol's Clean Development Mechanism (CDM) and the UN's Reducing Emissions from Deforestation and Forest Degradation (REDD) program, or generating new certification schemes, perhaps on a regional level (for example, accepting only biofuel feedstock from areas with forest protections in place). Some of these options face the same administrative and enforcement difficulties as other offset programs, including how to establish that the advances would not have taken place in the absence of the mitigation action (additionality) and do not prompt emissions elsewhere (leakage). Leakage could be dealt with most effectively by using an economy-wide carbon market across all potentially affected jurisdictions and sectors, but such policies may take a long time to implement, especially a globally implemented carbon market that reduces international leakage and iLUC emissions.
- **Finding situations where biofuel feedstock production can occur without displacing another land use** through, for example, changes or improvements in production system management. For example, a strengthened linkage between the biofuel and cattle-ranching production systems in Brazil could significantly reduce the risk of indirect land-use changes caused by biofuels.

While commercial development of low-iLUC biofuels lies largely in the future, there are indications the United States could produce large quantities at a reasonable cost given sustained and aggressive efforts to accelerate the development and penetration of low-carbon alternative fuels and technologies. To prevent iLUC and other unintended

> consequences, governments should also adopt enforceable, effective sustainability policies to prevent conversion of ecologically sensitive and high-carbon areas for biofuels or any other purpose; encourage appropriate use of fertilizers and other inputs for biofuels and other crops to reduce harmful environmental impacts from excess run-off; and work to improve access to food by the poor, especially if prices rise. These policies, not specifically aimed at biofuels, target the sweeping economy-wide changes needed to reduce the unwanted "leakage" effects from biofuel (or other) policies aiming to reduce GHG emissions.

Modeling Climate Impacts of Forest Management[22]

One way to reduce GHG emissions from iLUC is to use biofuel feedstocks that avoid or minimize land and resource competition, such as agricultural and forest wastes. However, the proposal to drastically increase the utilization of forest wastes for biofuel production has been met with strong criticisms and doubts about its actual climate benefits and sustainability impacts.

The use of wood biomass from forests has multiple effects on GHG balances. Biomass from forests can be used to produce energy and materials that offset the use of products derived from fossil sources, thereby reducing anthropogenic emissions. Forests sequester carbon through photosynthesis at varying rates influenced by tree age, stand conditions, rainfall, and other factors. Ecological disturbances such as wildfire, severe weather, pests, and disease have the potential to catastrophically alter the carbon dynamics of forests. Some studies suggest that producing biofuel and bioenergy from forest waste products considered to be uneconomical to harvest displaces significant well-to-tank GHG emissions from fossil resources. But comprehensive life-cycle modeling has not yet been done that would enable forest management decisions to be made based on maximizing GHG benefits.

Policies in California intended to increase the rate of sequestration in managed forests have resulted in changes to forestry practices on private lands in California. The Climate Action Reserve (http://www.climateactionreserve.org/) has registered 1.4 million tons of additional GHG sequestration by forests in California resulting from changes in forest practices. In parallel, several energy policy initiatives in California promote renewable energy by requiring more use of renewable sources including biomass produced from forests. These policies, though targeted at the electricity and transportation fuel sectors, will directly impact California's forests, which are already managed for a broad range of environmental, public interest, and market-driven objectives. As such, these new policies challenge the capacity of traditionally disparate research and policy communities to develop analysis and tools that address tightly coupled environmental, climate, and industrial wood and energy production systems.

There is also the fact that forests are valued for a range of public and economic products and services, and managing forests for maximal GHG benefit can have adverse impacts on other forest values. For example, in regions of high growth rate and where an efficient multi-product supply chain is in place, short-return, even-aged management may produce the greatest climate benefit. But silviculture of this type can reduce habitat diversity, alter hydrologic systems, and reduce the scenic and cultural value of forest ecosystems. In other regions, GHG management may be more in harmony with other forest values. Reconciling the range of environmental, ecological, social, and climate values present in forests while significantly increasing sequestration and offset of fossil energy sources through management is a significant policy and political challenge.

OBJECTIVES OF FOREST MANAGEMENT IN CALIFORNIA

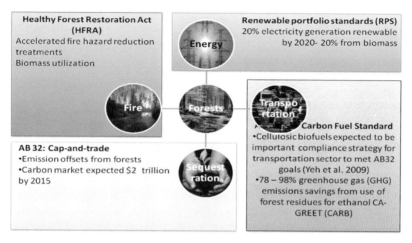

California's forests are already managed for a broad range of environmental, public interest, and market-driven objectives.

Tittmann and Yeh have proposed an integrated accounting framework that encompasses the dynamic interactions between carbon pools taking into account forest management practices, forest fire behavior, and the fate of forest biomass in debris, forest products, and energy production. Using a consistent framework like this for policy planning would maximize the overall benefits of GHG policy and would have a better chance of balancing the trade-offs and maximizing synergies between carbon management and sustainability goals.

A PROPOSED GHG ACCOUNTING FRAMEWORK FOR FOREST MANAGEMENT

Tittmann and Yeh propose this GHG accounting framework. The biofuel/bioenergy GHG balance system illustrates a biofuel production pathway, though a similar (but slightly more complex) flow diagram can also apply to bioenergy production. Because biofuel production affects the forest system, an accounting of the GHG impacts of utilizing forest wastes for biofuel/bioenergy production should also consider the impacts of GHG balance within an integrated forest system, especially changes in the fire behaviors, forest sinks, soil emissions, and other forest carbon pools.

Tittmann and Yeh suggest that in comparison with a no-action alternative, utilizing material from treated stands to offset the use of gasoline and diesel in the transportation sector could result in substantial systemwide GHG reduction. This initial analysis points to the need for more comprehensive statewide and regional modeling of risk-based forest management in order to maximize the net life-cycle carbon balance over the long term. A 2005 study commissioned by the California Energy Commission (CEC) estimated that 11.7 million bone-dry tons (BDT)/y of forest residue are available accounting for technical and administrative constraints and 2.7 million BDT/y could be generated from treating forests determined to be in critical need of Fire Threat reduction. Annual electricity generation from 11.7 million BDT/yr can reach 2,048 MWe and 15 million MWh/yr.[23]

Summary and Conclusions

- Key areas of scientific uncertainty exist about how to quantify the climate impacts of biofuel production. Policy makers need to acknowledge this and to create a robust policy framework that reflects evolving scientific understanding and provides a stable compliance environment while work is done to better understand and quantify these areas of uncertainty.

- More needs to be known about how to account for GHG emissions timing and other factors affecting measurement of GHG impacts. The NPV and TCF methods offer differing approaches. Policy makers may not be in the best position to decide between these approaches. Instead, conducting sensitivity analysis and testing the robustness of the results of different approaches may be the best way to ensure the policy choices are robust given uncertainties.

- Recent work reviewing iLUC modeling has highlighted the data uncertainties, modeling choices, and scenario dependencies inherent in iLUC modeling. These make it more difficult to argue that a single model or scenario of the future has sufficient scientific grounding to generate a single iLUC factor to serve as the basis for a policy decision with large social, economic, and technology implications. One approach to the uncertainty about iLUC emissions would be to establish the *range* of likely emissions consequences based on best scientific information (such as peer-reviewed modelling outcomes published to date) as an input for policymakers, to be updated as new scientific estimates become available.

- Policies should adopt, as much as possible, integrated frameworks for evaluating the GHG benefits of alternative fuels and should consider balancing the trade-offs and maximizing synergies between carbon management and sustainability goals of different policies. In the case of utilizing forest waste for biofuel production, conducting integrated analysis that takes into account the dynamic interactions between carbon pools and sustainability outcomes can maximize the overall benefits of GHG policy and sustainability goals.

Notes

1. This section and the next ("Accounting for Other Non-GHG Climate-Forcing Attributes") are condensed from M. A. Delucchi, *A Lifecycle Emissions Model (LEM): Lifecycle Emissions from Transportation Fuels, Motor Vehicles, Transportation Modes, Electricity Use, Heating and Cooking Fuels, and Materials*, UCD-ITS-RR-03-17 (Institute of Transportation Studies, University of California, Davis, 2003); M. A. Delucchi, *Lifecycle Analysis of Biofuels*, UCD-ITS-RR-06-08 (Institute of Transportation Studies, UC Davis, 2006); M. A. Delucchi, "Impacts of Biofuels on Climate, Land, and Water," *Annals of the New York Academy of Sciences* 1195 (2010): 28–45 (issue *The Year in Ecology and Conservation Biology*, ed. R. S. Ostfeld and W. H. Schlesinger); M. A. Delucchi, "A Conceptual Framework for Estimating Bioenergy-Related Land-Use Change and Its Impacts over Time," *Biomass and Bioenergy*, 35 (2011): 2337-2360; A. Kendall, B. Chang, and B. Sharpe, "Accounting for Time-Dependent Effects in Biofuel Life Cycle Greenhouse Gas Emissions Calculations," *Environmental Science and Technology* 43 (2009): 7142–47.
2. T. Searchinger, R. Heimlich, R. A. Houghton, F. Dong, A. Elobeid, J. Fabiosa, S. Tokgoz, D. Hayes, and T. H Yu, "Use of U.S. Croplands for Biofuels Increases Greenhouse Gases Through Emissions from Land-Use Change," *Science* 319 (2008): 1238. Accounting for emissions from advanced fuels and vehicles has historically been narrowly focused on summing the major GHGs identified by the Intergovernmental Panel on Climate Change (IPCC) in CO_2 equivalents averaged over 100 years.
3. Delucchi, *Lifecycle Emissions Model*.
4. A. Kendall, B. Chang, and B. Sharpe, "Accounting for Time-Dependent Effects in Biofuel Life Cycle Greenhouse Gas Emissions Calculations," *Environmental Science and Technology* 43 (2009): 7142–47.
5. Delucchi, "Conceptual Framework for Estimating Bioenergy-Related Land-Use Change."
6. R. Warren, C. Hope, M. Mastrandrea, R. Tol, N. Adger, and I. Lorenzoni, "Spotlighting Impacts Functions in Integrated Assessment; Research Report Prepared for the Stern Review on the Economics of Climate Change," Working Paper 91 (Tyndall Centre for Climate Change Research, University of East Anglia, United Kingdom, September 2006).
7. Intergovernmental Panel on Climate Change (IPCC), *Climate Change 2007: The Physical Science Basis*, Contribution of Working Group I to the Fourth Assessment Report of the IPCC, ed. by S. Solomon, D. Qin, et al. (Cambridge: Cambridge University Press, 2007).
8. Searchinger et al., "Use of U.S. Croplands."
9. http://rael.berkeley.edu/sites/default/files/EBAMM.

10. EBAMM specifies the time horizon of amortization for farm equipment at 10 years; however, the time horizon for other capital investments is not clear. We use an average TCF of 1.77 (the mean of TCFs calculated for time horizons between 10 and 50 years) and apply it to all amortized emissions for capital equipment in EBAMM.
11. A. E. Farrell, R. J. Plevin, B. T. Turner, A. D. Jones, M. O'Hare, and D. M. Kammen, "Ethanol Can Contribute to Energy and Environmental Goals," *Science* 311 (2006): 506–08.
12. Delucchi, *Lifecycle Emissions Model*; Delucchi, *Lifecycle Analysis of Biofuels*.
13. G. Bala, K. Caldeira, M. Wickett, T. J. Phillips, D. B. Lobell, C. Delire, and A. Mirin, "Combined Climate and Carbon-Cycle Effects of Large-Scale Deforestation," *Proceedings of the National Academy of Sciences* 104: 6550–6555 (2007).
14. This section is based on S. Yeh and J. Witcover, "Indirect Land-Use Change from Biofuels: Recent Developments in Modeling and Policy Landscapes," International Food and Agricultural Trade Policy Council policy brief, 2010; J. Witcover, "Biofuels GHG Emissions and Indirect Land Use Change: Surveying the Model Landscape," presentation to the American Chemical Society annual meeting, San Francisco, CA, March 22, 2010; S. Yeh, D. A. Sumner, S. R. Kaffka, J. M. Ogden, and B. M. Jenkins, *Implementing Performance-Based Sustainability Requirements for the Low Carbon Fuel Standard—Key Design Elements and Policy Considerations*, UCD-ITS-RR-09-42 (Institute of Transportation Studies, UC Davis, 2009); S. Yeh and D. Sperling, "Role of Low Carbon Fuel Standard in Reducing U.S. Transportation Emissions," in *Climate and Transportation Solutions: Findings from the 2009 Asilomar Conference on Transportation and Energy Policy*, ed. D. Sperling and J. Cannon (2010).
15. J. Fargione, J. Hill, D. Tilman, S. Polasky, and P. Hawthorne, "Land Clearing and the Biofuel Carbon Debt," *Science* 319 (2008): 1235–38; H. K. Gibbs, M. Johnston, J. A. Foley, T. Holloway, C. Monfreda, N. Ramankutty, and D. Zaks, "Carbon Payback Times for Crop-Based Biofuel Expansion in the Tropics: The Effects of Changing Yield and Technology," *Environmental Research Letters* 3 (2008): 034001.
16. California Air Resources Board (CARB), *Proposed Regulation to Implement the Low Carbon Fuel Standard, Volume 1* (2009), http://www.arb.ca.gov/fuels/lcfs/030409lcfs_isor_vol1.pdf; U.S. Environmental Protection Agency (EPA), "Regulation of Fuels and Fuel Additives: Changes to Renewable Fuel Standard Program," 40 CFR Part 80 (U.S. Environmental Protection Agency, 2009).
17. For a discussion of the policy justification and challenges surrounding iLUC, see M. Khanna, C. L. Crago, and M. Black, "Can biofuels be a solution to climate change? The implications of land use change-related emissions for policy," *Interface Focus* 1, no. 2 (April 6, 2011): 233-247. doi:10.1098/rsfs.2010.0016.
18. GTAP is a general equilibrium model addressing the short term (until about 2030). See T. W. Hertel, W. E. Tyner, and D. K. Birur, "The Global Impacts of Biofuel Mandates," *Energy Journal* 31 (2010): 75–100; and California Air Resources Board (CARB), *Proposed Regulation to Implement the Low Carbon Fuel Standard, Volume 1* (2009), http://www.arb.ca.gov/fuels/lcfs/030409lcfs_isor_vol1.pdf. Models addressing the long term (until 2100) include EPPA-TEM. See J. M. Melillo, J. M. Reilly, D. W. Kicklighter, A. C. Gurgel, T. W. Cronin, S. Paltsev, B. S. Felzer, X. Wang, A. P. Sokolov, and C. A. Schlosser, "Indirect Emissions from Biofuels: How Important?" Science 326 (2009): 1397–99.
19. FAPRI and FASOM are partial equilibrium models looking at the short term. Regarding FAPRI, see J. Dumortier, D. J. Hayes, M. Carriquiry, F. Dong, X. Du, A. Elobeid, J. F. Fabiosa, and S. Tokgoz, "Sensitivity of Carbon Emission Estimates from Indirect Land-Use Change," Working Paper 09-WP493 (Food and Agricultural Policy Research Institute, 2009); Searchinger et al., "Use of U.S. Croplands"; U.S. Environmental Protection Agency (EPA), Renewable Fuel Standard Program (RFS2) Regulatory Impact Analysis, 2010, http://www.epa.gov/otaq/fuels/renewablefuels/420410006.pdf. Regarding FASOM, see U.S. EPA, *Renewable Fuel Standard Program (RFS2) Regulatory Impact Analysis*. Models addressing the long term include MiniCAM. See M. Wise, K. Calvin, A. Thomson, L. Clarke, B. Bond-Lamberty, R. Sands, S. J. Smith, A. Janetos, and J. Edmonds, "Implications of Limiting CO2 Concentrations for Land Use and Energy," *Science* 324 (2009): 1183.
20. See, for example, Fritsche and Wiegmann (2008) as summarized in G. Fischer, E. Hizsnyik, S. Prieler, M. Shah, and H. van Velthuizen, *Biofuels and Food Security* (IIASA-report, Laxenburg, Austria, 2009); B. Dehue, J. van de Steeij, and J. Chalmers, "Mitigating Indirect Impacts of Biofuel Production: Case Studies and Methodology" (Ecofys, Winrock International, 2009), http://www.renewablefuelsagency.gov.uk/sites/rfa/files/_documents/Avoiding_indirect_land-use_change_-_Ecofys_for_RFA.pdf; E4Tech, "Causal-Descriptive Modelling of the Indirect Land Use Change Impacts of Biofuels: Introduction and Draft Methodology," 2009, http://www.ilucstudy.com/meetings.htm.

21. Adapted from Table 2 in S. Yeh and J. Witcover, "Indirect Land-Use Change from Biofuels: Recent Developments in Modeling and Policy Landscapes," International Food and Agricultural Trade Policy Council policy brief, 2010.
22. This sidebar is based on P. W. Tittmann and S. Yeh, "A Framework for Assessing the Environmental Performance of Forestry in an Era of Carbon Management," *Journal of Sustainable Forestry* (in press, 2010).
23. California Energy Commission, *Biomass Potentials from California Forest and Shrublands Including Fuel Reduction Potentials to Lessen Wildfire Threat* (2005).

Chapter 13:
Beyond Life-Cycle Analysis: Developing a Better Tool for Simulating Policy Impacts

Mark A. Delucchi

As mentioned in this book's introduction and illustrated in various chapters, life-cycle analysis (LCA) is a powerful method for evaluating and comparing fuel/vehicle pathways with respect to a set of sustainability metrics. For more than twenty years, analysts have used LCA to estimate emissions of greenhouse gases (GHGs) from the use of a wide range of transportation fuels. The distinguishing feature of LCA is that it considers all of the activities involved in producing, distributing, and using a product.

However, as commonly employed, LCA cannot accurately represent the impacts of complex systems, such as those involved in making and using biofuels for transportation. LCA generally is linear, static, highly simplified, and tightly circumscribed, and the real world, which LCA attempts to represent, is none of these. In order to better represent the impacts of complex systems such as those surrounding biofuels, we need a different tool, one that has the central features of LCA but not the limitations. If this tool is to be relevant to policy making, it must start with the specification of a policy or action and end with the impacts on environmental systems.

We propose as a successor to LCA a method of analysis that combines integrated assessment modeling, life-cycle analysis, and scenario analysis. We call this method integrated modeling systems and scenario analysis (IMSSA). This chapter describes the key features of IMSSA for transportation fuels. Because IMSSA is meant to be a better model of reality than is conventional LCA, our discussion of IMSSA is a discussion of what an ideal model of reality looks like and how this differs from conventional LCA. We frame our discussion around the climate impact of biofuels because this is a particularly complex problem that nicely illustrates the deficiencies of conventional LCA.

Background and General Critique of LCA

Current LCAs of transportation and climate change can be traced back to "net energy" analyses done in the late 1970s and early 1980s in response to the energy crises of the 1970s, which had motivated a search for alternatives to petroleum. These were relatively straightforward, generic, partial engineering analyses of the amount of energy required to produce and distribute energy feedstocks and finished fuels. Their objective was to compare alternatives to conventional gasoline and diesel fuel according to total life-cycle use of energy, fossil fuels, or petroleum.

In the late 1980s, analysts, policy makers, and the public began to worry that burning coal, oil, and gas would affect the global climate. Interest in alternative transportation fuels, which had subsided with the low oil prices of the mid-1980s, was renewed. Motivated now by global (and also local) environmental concerns, engineers again analyzed alternative transportation life cycles. Unsurprisingly, they adopted the methods of their net-energy engineering predecessors, except that they took the additional step of estimating net carbon dioxide (CO_2) emissions based on the carbon content of fuels.

By the early 1990s, analysts had added two other GHGs, methane (CH_4) and nitrous oxide (N_2O), weighted by their "global warming potential" (GWP), to come up with life-cycle CO_2-equivalent emissions for alternative transportation fuels. Today, most LCAs of transportation and global climate are not appreciably different in general method from those analyses done in the early 1990s.[1] And although different analysts have made different assumptions and used slightly different specific estimation methods, and as a result have come up with different answers, only recently have some analysts begun to question the validity of the general method that has been handed down to them.

In principle, LCAs of transportation and climate are much broader than the net-energy analyses from which they were derived, and hence they have all of the shortcomings of net-energy analyses plus many more. For example, if the original net-energy analyses of the 1970s and 1980s could be criticized for failing to include economic variables on the grounds that any alternative-energy policy would affect prices and hence uses of all major sources of energy, the life-cycle GHG analyses that followed can be criticized on the same grounds but even more deeply because in the case of life-cycle GHG analyses we care about any economic effect anywhere in the world, whereas in the case of net-energy analyses we care about economic effects only insofar as they affect the country of interest. Beyond this, life-cycle GHG analysis in principle encompasses additional areas of data (such as emission factors) and, more importantly, additional large and complex systems (such as the nitrogen cycle, the hydrologic cycle, and global climate), all of which introduce considerable additional uncertainty.

The upshot is that traditional or conventional LCAs of transportation and climate are *not* built on a carefully derived, broad, theoretically solid foundation but rather are an ad-hoc extension of a method—net-energy analysis—that was itself too incomplete and theoretically ungrounded to be valid on its own terms and that could not reasonably be extended to the considerably broader and more complex problem of global climate change. And although recent LCAs of transportation and climate have been made to be consistent with LCA guidelines established by the International Organization for Standardization (ISO),[2] the ISO guidelines have only recently properly addressed a few of the issues discussed here and have not yet developed a proper policy/economic conceptual framework.

The broader LCA community is beginning to recognize this need for a more comprehensive, integrated modeling approach to traditional LCA problems. In this respect, researchers have discussed "system-wide accounting,"[3] "consequential environmental systems analysis,"[4] and "environmental systems analysis using life cycle methodology."[5] At a general conceptual level, all of these approaches, and our own, are a version of the well-established field of integrated assessment modeling (IAM).[6] We are proposing something similar to IAM but with more emphasis on the systems integration and scenario analysis; hence, we suggest the term *integrated modeling systems scenario analysis* (IMSSA).

Comparison of Conventional LCA with an Ideal Model

When we begin to examine the development and application of conventional life-cycle models for transportation we immediately run into a major problem: it is not clear what precise questions the models are supposed to answer. This is a serious flaw, because if we don't know what question a model is meant to answer, we cannot comprehend the answers (outputs) the model provides. In the case of conventional LCAs of transportation and global climate, we are forced often to infer a question from the nature of the outputs and the methods used. What we find, generally, is an unrealistic and irrelevant research question and a limited modeling method.

The weaknesses of conventional LCAs applied to transportation can best be seen by comparing current practice with an ideal model, which would replicate reality. In conventional LCA, a series of production and consumption activities are linked in fixed input-to-output ratios, with emissions per unit of input or output quantified for each activity. The total emissions are added up and expressed per unit of final product or service output. The linkages can be extensive and interrelated, but conventional LCA cannot be made to adequately represent reality simply by multiplying the number of linkages within the same static, circumscribed, linear framework. To see this better, we turn now to an ideal model of reality and compare this with conventional LCA.

HOW CONVENTIONAL LCA IS APPLIED TO TRANSPORTATION

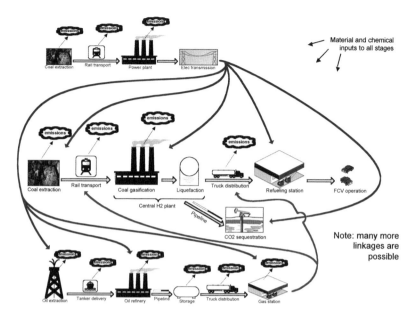

A conventional life-cycle analysis (LCA) links a series of production and consumption activities in fixed input-to-output ratios, with emissions per unit of input or output quantified for each activity. The total emissions are added up and expressed per unit of final product or service output. The linkages can be extensive. Here, for example, the coal life cycle is connected to the electricity life cycle, which is connected to the petroleum life cycle, which is connected back to the coal life cycle.

In principle, LCAs of transportation and climate change are meant to help us understand the impact on global climate of some proposed transportation action. Let us call this a policy/action and refer generally to the impacts of the policy/action on environmental systems. Hence, the ideal model starts with the specification of a policy/action and ends with the impacts on environmental systems. In between are a series of steps that constitute the conceptual components of our model of reality.

HOW CONVENTIONAL LCA COMPARES WITH AN IDEAL MODEL

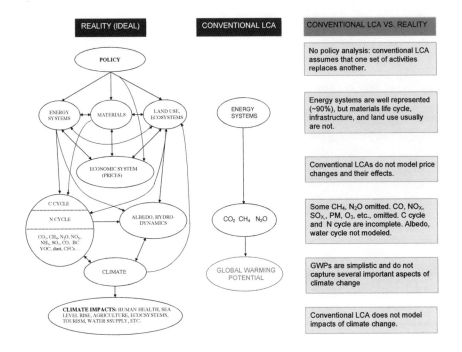

This conceptual flowchart of an ideal model, on the left, shows that it replicates reality as well as possible. Arrows show the relationships between various components. Next to the ideal model is a comparable conceptualization of conventional LCA. Across from each component, on the right side, is a yellow box that discusses whether and how the component is treated in conventional LCA.

In reviewing these components, it is easiest to work backward from the output of interest, the impact on environmental systems. The impact of climate change—the ultimate output of interest—is determined by the dynamic state of the climate system. The climate system is influenced by a wide range of emissions other than the three commonly considered in transportation LCAs (CO_2, CH_4, and N_2O) and by other factors, such as albedo. Emissions and nonemission factors, in turn, are affected by energy systems, material systems, and land use and ecosystems. All of these are affected by, and in some cases in turn affect, policies and economic systems. Indeed, in reality and hence in an ideal model, there are many important feedbacks, especially among energy systems, material systems, land use and ecosystems, economic systems, nonemission factors, and climate systems.

By contrast, conventional LCA generally represents a simplistic, one-way system from energy use to emissions of three GHGs to a simplified measure of climate, the global warming potential (GWP). Some LCAs also include the life cycle of materials, and recently many LCAs have added a simple, partial treatment of land-use change (LUC). Thus, conventional LCA lacks altogether explicit representations of policy, economic systems, and climate impacts, and offers simplified or incomplete treatments of the nitrogen cycle, land use and ecosystems, the climate system, and GHGs other than CO_2, CH_4, and N_2O.[7]

We will now examine in more detail the major deficiencies of conventional LCA compared with an ideal model (our integrated modeling systems and scenario analysis or IMSSA).

LCA Deficiency 1: Inability to Analyze a Specific Policy/Action

Conventional LCAs of transportation and climate change typically do not analyze a specific policy. Indeed, conventional LCAs typically do not even posit a specific question for analysis. The implicit questions of conventional LCA must be inferred from the conclusory statements and the methods of analysis. In transportation, the *conclusory statements* of life-cycle analysis typically are of this sort: "The use of fuel F in light-duty vehicles results in x% greater [or fewer] emissions of CO_2-equivalent GHGs per mile than does the use of gasoline in light-duty vehicles." The *method of analysis* is usually a limited input-output representation of energy use and emissions for a relatively small number of activities linked together to make a life cycle, with no parameters for policies or the function of markets, and no or limited representation of environmental and climate systems.

Given that CO_2-equivalent emissions (which typically are part of the conclusory statements) are equal to emissions of CO_2 plus equivalency-weighted emissions of non-CO_2 gases, where the equivalency weighting usually is done with respect to radiative forcing over a 100-year time period, we can infer that the question being addressed by most conventional LCAs of GHG emissions in transportation is something like this:

> *What would happen to radiative forcing over the next 100 years if we simply replaced the set of activities that we have defined to be the gasoline life cycle with the set of activities that we have defined to be the fuel-F life cycle, with no other changes occurring in the world?*

The problem here is that this question is irrelevant, because we don't care about radiative forcing per se, and because no action that anyone can take in the real world will have the net effect of just replacing the narrowly defined set of gasoline activities with the narrowly defined set of fuel-F activities. Any policy/action that involves fuel F will have complex effects on production and consumption activities throughout the world via global political and economic linkages. These effects *will* occur and a priori cannot be dismissed as insignificant. Because conventional LCAs do not evaluate specific policies but rather evaluate implicit, unrealistic questions, it is difficult if not impossible to relate the results of conventional LCAs to any actual policies/actions in the real world.

The details of the specification of the policy/action are important because different policies will have different climate-change impacts. For example, considering the case of ethanol from corn, a policy to increase (or eliminate) the ethanol subsidy will have a different impact on climate than will a policy to mandate ethanol vehicles, mainly because different policies affect people, prices, and choices differently. In order to analyze the impacts of a particular policy, or indeed of any conceivable policy, one must include all of the variables affected directly by the policy. Many of these are economic variables, which are conspicuously absent from virtually all conventional transportation LCAs.

A related deficiency of conventional LCA is the failure to specify clearly the alternative world with which a specific policy scenario (say, a specific policy regarding ethanol) is being compared. It is conceptually impossible to evaluate a fuel such as ethanol by itself; rather, we must estimate the difference between one course of action involving ethanol and another course of action. These differences between alternative worlds are a function of the initial conditions in each world, the initial perturbations (or changes), and dynamic economic, political, social, and physical/environmental forces. Yet very few transportation life-cycle studies, old or new, have any sort of serviceably modeled alternative world—most likely because such a model requires something like general economic equilibrium analysis and integrated assessment modeling, and most life-cycle analysts are not familiar with these.

LCA Deficiency 2: Failure to Account for Price Effects

All energy and environmental policies affect prices. Changes in prices affect consumption and hence output. Changes in consumption and output change emissions. In the real world, price effects are ubiquitous and often important. They occur in every market affected directly or indirectly by transportation fuels—the markets for agricultural commodities, fertilizer, oil, steel, electricity, and new cars. An ideal model should account for them.

The LCA community is beginning to incorporate economic modeling into LCAs in order to account for price effects. As discussed below, a few LCAs have estimated how changes in biofuel production change the prices of agricultural commodities and thereby change the use of land, which leads to emission or sequestration of carbon. Researchers have also begun to examine some aspects of one of the most important potential price effects: the impact of any nonpetroleum alternative on the price of oil.

Price effects related to oil use

In general, the substitution of any nonpetroleum fuel for gasoline will contract demand for gasoline, which in turn will contract demand for crude oil, which will probably reduce the price of crude oil. This reduction in the world price of oil will stimulate increased consumption of petroleum products, for *all* end uses, *worldwide*. The increased use of petroleum products will increase all of the energy and environmental impacts of petroleum use, including climate change impacts. Hence, the use of nonpetroleum alternative fuels can cause increases in GHG emissions in the petroleum sector via price feedback effects.

Economic theory suggests that the web is even more complex. For example, a large price subsidy, such as corn ethanol enjoys, ultimately causes a deadweight loss of social welfare because output is suppressed below optimal levels by the inefficient use of (tax) resources. This loss of output probably is associated ultimately with lower GHG emissions. Thus, in this case, a subsidy policy may have countervailing effects: on the one hand, there will be an increase in GHG emissions caused by increased use of petroleum due to the lower price of oil due to the substitution of ethanol, but on the other, there will be a decrease in GHG emissions due to the reduction in output caused by the economic deadweight loss from the subsidy. By contrast, a research-and-development policy that succeeds in bringing to market a new low-social-cost fuel will because of the more efficient use of energy resources unambiguously improve social welfare.

Research on price effects related to oil use is relatively recent. Elsewhere I have detailed a formal scheme for incorporating price effects into existing conventional LCA models.[8] Dixon et al. use a dynamic computable-general-equilibrium (CGE) model of the U.S. economy to quantify the economy-wide effects of partial replacement of crude petroleum with biomass and conclude that there is "a noticeable damping effect on world demand for crude petroleum, generating a reduction in its price" (p. 716).[9] Kretschmer and Peterson survey approaches to incorporating biofuels into CGEs.[10] Zhang et al. use a theoretical model to examine the effects on fossil-fuel use of increased use of ethanol as a blend fuel and find that "making higher ethanol fuel blends available for all vehicles potentially has the adverse spillover effect of reducing the demand for flex-fuel vehicles [using 85 percent ethanol]" (p. 3429), thereby increasing the use of fossil fuels.[11] Rajagopal et al. estimate the "indirect fuel use change" (IFUC) effect of biofuel policies on petroleum consumption, where IFUC is the fuel-use analog of indirect land-use change (ILUC). They note that "the adoption of renewable fuels will affect the price of fuel and therefore affect total fuel consumption, which may increase or decrease depending on the policy regime and market conditions" (p. 228).[12] Finally, and most pertinently, Hochman et al. quantify the effects of biofuels on global crude oil markets and find that the introduction of biofuel reduces international fuel prices by between 1.07 and 1.10 percent and increases global fuel consumption by 1.5 to 1.6 percent (p. 112).[13]

Prices in the context of "joint production"

Price effects also are likely to be important in cases of joint production, where one process and one set of inputs inseparably produce more than one marketed output. It is well known that corn-ethanol plants, for example, produce commodities other than ethanol. A policy promoting ethanol therefore is likely to result in more output of these other goods as well as more production of ethanol. What is the impact on climate of the production of the other goods? The only way to answer this question is to model the market for the other goods to see, in the final equilibrium, what changes in consumption and production, mediated by price changes, occur in the world with the ethanol policy. The same issue of joint production also arises in petroleum refineries and in other processes in fuel life cycles. Economic models are needed to analyze these effects.

Other price effects

As mentioned earlier, price effects occur in every market, from the market for steel to the market for new fuels. For example, an economist might argue that price effects might eliminate and even reverse the environmental benefits of electric vehicles (EVs) as estimated in simple life-cycle analyses because if EVs are mandated but are quite costly, car buyers might delay purchase of new, clean, efficient vehicles to the possible detriment of the local and global environment.

Price changes can have a practically infinite number of what are likely to be relatively minor effects. For example, different life cycles use different amounts of steel and hence have different effects on the price and thereby the use of steel in other sectors. The same can be said of any material, or of any process fuel, such as coal used to generate electricity used anywhere in a life cycle. It might be reasonable to presume that in these cases the associated differences in emissions of GHGs are a second-order effect on a second-order process (for example, that the price effect of steel use is no more than 10 percent of the first-order or direct effect of using steel, which itself probably is much less than 10 percent of life-cycle CO_2-equivalent emissions) and hence relatively small. On the other hand, we might be surprised, and sometimes many individually quite small effects add up (rather than cancel each other). For these reasons, it would be ideal for life-cycle analysts to investigate a few classes of these apparently minor price effects.

LCA Deficiency 3: Incomplete Treatment of Land-Use Change (LUC)

Changes in land use and associated changes in climate impacts are another part of the complex web that links bioenergy policies with climate change. As touched on in Chapters 7 and 12, changes in land use can affect climate in several ways:
- by affecting the flows of carbon between the atmosphere and soil and plants
- by affecting climate-relevant physical properties of land, such as its albedo
- by affecting the nitrogen cycle, which in turn can affect climate in several ways—for example, via production of N_2O or by affecting the growth of plants, which in turn affects carbon-CO_2 removal from the atmosphere via photosynthesis
- by affecting the hydrologic cycle, which again affects climate in several ways—for example, via the direct radiative forcing of water vapor, via evapotranspirative cooling, via cloud formation, or via rainfall and thus the growth of and hence carbon sequestration in plants[14]
- by affecting the fluxes of other pollutants that can affect climate, such as CH_4, volatile organic compounds, and aerosols

CO_2 emissions from land-use change

As just indicated, CO_2 emissions from plants and soils due to LUC is just one of several ways that LUC can affect climate, and LUC, in turn, is just one of several consequences of bioenergy policies that can affect climate. However, this does not mean that the climate impact of CO_2 emissions from LUC is small; indeed, several analyses have suggested that CO_2 emissions from LUC could be a large fraction of total CO_2-equivalent GHG emissions from the entire life cycle of biofuels.[15]

Conceptually, an ideal model of the climate impact of changes in carbon emissions due to LUC caused by bioenergy policies would have several components. Emissions of CO_2 from LUC would be estimated based on the difference, over time, between ecosystem carbon content in a "no bioenergy program" baseline case compared with ecosystem carbon content in a "with bioenergy program" case, where "bioenergy program" refers to a specific program and need not encompass all bioenergy in the world. To represent this, one would create an economic/land-use model with dynamic, price-endogenous supply and demand functions, with land supply treated explicitly, and with yields determined as a function of endogenous parameters (such as price) and exogenous parameters (such as government R&D policy). One would run this model once with no bioenergy program to establish a dynamic "no bioenergy program" land-use baseline (that is, one in which prices, yields, supply curves, and land uses change year by year) and then run it again for a "with bioenergy program" case, simulated by an outward shift of demand at time zero and then demand contractions following the end of the program.

One would then compare land uses between the two cases year by year for as long as differences remain between the two cases (stream #1). For each year that there was a difference in land use, one would estimate the change in carbon stocks and emissions (stream #2) and then the change in atmospheric CO_2 (stream #3), the change in radiative forcing and climate (stream #4), and the change in climate impacts (stream #5). One would then track these changes in carbon stocks and climate for every land-use category every year. The *impacts* of climate change in each year would then be expressed in the values of a reference year (stream #6); in any cost-benefit or economic framework, this would be done by discounting the impacts to their present value. The sum of the reference-year values of each stream of the impacts of climate change—associated ultimately with the year-by-year differences in land uses between the "no bioenergy program" and "with bioenergy program" cases—would represent the climate-change impact of CO_2 emissions from LUC resulting from a bioenergy program.

HOW STREAMS IN THE REAL WORLD ARE TREATED IN AN IDEAL MODEL

An ideal model of the climate impact of changes in carbon emissions due to LUC caused by bioenergy policies would have several components. This table shows the hierarchy of streams in the real world that would be represented.

Stream in the Real World	Treatment in an Ideal Model (IMSSA)
1. Program actions. Prices, yields, supply curves, and land uses can change over time, year by year, in the "with bioenergy program" case compared to the "no bioenergy program" case. These changes occur at the end of the program as well as at the beginning.	Socioeconomic model of the relationship between changes in bioenergy production and changes in land use
2. Emissions. Then, each change in land use (in each year) generates its own time series of changes in carbon emissions; for example, a change in land use in any year T initiates a process of carbon emission or sequestration that can continue for many years after T. These emission streams occur at the end of the program as well as at the beginning.	Soil and plant carbon database; explicit representation of duration and shape of soil-carbon and plant-carbon emission streams, including post-program ("reversion") streams
3. Concentration and radiative forcing. Next, each change in carbon emission or sequestration (in each year) generates its own time series of changes in CO_2 concentration (atmospheric carbon stocks) and radiative forcing; for example, an emission of carbon from soils in year T+x (due ultimately to LUC in year T) will generate an atmospheric CO_2 concentration and decay profile and associated radiative-forcing effects that extend for many decades beyond T+x.	Simplified but realistic climate model showing CO_2 decay and radiative forcing
4. Climate (temperature) change. Next, any change in radiative forcing in any year will generate a stream of climate changes, with the lag between radiative forcing (stream #3) and climate change (stream #4) being due mainly to the thermal inertia of the oceans.	Explicit representation of the thermal inertia lag between radiative forcing and climate change
5. Impacts. Finally, any change in climate in any year (stream #4) can impact people and ecosystems for many years (for example, by changing the incidence of chronic diseases).	Comprehensive assessment of damages of climate change in a present-value/annualization framework

Ideally, this modeling would be part of a comprehensive analysis of the climate impacts of bioenergy programs, which would include, in addition to the impacts of CO_2 emissions from LUC just described, two other general kinds of impacts: the climate impacts of LUC *other* than those resulting from CO_2 emissions (for example, changes in albedo) and the climate impacts from the rest of the bioenergy production-and-use chain. The value of all of these other impacts would be added to the value of the impacts of the CO_2 emissions from LUC to produce a comprehensive measure of the climate impact of a bioenergy program.

Note that reality and hence the ideal representation comprise a hierarchy of several separate streams over time: policy/action streams generate LUC streams, which generate soil-carbon and plant-carbon change streams, which generate CO_2-concentration-change streams, which generate climate-change streams, which finally generate climate-impact-change streams. An accurate representation of the climate impacts of a bioenergy program should have an explicit treatment of these streams and a method for making impact streams with different time profiles commensurate.

My 2011 review of the literature[16] shows that while a few recent LCA studies have addressed economic modeling of LUC,[17] the treatment of this component is incomplete, and no published, peer-reviewed LCA study has addressed the other four components properly or at all. Most important, no LCA work apart from my 2011 review has a conceptual framework that properly represents the reversion of land uses at the end of the biofuels program, the actual behavior of emissions and climate over time, and the treatment of future climate-change impacts relative to present impacts.

Biogeophysical impacts of land-use change

Changes in land use and vegetation can change physical parameters, such as albedo (reflectivity) and evapotranspiration rates, that directly affect the absorption and disposition of energy at the surface of the earth and thereby affect local and regional temperatures.[18] Changes in temperature and evapotranspiration can affect the hydrologic cycle,[19] which in turn can affect ecosystems and climate in several ways—for example, via the direct radiative forcing of water vapor, via evapotranspirative cooling, via cloud formation, or via rainfall, affecting the growth and hence carbon sequestration by plants.

In some cases, the climate impacts of changes in albedo and evapotranspiration due to LUC appear to be of the same order of magnitude but of the opposite sign as the climate impacts that result from the associated changes in carbon stocks in soil and biomass due to LUC. For example, Bala et al. find that "the climate effects of CO_2 storage in forests are offset by albedo changes at high latitudes, so that from a climate change mitigation perspective, projects promoting large-scale afforestation projects are likely to be counterproductive in these regions" (p. 6553).[20] This suggests that the incorporation of these biogeophysical impacts into biofuel LCAs could significantly change the estimated climate impact of biofuel policies.

Interactive and feedback effects between climate change, land use, and water use

Climate change can affect water use and land use. For example, changes in precipitation and evapotranspiration (due to climate change) will affect groundwater levels[21] and cropping patterns, which in turn will give rise to other environmental impacts, including feedback effects on climate change. People in less wealthy countries may be most vulnerable to these changes because they have less capacity to mitigate or adapt to impacts on groundwater. These sorts of feedback interrelationships further complicate analyses of the impacts of biofuels on climate change, water use, and land use.

LCA Deficiency 4: Neglect of the Nitrogen Cycle

Anthropogenic inputs of nitrogen to the environment, such as from the use of fertilizer or the combustion of fuels, can disturb aspects of the global nitrogen cycle. These disturbances ultimately have a wide range of environmental impacts, including eutrophication of lakes and coastal regions, fertilization of terrestrial ecosystems, acidification of soils and water bodies, changes in biodiversity, respiratory disease in humans, ozone damages to crops, and changes to global climate.[22] Galloway et al. depict this as a "nitrogen cascade," in which "the same atom of Nr [reactive N, such as in NO_x, NH_3, or NH_4^+] can cause multiple effects in the atmosphere, in terrestrial ecosystems, in freshwater and marine systems, and on human health" (p. 341).[23]

THE NITROGEN CASCADE

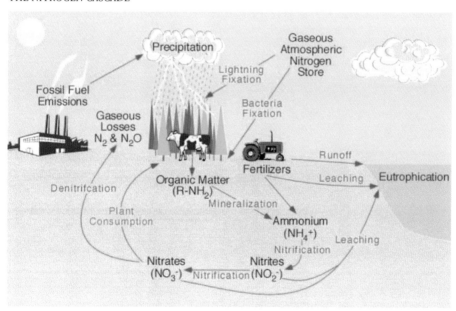

In what has been termed a nitrogen cascade, the same atom of reactive nitrogen can cause multiple effects in the atmosphere, in terrestrial ecosystems, in freshwater and marine systems, and on human health.

Nitrogen emissions to the atmosphere—as nitrogen oxide (NO_x), ammonia (NH_3), ammonium (NH_4^+), or N_2O—can contribute to climate change through a number of complex physical and chemical pathways that affect the concentration of ozone, methane, nitrous oxide, carbon dioxide, and aerosols:

1. NO_X participates in a series of atmospheric chemical reactions involving CO, nonmethane hydrocarbons (NMHCs), H_2O, OH-, O_2, and other species that affect the production of tropospheric ozone, a powerful GHG as well as an urban air pollutant.
2. In the atmospheric chemistry mentioned in (1), NO_X affects the production of the hydroxyl radical, OH, which oxidizes and thereby affects the lifetime of methane, another powerful GHG.
3. In the atmospheric chemistry mentioned in (1), NO_X affects the production of sulfate aerosol, which as an aerosol has, on the one hand, a net *negative* radiative forcing (and thereby a beneficial effect on climate[24]) but on the other hand adversely affects human health.
4. NH_Y (NH_3 or NH_4+) and nitrate from NO_X deposit onto soils and oceans and then eventually re-emit N as N_2O, NO_X, or NH_Y. Nitrate deposition also affects soil emissions of CH_4.
5. NH_Y and nitrate from NO_X fertilize terrestrial and marine ecosystems and thereby stimulate plant growth and sequester carbon in nitrogen-limited ecosystems.
6. NH_Y and nitrate from NO_X form ammonium nitrate, which as an aerosol has, on the one hand, a net *negative* radiative forcing (and thereby a beneficial effect on climate[25]) but on the other hand adversely affects human health.
7. As deposited nitrate, N from NO_X can increase acidity and harm plants and thereby reduce $C\text{-}CO_2$ sequestration.

Even though the development of many kinds of biofuels will lead to large emissions of NO_X, N_2O, and NH_Y, virtually all LCAs of CO_2-equivalent GHG emissions from biofuels ignore all N emissions and the associated climate effects except for the effect of N fertilizer on N_2O emissions. Some preliminary, more comprehensive estimates are provided in work I published in 2003 and 2006.[26] Even in the broader literature on climate change, relatively little analysis of the climate impacts of N emissions has been done, because as Fuglestvedt et al. note, "GWPs for nitrogen oxides (NO_X) are amongst the most challenging and controversial" (p. 324).[27] Shine et al. estimate the global warming impacts of the effect of NO_X on O_3 and CH_4, focusing on regional differences (1 and 2 above), but they merely mention and do not quantify the effect of NO_X on nitrate aerosols (6 above) and do not mention the other impacts.[28] Prinn et al. and Brakkee et al. estimate effects 1 and 2.[29] These studies, along with my preliminary work, suggest that the climate impacts of perturbations to the N cycle by the production and use of biofuels could be comparable to the impacts of LUC.

LCA Deficiency 5: Omission of Climate-Impact Modeling Steps and Climate-Relevant Pollutants

The ultimate objective of LCAs of GHG emissions in transportation is to determine the effect of a particular policy on global climate and the impact of global climate change on quantities of interest (such as human welfare). This requires a number of modeling steps beyond the economic and environmental modeling discussed above. These steps involve estimating relationships between policies and emissions, emissions and concentration, concentration and radiative forcing, radiative forcing and temperature change, and temperature change and climate impacts for all climate-relevant pollutants. Conventional LCAs omit or characterize poorly most of these steps and omit most climate-relevant pollutants.

Conventional LCAs do not estimate the climate-change impacts of emissions of GHGs from transportation fuels but rather use a quantity called the global warming potential (GWP) to convert emissions of CH_4, N_2O, and CO_2 into a common index of temperature change. GWPs tell us the grams of CH_4 or N_2O that produce the same integrated radiative forcing, over a specified period of time, as one gram of CO_2, given a single pulse of emissions of each gas.[30] Typically, analysts use GWPs for a 100-year time horizon.

There are several problems with this method.[31] First, we care about the impacts of climate change, not about radiative forcing per se, and changes in radiative forcing are not simply (linearly) correlated with changes in climate impacts. Second, the method for calculating the GWPs involves several unrealistic simplifying assumptions, which can be avoided relatively easily in a more realistic, comprehensive CO_2-equivalency factor (CEF). Third, by integrating radiative forcing from the present day to 100 years hence, the GWPs in effect give a weight of 1.0 to every year between now and 100 and a weight of 0.0 to every year beyond 100, which certainly does not reflect how society makes trade-offs over time (a more realistic treatment would use continuous discounting[32]). Fourth, the conventional method omits several gases and aerosols that are emitted in significant quantities from biofuel life cycles and can have a significant impact on climate, such as ozone precursors—(for example, volatile organic compounds (VOCs), carbon monoxide and nitrogen oxides), ammonia, sulfur oxides, black carbon, and other aerosols.

A better approach is to use CEFs that equilibrate the present-dollar value of the impacts of climate change from a unit emission of gas X with the present-dollar value of the impacts of climate change from a unit emission of CO_2. Ideally, these present-value CEFs would be derived from runs of climate-change models for generic but explicitly delineated policy scenarios.

Toward a More Comprehensive Model: IMSSA

Thus far this chapter has identified major deficiencies in the development and application of conventional LCAs of transportation and climate. This concluding section briefly synthesizes the findings and delineates a more comprehensive and accurate model. Such a model can be built from scratch or developed by expanding an existing LCA model or IAM. At ITS-Davis we are currently exploring all of these options.

If we want the results of analysis of the climate-change impacts of transportation policies to be interpretable and relevant, our models must be designed to address clear and realistic questions. In the case of LCA comparing the energy and environmental impacts of different transportation fuels and vehicles, the questions must be of the sort "What would happen to [some measure of energy-use or environmental impacts] if somebody did X instead of Y?" where X and Y are specific and realistic alternative courses of action. These alternative courses of action may be related to public policies or to private-sector market decisions, or to both. Then the model must be able to properly trace out all of the differences—political, economic, technological, environmental—between the world with X and the world with Y. So rather than ask what would happen (to some marginally relevant metric such as radiative forcing) if we replaced one very narrowly defined set of activities with another and then use a technology life-cycle model to answer this (misplaced) question, we instead should ask what would happen in the world were we to take one realistic course of action rather than another, and then we should use an integrated economic, environmental, and engineering model—IMSSA—to answer the question.

Given the tremendous uncertainty in data, methods, and model scope and structure, IMSSA emphasizes *scenario analysis* rather than simple point estimates. IMSSA results thus would be described with nuanced statements of this sort: "Under conditions A, B, and C, the distribution of climate-impact damages for policy option 1 tends to be shifted toward lower values than the distribution for policy option 2, but option 1 also tends to result in fewer vehicle miles of travel and lower GNP."

SUMMARY OF THE DIFFERENCES BETWEEN CONVENTIONAL LCA AND AN IDEAL MODEL (IMSSA)

	Ideal Modeling Approach (IMSSA)	**Conventional LCA Approach**
Aim of the analysis	Evaluate impacts (worldwide if necessary) of one realistic action compared with another	Evaluate impacts of replacing one limited set of "engineering" activities with another
Scope of the analysis	All energy, materials, and economic, social, technological, ecological, and climate systems, globally	Narrowly defined chain of energy and material production and use activities
Method of analysis	Dynamic, nonlinear, interrelated, feedback-modulated representations of all relevant systems	Simplified, static, often linear energy-and-materials-in/emissions-out representation of technology
What is evaluated	Ideally, physical and economic impacts of direct interest to society (for example, damages from climate change)	Emissions aggregated by some relatively simple weighting factors (for example, global warming potentials, ozone-forming potential)
How results are expressed	Distribution of results for a range of scenarios	Point estimates

As mentioned above, I have framed the discussion of IMSSA around the climate impact of biofuels because this is a particularly complex problem that nicely illustrates the deficiencies of conventional LCA. But might conventional LCA be acceptable for much less complex problems? In general, the more an energy alternative perturbs technological, economic, and environmental systems, the less suitable is conventional LCA. This suggests that, in principle, conventional LCA might be almost as accurate as IMSSA in estimating the impacts of alternatives that do not appreciably affect technological, economic, and environmental systems. The problem, however, is that often it is difficult to identify low-perturbation alternatives without using relatively complex models to determine the impacts. This difficulty is compounded by our experience that the harder we look, the more impacts we find, for any system. Even alternatives that at first glance seem to have very small impacts (e.g., wind, water, and solar power) can, upon further inspection, turn out to have potentially nontrivial impacts not covered by conventional LCA. For example, the deployment of wind turbines over the ocean may cause local surface cooling due to enhanced heat latent flux driven by an increase in turbulent mixing caused by the turbines.[33] Large-scale photovoltaic arrays in deserts can alter surface albedo, affecting local temperature and wind patterns, with the sign of the temperature effect depending on the efficiency of the photovoltaic system relative to the background albedo (very efficient PV systems will cause local cooling).[34]

Nevertheless, resources for research are limited, and we cannot research everything forever. Ideally we want to concentrate our efforts on problems that are important, uncertain, and tractable. (If a problem is unimportant, or well understood, or intractable, it is not worth a great deal of attention. Thus, it is beside the point to argue that conventional LCA might be suitable for analyzing the impacts of policies that are intended to make only inconsequential changes in energy use, because there is no need to analyze such policies in the first place.) Given this, the most sensible approach is to evaluate periodically the state of our knowledge so that we can continue to target important, uncertain, and tractable problems. Unfortunately, at the beginning of this process, we need fairly comprehensive tools in order to do any kind of screening at all. Thus, we should develop at least rudimentary IMSSA as quickly as possible in order to guide the evolution of our analyses.

Summary and Conclusions

- As commonly employed, life-cycle analysis (LCA) cannot accurately represent the impacts of complex systems, such as those involved in making and using biofuels for transportation. LCA generally is linear, static, highly simplified, and tightly circumscribed, and the real world, which LCA attempts to represent, is none of these.

- Among LCA's major deficiencies are its inability to analyze a specific policy or action, its failure to account for price effects, its incomplete treatment of land-use change, its neglect of the nitrogen cycle, and its omission of climate-impact modeling steps and climate-relevant pollutants.

- In order to better represent the impacts of complex systems such as those surrounding biofuels, we need a different tool, one that has the central features of LCA but not the limitations. We propose as a successor to LCA a method of analysis that combines integrated assessment modeling, life-cycle analysis, and scenario analysis. We call this method integrated modeling systems and scenario analysis (IMSSA).

- IMSSA uses dynamic, nonlinear, feedback-modulated representations of energy, economic, ecological, and technological systems in order to estimate the physical and economic impacts of particular policies or actions. IMSSA can be built from scratch or developed by expanding an existing LCA model or IAM. We are currently exploring all of these options.

Notes

1. For a review of early transportation LCAs, see M. A. DeLuchi, *Emissions of Greenhouse Gases from the Use of Transportation Fuels and Electricity, Volume 1: Summary*, ANL/ESD/TM-22 (Center for Transportation Research, Argonne National Laboratory, November 1991), http://www.its.ucdavis.edu/publications/1991/UCD-ITS-RP-91-30.pdf.
2. See the ISO website, http://www.iso.ch/iso/en/iso9000-14000/iso14000index.html.

3. H. Feng, O. D. Rubin, and B. A. Babcock, *Greenhouse Gas Impacts of Ethanol from Iowa Corn: Life Cycle Analysis versus System-wide Accounting*, Working Paper 08-WP 461 (Center for Agricultural and Rural Development, Iowa State University, 2008), http://www.card.iastate.edu/publications/DBS/PDFFiles/08wp461.pdf.
4. M. Pehnt, M. Oeser, and D. J. Swider, "Consequential Environmental System Analysis of Expected Offshore Wind Electricity Production in Germany," *Energy* 33 (2008): 747–59.
5. G. R. Finnveden, M. Z. Hauschild, T. Ekvall, J. Guinée, R. Heijungs, S. Hellweg, A. Koehler, D. Pennington and S. Suh, "Recent Developments in Life Cycle Assessment," *Journal of Environmental Management* 91(2009): 1–21.
6. For example, E. A. Parson and K. Fisher-Vanden, "Integrated Assessment Models of Global Climate Change," *Annual Review of Energy and the Environment* 22 (1997): 589–628. See also J. B. Guinée, R. Heijungs, G. Huppes, A. Zamagni, P. Masoni, R. Buonamici, T. Ekvall, and T. Rydberg, "Life Cycle Assessment: Past, Present, and Future," *Environmental Science and Technology* 45 (2011): 90–96; and B. Weidema and T. Ekvall, *Guidelines for Application of Deepened and Broadened LCA*, CALCAS D18 (Co-ordination Action for Innovation in Life-Cycle Analysis for Sustainability, July 2009), http://www.leidenuniv.nl/cml/ssp/publications/calcas_report_d18.pdf.
7. This applies to methods that use economic input-output (I-O) analysis, such as hybrid IO-LCA methods. (See M. Lenzen, "A Guide for Compiling Inventories in Hybrid Life-Cycle Assessments: Some Australian Results," *Journal of Cleaner Production* 10: 545-572 (2002).) Hybrid IO-LCA merely expands the "energy systems" component of conventional LCA. (See figure comparing conventional LCA with an ideal model.)
8. M. A. Delucchi, *Incorporating the Effect of Price Changes on CO_2-Equivalent Emissions from Alternative-Fuel Lifecycles: Scoping the Issues*, for Oak Ridge National Laboratory, UCD-ITS-RR-05-19 (Institute of Transportation Studies, University of California, Davis, 2005), http://www.its.ucdavis/people/faculty/delucchi.
9. P. B. Dixon, S. Osborne, and M. T. Rimmer, "The Economy-Wide Effects in the United States of Replacing Crude Petroleum with Biomass," *Energy and Environment* 18 (2007): 709–22.
10. B. Kretschmer and S. Peterson, "Integrating Bioenergy into Computable General Equilibrium Models—A Survey," *Energy Economics* 32 (2010): 673–86.
11. Z. Zhang, C. Qiu, and M. Wetzstein, "Blend-Wall Economics: Relaxing U.S. Ethanol Regulations Can Lead to Increased Use of Fossil Fuels," *Energy Policy* 38 (2010): 3426–30.
12. D. Rajagopal, G. Hochman, and D. Zilberman, "Indirect Fuel Use Change (IFUC) and the Lifecycle Environmental Impact of Biofuel Policies," *Energy Policy* 39 (2011): 228–33.
13. G. Hochman, D. Rajagopal, and D. Zilberman, "The Effect of Biofuels on Crude Oil Markets," *AgBioForum* 13 (2010): 112–18.
14. G. Bala, K. Caldeira, M. Wickett, T. J. Phillips, D. B. Lobell, C. Delire, and A. Mirin, "Combined Climate and Carbon-Cycle Effects of Large-Scale Deforestation," *Proceedings of the National Academy of Sciences* 104 (2007): 6550–55; G. Marland et al., "The Climatic Impacts of Land Surface Change and Carbon Management, and the Implications for Climate-Change Mitigation Policy," *Climate Policy* 3 (2003): 149–57; R. A. Pielke, "Land Use and Climate Change," *Science* 310 (2005): 1625–26.
15. T. Searchinger, R. Heimlich, R. A. Houghton, F. Dong, A. Elobeid, J. Fabiosa, S. Tokgoz, D. Hayes, and T.-H. Yu, "Use of U.S. Croplands for Biofuels Increases Greenhouse Gases Through Emissions from Land Use Change," *Science* 319 (2008): 1238–40; R. J. Plevin, M. O'Hare, A. D. Jones, M. S. Torn, and H. K. Gibbs, "Greenhouse Gas Emissions from Biofuels' Indirect Land-Use Change Are Uncertain but May Be Much Greater than Previously Estimated," *Environmental Science and Technology* 44 (2010): 8015–21; T. W. Hertel, A. A. Golub, A. D. Jones, M. O'Hare, R. J. Plevin, and D. M. Kammen, "Effects of U.S. Maize Ethanol on Global Land Use and Greenhouse Gas Emissions: Estimating Market-Mediated Responses," *BioScience* 60 (2010): 223–31.
16. M. A. Delucchi, "A Conceptual Framework for Estimating the Climate Impacts of Land-Use Change Due to Energy-Crop Programs," *Biomass and Bioenergy* 35 (2011): 2337–60.
17. For example, Searchinger et al., "Use of U.S. Croplands for Biofuels"; Hertel et al., "Effects of U.S. Maize Ethanol"; T. W. Hertel, W. E. Tyner, and D. K. Birur, "The Global Impacts of Biofuel Mandates," *The Energy Journal* 31 (2010): 75–100.
18. Bala et al., "Combined Climate and Carbon-Cycle Effects"; J. J. Feddema, K. W. Oleson, G. B. Bonan, L. O. Mearns, L. W. Buja, G. A. Meehl, and W. M. Washington, "The Importance of Land-Cover Change in Simulating Future Climates," *Science* 310 (2005): 1674–78; C. R. Pyke and S. J. Andelman, "Land Use and Land Cover Tools for Climate Adaptation," *Climatic Change* 80 (2007): 239–51; M. Notaro, S. Vavrus, and Z. Liu, "Global Vegetation and Climate Change Due to Future Increases in CO_2 as Projected by a Fully Coupled Model with Dynamic Vegetation," *Journal of Climate* 20 (2007): 70–90; Marland et al., "Climatic Impacts of Land Surface Change"; D. B. Lobell, G. Bala, and P. B. Duffy, "Biogeophysical Impacts of Cropland Management Changes on Climate," *Geophysical Research Letters* 33 (2006), L06708.

19. M. Georgescu, D. B. Lobell, and C. B. Field, "Potential Impact of U.S. Biofuels on Regional Climate," *Geophysical Research Letters* 36 (2009), L21806.
20. Bala et al., "Combined Climate and Carbon-Cycle Effects."
21. C. I. Bovolo, G. Parkin, and M. Sophocleous, "Groundwater Resources, Climate, and Vulnerability," *Environmental Research Letters* 4 (2009), 035001.
22. J. N. Galloway, J. D. Aber, J. W. Erisman, S. P. Seitzinger, R. W. Howarth, E. B. Cowling, and B. J. Cosby, "The Nitrogen Cascade," *BioScience* 53 (2003): 341–56; A. R. Mosier, M. A. Bleken, P. Chaiwanakupt, E. C. Ellis, J. R. Freney, R. B. Howarth, P. A. Matson, K. Minami, R. Naylor, K. N. Weeks, and Z-L Zhu, "Policy Implications of Human-Accelerated Nitrogen Cycling," *Biogeochemistry* 57/58 (2002): 477–516; D. W. Jenkinson, "The Impact of Humans on the Nitrogen Cycle, with Focus on Temperate Arable Agriculture," *Plant and Soil* 228 (2001): 3–15; J. N. Galloway, "The Global Nitrogen Cycle: Changes and Consequences," *Environmental Pollution* 102, S1 (1998): 15–24; P. M. Vitousek, J. D. Aber, R. W. Howarth, G. E. Likens, P. A. Matson, D. W. Schindler, W. H. Schlesinger, and D. G. Tilman, "Technical Report: Human Alteration of the Global Nitrogen Cycle: Sources and Consequences," *Ecological Applications* 7 (1997): 737–50.
23. Galloway et al., "Nitrogen Cascade."
24. Intergovernmental Panel on Climate Change, *Climate Change 2007: The Physical Science Basis*, Contribution of Working Group I to the Fourth Assessment Report of the IPCC, ed. S. Solomon, D. Qin, M. Manning, Z. Chen, M. Marquis, K. B. Averyt, M. Tignor, and H. L. Miller (Cambridge, UK: Cambridge University Press, 2007), http://www.ipcc.ch/ipccreports/ar4-wg1.htm.
25. Ibid.
26. M. A. Delucchi et al., *A Lifecycle Emissions Model (LEM): Lifecycle Emissions from Transportation Fuels, Motor Vehicles, Transportation Modes, Electricity Use, Heating and Cooking Fuels, and Materials*, UCD-ITS-RR-03-17 (Institute of Transportation Studies, University of California, Davis, December 2003) and M. A. Delucchi, *Lifecycle Analysis of Biofuels*, UCD-ITS-RR-06-08 (Institute of Transportation Studies, University of California, Davis, May 2006).
27. J. S. Fuglestvedt, T. K. Bernsten, O. Godal, R. Sausen, K. P. Shine, and T. Skodvin, "Metrics of Climate Change: Assessing Radiative Forcing and Emission Indices," *Climatic Change* 58 (2003): 267–331.
28. K. P. Shine, T. K. Bernsten, J. S. Fuglestvedt, and R. Sausen, "Scientific Issues in the Design of Metrics for Inclusion of Oxides of Nitrogen in Global Climate Agreements," *Proceedings of the National Academy of Sciences* 102 (2005): 15768–73.
29. R. G. Prinn, J. Reilly, M. Sarofim, C. Wang, and B. Felzer, *Effects of Air Pollution Control on Climate*, Report No. 118 (MIT Joint Program on the Science and Policy of Global Change, Massachusetts Institute of Technology, January 2005), http://web.mit.edu/globalchange/www/MITJPSPGC_Rpt118.pdf; K. W. Brakkee, M.A.J. Huijbregts, B. Eickhout, A. J. Hendriks, and D. van de Meent, "Characterisation Factors for Greenhouse Gases at a Midpoint Level Including Indirect Effects Based on Calculations with the IMAGE Model," *International Journal of Lifecycle Analysis* 13 (2008): 191–201.
30. Intergovernmental Panel on Climate Change, *Climate Change 2007: The Physical Science Basis*.
31. Ibid.; T.M.L. Wigley, "The Climate Change Commitment," *Science* 307 (2005): 1766–69; Fuglestvedt et al., "Metrics of Climate Change"; B. C. O'Neill, "Economics, Natural Science, and the Costs of Global Warming Potentials," *Climatic Change* 58 (2003): 251–60; O. Godal, "The IPCC's Assessment of Multidisciplinary Issues: The Case of Greenhouse Gas Indices," *Climatic Change* 58 (2003): 243–49; D. F. Bradford, "Time, Money, and Tradeoffs," *Nature* 410 (2001): 649–50; S. Manne and R. G. Richels, "An Alternative Approach to Establishing Tradeoffs Among Greenhouse Gases," *Nature* 410 (2001): 675–77; J. Reilly, M. Babiker, and M. Mayer, *Comparing Greenhouse Gases*, Report No. 77 (MIT Joint Program on the Science and Policy of Global Change, Massachusetts Institute of Technology, July 2001), http://globalchange.mit.edu/files/document/MITJPSPGC_Rpt77.pdf.
32. Bradford, "Time, Money, and Tradeoffs"; Delucchi, "Conceptual Framework for Estimating the Climate Impacts of Land-Use Change."
33. C. Wang and R. G. Prinn, "Potential Climatic Impacts and Reliability of Large-Scale Offshore Wind Farms," *Environmental Research Letters* 6, 025101, doi:10.1088/1748-9326/6/2/025101 (2011).
34. D. Millstein and S. Menon, "Regional Climate Consequences of Large-Scale Cool Roof and Photovoltaic Deployment," *Environmental Research Letters* 6, 034001, doi:10.1088/1748-9326/6/3/034001 (2011).

Conclusion: Key Findings and Paths Forward

Revolutionary changes in transportation will be required to meet societal goals for climate, environment and energy security. Through STEPS we explored the role of alternative fuels and vehicles in this revolution, focusing on biofuels, electricity and hydrogen. We assessed the prospects for new fuels and vehicles, and compared their characteristics, costs, and benefits across multiple dimensions. From this knowledge base, we have begun to develop scenarios for how these new technologies might transform the transportation sector over the next several decades and what policies would be needed to support the transition. In this conclusion, we take stock of what we have learned so far, and identify critical questions going forward.

Summary of Key Findings

1. Insights about Individual Fuel/Vehicle Pathways

Biofuels

- There are a large number of pathways for biofuels production. The costs and benefits of biofuels vary greatly, depending on the specific pathway taken.

- With current (or "first generation") biofuels production technology, the lowest-cost biofuels do not provide major environmental benefits. Some represent marginal improvements over petroleum while others are actually worse than petroleum fuels in terms of environmental impacts.

- Advanced biofuels now under development could provide significant environmental benefits. The first commercial-scale biorefineries are expected to produce large quantities of advanced biofuels by 2015. If these technologies prove to be viable, rapid expansion could take place in the United States to meet the 2022 requirements of the Renewable Fuel Standard. Advanced biofuels are expected to have small greenhouse gas footprints, but face some of the same indirect land-use change challenges as conventional biofuels if cultivating their feedstocks displaces food crops.

- Biofuels can be blended with gasoline or diesel and used in existing vehicles, which eases their introduction into the transportation system. Advanced liquid biofuels require few vehicle changes, and some biofuels (so-called "drop-in" biofuels) can be compatible with existing petroleum infrastructure. Liquid biofuels have an advantage over other petroleum alternatives (hydrogen and electricity) in serving sectors such as aviation and freight that require easily transportable, energy-dense fuels.

- Biofuels can make limited but significant contributions to a sustainable transportation energy supply. STEPS research on the supply potential of biofuels shows that advanced biofuels from waste, residues, and energy crops grown on marginal land could provide between 2 percent and 16 percent of transportation energy in the United States in the next decade. (These biomass sources would avoid potential negative impacts of energy crops grown on agricultural land.) An additional 5 percent could come from conventional corn and soy-based biofuels. In total, we estimate it would be possible to meet 6.5–22 percent of U.S. transportation fuel demand (15–45 billion gallons gasoline equivalent per year) with biofuels costing $3–4 per gallon gasoline equivalent (gge). Biofuel costs would increase sharply above this level of demand because of biomass supply constraints. This result depends on advancements in conversion technologies, the development of reliable feedstock supply chains, and the participation of potential biomass suppliers.

- Balancing sustainability with increasing biofuel production requires the consideration of many factors. Capturing all these factors within a policy and regulatory framework will be challenging.

Electricity

- While plug-in electric vehicles (PEVs) offer significant long term potential for environmental benefits and oil displacement, they also present a radical departure from conventional vehicles in terms of efficiency, range, utility, flexibility, and the refueling experience. There is a range of possible configurations for plug-in electric vehicles including pure battery cars, and plug-in hybrids that rely partly on batteries and partly on engines using fuels such as gasoline or biofuels. STEPS research on PEVs has attempted to better understand different electric vehicle designs and their resource utilization and emissions impacts, especially when in the hands of consumers, as driving and charging behavior influence the potential benefits of PEVs.

- Costs of batteries are a key issue for adoption of electric vehicles, although these costs are coming down. Automakers are making major commitments to plug-in hybrid electric vehicles (PHEVs) and battery electric vehicles (BEVs), and models are entering the market. For PHEVs, there is a trade-off between vehicle cost, which is higher for larger battery models, and fraction of miles run on electricity. The high cost of batteries may encourage use of small-battery PHEVs even beyond early markets.

- Our work indicates that most drivers will charge at home at night. This requires home chargers (which can be built as needed) as part of the larger electric power system. As many as 50% of U.S. consumers may have access to plug in at home and even more if charging at work is an option.

- The existing grid is capable of sustaining projected increases in electric vehicles for decades to come. Further, our studies suggest that lack of public charging infrastructure will not impede the market for PEVs in its initial years.

- Electricity offers a huge low-carbon resource base and large potential benefits in terms of reducing GHG emissions and air pollutants and displacing oil. With the current average U.S. grid mix, there is relatively little GHG benefit for PEVs compared to gasoline hybrids. To realize potential GHG benefits of PEVs it is necessary to substantially decarbonize the electricity supply over time by incorporating renewables and fossil electricity with carbon capture and sequestration.

Hydrogen and fuel cell vehicles

- Hydrogen and fuel cell vehicle technologies are progressing rapidly and could be commercially ready by about 2015. Hydrogen fuel cell vehicles offer high efficiency, good performance, a greater-than-300-mile range, and a fast refueling time. Larger vehicles could be powered by fuel cells, and the consumer could use FCVs much like today's gasoline vehicles. Many major automakers have committed to introducing fuel cell vehicles. Remaining issues are fuel cell system cost and durability and hydrogen storage cost.

- Hydrogen will require a new fueling infrastructure and infrastructure build-out is currently the rate limiting factor for introducing hydrogen vehicles. Early infrastructure "cluster" strategies that co-locate vehicles and stations in lighthouse cities could allow good fuel access for consumers even with a sparse, relatively low-cost early fueling network. Although hydrogen will be more costly than gasoline initially, costs will become competitive as demand grows and the system scales up.

- In the near term, most hydrogen will probably come from natural gas. Many very low-carbon supply pathways are available for future hydrogen supply including renewable hydrogen and fossil hydrogen with carbon capture and sequestration. In the long term, low-carbon hydrogen could cost $3–4/gge, competing with gasoline at $2–3/gallon on a cents-per-mile basis.

- Hydrogen fuel cell vehicles could play a major role in a future light-duty vehicle market beyond 2025, but realizing this will require strong stakeholder coordination and policy and consistent support during an initial transition period.

- Hydrogen offers a huge low-carbon resource base and large potential benefits in terms of reducing GHG emissions and air pollutants and displacing oil. With the current hydrogen production (mostly from natural gas), well-to-wheels GHG emissions would be about half those of a conventional gasoline ICEV; there is a modest GHG benefit for FCVs compared to gasoline hybrids. To realize the full potential GHG benefits of FCVs it is necessary produce hydrogen from low-carbon sources such as renewables or fossil with carbon capture and sequestration.

2. Pathway Comparisons

There are many promising fuel/vehicle pathways but no single clear winner among electricity, hydrogen, and biofuels. Each fuel faces unique challenges, and each could play a role for different consumer needs, regions, and transportation sectors. Rather than down-selecting now among electricity, hydrogen, and biofuels, we see potential roles for each, and a need for flexibility to keep multiple pathway options open.

Technology status and timing: Vehicles

- Several promising new types of vehicles will be ready for initial deployment over the next few years. These include battery electric vehicles, which are starting to appear now as plug-in hybrids and full battery cars, and hydrogen fuel cell vehicles, which are slated for market introduction by about 2015. Our assessments suggest that fuel cell vehicle technology is about as mature as battery electric vehicle technology, and commercial-ready fuel cell vehicles will lag electric vehicles by only a few years, not by several decades. Most automakers see roles for both battery and fuel cell technologies in a future electrified light duty vehicle fleet. It is important to note that improved efficiency of gasoline internal combustion engine vehicles (ICEVs) and hybridized drivetrains could significantly reduce GHG emissions and oil use while alternative vehicle and fuel technologies are developed, and that current ICEV technologies could utilize biofuels with relatively minor changes.

Technology status and timing: Fuel infrastructure

It is technically feasible to build new fuel infrastructures for biofuels, electricity, or hydrogen. Each faces infrastructure challenges that differ among fuels and conversion pathways.

- For biofuels, the main infrastructure issue is developing advanced biorefineries that can produce biofuels at large scale with competitive costs and low net carbon emissions. New biomass delivery systems will be needed to collect biomass and bring it to biorefineries, but the technologies for biomass harvesting and transport are well known. Liquid biofuels are relatively easy to store and transport, and require few vehicle changes to implement. Some biofuels may be at least partly compatible with the existing petroleum delivery and refueling infrastructure.

- Electricity is already widely available to consumers, and it is unlikely that battery electric vehicles will have a major impact on the grid for several decades. (A large number of PEVs will need to be driven in a region before power plants are operated differently or new ones are required.) The main near-term infrastructure needs are new in-home chargers plus some public fast chargers to facilitate longer-distance travel. (Availability of secure charging sites at home or work will impact ultimate market penetration of EVs.) In the longer term, integration of charging demands (via smart grid concepts) will need to occur as part of the larger evolving electric power system, and a low-carbon electricity supply will be needed.

- Hydrogen requires infrastructure changes throughout the supply chain: new hydrogen production and delivery systems and a network of refueling stations. Successful introduction will require close coordination of vehicle and infrastructure deployments in carefully chosen geographic areas or lighthouse cities to finesse the "chicken or egg" problem. The largest near-term infrastructure issue is more logistical than technical: finding strategies for low-cost build-out until hydrogen demand is large enough to exploit economies of scale. Before 2025, hydrogen fuel will likely be produced from natural gas via distributed production at refueling stations, or, where available, excess industrial or refinery hydrogen. Beyond 2025, central production plants with pipeline delivery will become economically viable in urban areas and regionally. For the long term, low-cost, low-carbon hydrogen production technology will be needed.

Consumer behavior

Understanding consumer behavior is important for market introduction of alternative-fuel vehicles and infrastructure, and to realize maximum benefits from these technologies.

- For battery electric vehicles, it is critically important to understand the trade-offs among battery size, vehicle cost, and consumer travel and recharging behavior. Our study of vehicle recharging behavior showed that more new vehicle buyers may be pre-adapted for vehicle recharging than estimated in previous analyses (about half have access to charging when parked at home) and that the success of EVs in meeting energy and emission goals depends on users' recharging and driving behavior as much as or more than on vehicle design.

- For hydrogen, early station placement is an important factor influencing refueling convenience and consumer acceptance. We found that strategic co-location of early hydrogen vehicles and stations in clusters can greatly improve fuel accessibility for early consumers while reducing initial infrastructure costs. Initially a sparse network of less than 1% of gasoline stations may be enough to assure fuel availability.

Regional transition issues

- Geography and regional issues such as availability of primary resources and size and spatial density of demand are key factors influencing fuel and pathway choice. Unlike in the current petroleum-based system, we might see a variety of primary sources being used to make a diverse set of transportation fuels.

Costs

- When mass produced, advanced vehicles are likely to cost $3,000–10,000 more than comparable gasoline internal combustion engine vehicles. PHEVs are estimated to cost $4,500–7,000 more depending on the battery size, FCVs $3,600–6,000 more, and pure battery cars $9,000–20,000 more.

- Fuel costs from early hydrogen refueling systems will be higher than gasoline. Once the infrastructure reaches full scale, it will be possible to supply large amounts of low-carbon hydrogen from biomass or fossil with CCS at $3–4/gge (electrolytic hydrogen would likely cost more). With projected large-scale advanced biorefinery technology, biofuels could become competitive with other liquid fuels at $3–4/gge. Biofuel costs increase sharply above a certain level of demand because of biomass supply constraints.

- A variety of fuel/vehicle options (including plug-in hybrids and hydrogen fuel cell vehicles) could become cost competitive on a life-cycle cost basis with gasoline internal combustion engine vehicles between 2020 and 2030. However, it will be more difficult for pure battery cars to compete even if batteries reach their cost goals. This assumes there is continued technical progress, the vehicle technology is mass produced, the fuel infrastructure is scaled up, and gasoline is priced at $2.5–4/gallon. (The life-cycle cost includes vehicle, fuel, and other operating costs.)

- During a 10- to 20-year transition period, new vehicles and fuels will be more expensive than incumbents, and significant investments will be required to bring them to the point of breaking even with gasoline vehicles. Total investment costs to get to the break-even point are estimated to be tens to hundreds of billions of dollars spent over the next 10 to 20 years.

- When external costs such as air pollution damage, climate change, and energy supply insecurity are taken into account in a social life-cycle cost framework, alternative-fuel vehicles become more competitive with gasoline vehicles.

Primary resources

- Ultimately, the availability of low-cost, sustainable biomass, and competing uses in other parts of the economy, will limit how much biofuel will be deployed in transportation. Because of these limitations, biofuels might be used in sectors like air and marine transport where a liquid fuel is preferred.

- At present most electricity and hydrogen are produced from fossil sources. Demand for electricity and hydrogen for vehicles will be small before 2025, because of the small numbers of electric and hydrogen vehicles in the fleet, and will have relatively small impact on primary resource use. In theory, both electricity and hydrogen could utilize vast low-carbon resources, including biomass, hydro, geothermal, and intermittent renewable energies like wind and solar and fossil energy with carbon capture and sequestration. Primary energy availability should not be a major constraint for either electricity or hydrogen, but continued development of low-cost, low-carbon conversion technologies is needed.

Environmental impacts

- There is considerable scope for reducing GHG emissions compared to today's gasoline vehicles. Well-to-wheels GHG emissions depend sensitively on the particular conversion pathway for biofuels, electricity, and hydrogen. The GHG signature of biofuels is sensitive to the conversion process and to indirect land-use considerations. There are many low-carbon pathways for electricity and hydrogen that rely on renewables or fossil fuels with carbon capture and sequestration (CCS). Unless CCS becomes a reality, though, using electric or hydrogen vehicles in conjunction with a heavily coal-based fuel supply offers little or no benefit compared to gasoline hybrids.

- Plug-in electric vehicles and hydrogen fuel cell vehicles would have significant ancillary benefits in terms of reduced air pollutant emissions and oil use.

- Sustainability issues associated with land, water, and materials impacts of alternative fuel pathways compared with petroleum-based gasoline and diesel are important, but much work remains to be done on understanding and measuring these impacts.

- It is unlikely that material use will impose serious constraints on vehicle technology development in the long term. However, short-term material price volatility and sustainability impacts due to extraction activities need to be considered and mitigated whenever appropriate. Constraints on platinum supplies for fuel cells or lithium for batteries are not likely be long-term "show stoppers," assuming continued progress in reducing materials use and expanded recycling.

- The sustainability impacts of fuel production on water resources need to be compared at the local and regional levels. Concerns about local impacts on water availability, water quality, and ecosystem health should be carefully evaluated. The relative importance of water aspects compared to other aspects of the shift to a new transportation energy system—such as effects on GHG emissions, soil quality, biodiversity, and economic sustainability—must be weighed.

3. Reaching Societal Goals for Sustainable Transportation: Scenarios and Transition Issues

Building on the knowledge gained about individual pathways, STEPS researchers used a variety of analytic approaches to develop scenarios for low-carbon transportation futures. STEPS research has shown that emerging vehicle and fuel technologies could greatly reduce GHG emissions and oil use in transportation by 2050, as part of portfolio approach. These changes will take several decades, but must start now because of the long time constants inherent in changing the energy system.[1]

Future sustainable transportation systems

- A portfolio approach is essential to meet stringent long-term goals for transportation-related GHG emissions reduction and energy security. To achieve deep reductions in GHG emissions and oil use, alternative fuels should be pursued in coordination with improved vehicle efficiency and reduced travel demand. It appears that no one fuel or pathway could meet long-term goals by itself, but combinations of improved vehicle efficiency (most likely relying on increased use of electric drivetrains), decarbonized fuels from diverse low-net-carbon primary sources, and reduced travel demand could.

- There is more than one route to deep reductions in GHGs and oil use by 2050. STEPS researchers identified several distinct "portfolio" scenarios combining multiple strategies that could reach an 80% reduction in transportation-related GHG emissions by 2050. These scenarios differ, but all are characterized by a 2050 light-duty vehicle fleet that relies on highly efficient vehicles, some degree of electrification of drivetrains, and decarbonized transportation fuels. The availability of low-carbon biofuels and the amount of travel demand reduction are "swing factors" that will impact the degree of electrification required to meet GHG reduction goals.

- Unlike today's transportation system, which is 97-percent dependent on petroleum, the transportation system in 2050 might feature a mix of different fuels that could vary by region and transport sector. For example, we might see an electrified light-duty sector and reliance on liquid fuels in the heavy-duty, air, and marine sectors.

- The relative success of several critical technologies could influence the mix of future fuels and vehicles. These include electric batteries, hydrogen fuel cells, biomass conversion technologies, carbon capture and sequestration, and renewable conversion (solar and wind to electricity or hydrogen). The long-term performance and cost of these technologies is still uncertain. This highlights the need for broad and consistent RD&D support to assure rapid progress across a broad front of crucial technologies.

Transition issues and timing

- Making a transition to a low-carbon transportation system is a complex undertaking with multiple actors: consumers, energy suppliers, vehicle manufacturers, and policymakers. Consumer behavior will have a strong influence on which types of vehicles are adopted and what type of infrastructure is needed. Some fuels, notably hydrogen, will require close stakeholder coordination to introduce fuels and vehicles together in particular locations. And most decarbonized fuels will require development of new primary supply chains that could interact with other sectors of the economy (electricity, food, land use). One of the key challenges for policymakers is mitigating the stakeholder risks inherent in introducing new technologies.

- The time required for fuel and vehicle transitions is long. Although electric vehicles and biofuels are beginning to enter the market (and hydrogen fuel cell vehicles could enter by about 2015), it will take several decades for any alternative fuel pathway to make a major difference in GHG emissions or oil use because of the time required for market penetration, vehicle stock turnover and fuel supply development.

- The transitions in vehicle fleets and energy supply systems necessary to reach low-carbon scenarios for 2050 must begin soon and progress rapidly, with rates of market penetration and change near feasible limits. To reach major market penetrations by 2050, new vehicles and fuels need support during early commercialization, as manufacturing and fuel supply systems are scaled up.

Transitions in the light duty vehicle fleet

STEPS researchers developed transition models for the U.S. light-duty vehicle sector to 2050 in support of the National Academies' assessment of the investments needed to bring hydrogen fuel cell vehicles (HFCVs) and plug-in hybrid vehicles (PHEVs) to cost competitiveness with gasoline vehicles. A variety of scenarios were explored that stressed (1) more efficient internal combustion engine technologies, (2) biofuels, (3) hydrogen and fuel cells, (4) plug-in hybrids, and (5) combinations of technologies. Dynamics and costs for vehicle technology learning and infrastructure development were included to find a break-even year when each technology becomes competitive. We also assessed the potential for GHG emissions reduction and oil displacement over time. These studies confirmed the importance of a portfolio approach to oil displacement and GHG emissions reduction. Major findings on transition costs are shown below.

- Transition costs are similar for PHEVs and HFCVs and are in the range of tens to hundreds of billions of dollars. In each case, it will take 15 to 20 years and 5 to 10 million vehicles for the new technology to break even with initial purchase and fuel supply costs for a reference gasoline car. For radically new types of vehicles like FCVs or PHEVs, there is a need to buy down the cost of the vehicle through improvements in technology and scale-up of manufacturing. (Vehicle buy-down costs are typically 80 percent of the total transition cost, and infrastructure costs 20 percent for both PHEVs and HFCVs.) For hydrogen, the fuel cost is initially high and comes down by focusing scaled-up development in lighthouse regions.

- The main transition cost issues for biofuels are to improve biorefinery technology and scale up the supply chain to the point where biofuel competes with other liquid fuels. In the United States, the total investment needed to meet the RFS standard is estimated by various studies to be $100–360 billion for biorefineries, fuel storage terminals, feedstock, and fuel transport to provide enough fuel for 30 to 60 million cars.

- Infrastructure investment costs during a transition are significant but are still relatively small compared to the investment and money flows in the current petroleum system. Maintaining and expanding the existing petroleum infrastructure is projected to cost about $1 trillion in North America alone between 2007 and 2030. Perhaps 20 percent of this capital is for building refineries and fuel transport; the remainder is for exploration and production. By contrast, the cumulative infrastructure capital costs during a 15–20 year transition to competitive hydrogen FCVs or PHEVs would be $10–20 billion (or $1,000–2,000 per vehicle served). In the United States, we estimate that building sufficient biorefinery capacity to meet the Renewable Fuel Standard over the next decade might require $100–360 billion.

4. Policy Needed to Support a Transition

We have begun to explore what kinds of policies would be needed to move toward a transportation system that is more efficient, uses lower-carbon fuels, and employs new types of vehicles. Since a portfolio approach is required and there is considerable uncertainty about the adoption, technology cost, and performance of new vehicles and fuels, one of the challenges for policy is reducing risk.

- Innovative policies will be needed if transportation and alternative fuels are to play a major role in meeting societal goals.

- Policy analyses suggest that economy-wide energy use and GHG emissions can be reduced with strong pricing policy instruments such as a tax on fuel or carbon. But in the transport sector, the evidence suggests that such an approach would not be effective, partly because consumer demand and industry supply responses to such market instruments are highly inelastic and partly because increased fuel and carbon taxes are political anathema at this time in most countries.

- A different—or at least complementary—approach is needed. This other approach might include a mix of market and regulatory policy instruments but by definition will be more fragmented and more targeted at specific technologies and activities. Specific measures could include policies to improve vehicle efficiencies (such as the CAFE standard or the vehicle GHG emission standard), to encourage the reduction of fuel carbon intensities (such as the low-carbon fuel standard), and to encourage the production and adoption of advanced vehicle technologies that both reduce fuel GHG intensity and increase vehicle efficiency (such as the zero-emission vehicle or ZEV program). Policies will almost certainly be required to mitigate stakeholder risk in the early stages of a transition to new fuels and vehicles.

- Key areas of scientific uncertainty exist about how to quantify the social and environmental impact of alternative fuels and advanced vehicles. Policy needs to acknowledge and work around these areas of uncertainty. The question is how to create a robust policy framework that reflects evolving scientific understanding and provides a stable compliance environment.

- More needs to be known about how to account for GHG emissions timing and other factors affecting measurement of GHG impacts. We need a better understanding of how to model the climate impact of land-use change and of forest management. Equally challenging are the sustainability issues associated with market-mediated effects at the system level, such as food prices, indirect land-use change (iLUC), and cumulative environmental impacts.

- So far, there has been limited experience in implementing sustainability standards over large geographical and political regions. Many technical, policy, and implementation issues remain to be tested. Continued improvement in the underlying science and models will pave the way for more effective policy in the future.

Paths Forward

We have found that there is no single transportation fuel or vehicle of the future. Just a few years ago the policy discussion about alternative fuels was framed around finding a single "silver bullet" replacement for petroleum. In light of STEPS research and other recent studies, we now believe that the future is unlikely to be a winner-take-all competition among biofuels, electricity, hydrogen, and petroleum. Instead, the path toward sustainable transportation will be paved by a long series of actions taken together across many fronts over the next decades to improve vehicle efficiency and reduce travel demand while developing new types of vehicles and building new fuel systems tapping into low-carbon primary supplies. By 2050, we will probably see a diverse mix of low-carbon fuels and efficient vehicles in different transportation sectors and regions.

A portfolio approach is essential if we want to achieve deep cuts in transportation GHG emissions and oil use by 2050. The long-term performance and cost projections for key technologies like electric batteries, fuel cells, and advanced biofuels are promising but still uncertain and it will take at least a decade bring these technologies to scale. If we down-select too soon, we run the risk of cutting needed options. This suggests that we need to nurture a range of options over the next decade or so with strong, consistent policy to improve our overall chance of long-term success. A successful portfolio strategy will require a new approach to alternative fuel policy, one that recognizes the uncertainties and long time horizon for change.[2]

Despite the uncertainties, there are clearly measures that could be taken now with a high degree of confidence to reduce GHG emissions and oil use. These include increasing the efficiency of internal combustion engine vehicles (including hybridizing drivetrains) and bringing lower-carbon biofuels into use. In parallel, we need a strong program of support to nurture emerging electric drive transportation technologies (batteries and fuel cells) so that they can be commercialized soon enough to bring deep cuts by 2050. And we need ongoing science to assess the impacts of choices with respect to GHG emissions, oil use, and water, land, air, and materials.

Fortunately, it appears that staying in the game to commercialize multiple fuel/vehicle options would have a relatively low cost compared to the money flows in the current transportation fuel system, although it is more expensive than traditional government spending on research and is risky for individual industries. It will be challenging to craft policies that can support a range of new technologies and are flexible enough to not pick winners. At the same time, we will need measures of success for different options over time, and the ability to stage public support in a timely way.

We are moving into a creative new era for the transportation energy system. As we did 100 years ago at the dawn of the automobile and oil age, we are rethinking our energy system's design and structure; new fuels and vehicles are a critical piece of the picture. This analysis and most others are reaching toward the future from the perspective of our current system. But ultimately the shape of our transportation system may be quite different as we design within the constraints of not just energy and climate but also land, water, air, and materials. Technology and policy are evolving rapidly, with decision makers facing a dynamic future playing field. Perhaps the greatest need now is for strong and consistent policies, and roadmaps and strategies showing how stakeholders can coordinate to take feasible steps toward a sustainable transportation future. STEPS has helped provide solid information and analyses to illuminate paths within the coming revolution. The ongoing challenge is putting the pieces together into realistic visions that can inspire action.

QUESTIONS FOR FUTURE RESEARCH: UNDERSTANDING TRANSITIONS TO A SUSTAINABLE TRANSPORTATION SYSTEM

STEPS researchers have identified a number of critical issues where new research is needed to understand transition paths toward a more sustainable transportation system. These are important but are not well understood and are generally not included in existing energy-economic models that guide decision making in industry and government.

Understand the underlying dynamics of transitions. In most analyses of transportation futures, many assumptions are made about the rate of adoption of new vehicles and fuels, but the underlying factors that govern transition dynamics are not well understood.
- Investigate consumer values and behavior to understand how drivers utilize new technologies and value attributes such as vehicle range and refueling time, especially during a transition.
- Examine possible commercialization pathways for critical technologies such as batteries, fuel cells, and advanced biofuels.
- Explore the roles of different stakeholders during a transition. (For example, under what conditions would automakers and energy companies coordinate to introduce new fuels? What are the pros and cons of policies to stimulate investments in energy infrastructure?)

Improve tools for modeling and technology assessment.
- Develop life-cycle analysis (LCA) tools to better understand societal costs and benefits of different fuels and vehicles during a transition. Integrate sustainability concerns to investigate whether water, land, or materials constraints will be "showstoppers" for clean transportation technologies.
- Expand LCA methodology to assess a wide range of sustainability issues—including GHG emissions, primary energy use, air pollution, energy system reliability/resilience, and water, land, and materials use—in a holistic framework.
- Study the potential interplay between new fuel/vehicle technologies and the design of the future energy system. (For example, would a smart grid and extensive use of intermittent renewables help enable electric vehicles?)
- Use optimization tools to design reliable clean-energy systems. Analyze the reliability and resilience of future renewable energy systems.
- Use geographic information systems and optimization tools to consider regional solutions for building low-cost clean-fuel supply systems.

Develop realistic scenarios and transition strategies to inform industry planning and government policy. Develop visions of the future accounting for engineering design, resource and environmental impacts, policy constraints, and what we know about consumer behavior and economics.
- Elaborate strategies for renewable-intensive transportation, expanded use of natural gas fuels, "smart growth" land use, and low-carbon options for heavy-duty trucks and air and marine transport.
- Analyze how future sustainable transportation systems vary in different regions and for different transportation applications.
- Develop region-specific transition scenarios for the United States, China, and Europe.

Analyze policy approaches for reducing GHG emissions and meeting other sustainability goals. Assess the feasibility and effectiveness of:
- Broad market instruments such as fuel and carbon taxes and cap and trade;
- Fuel policies such as a low-carbon fuel standard, fuel-specific rules, fuel infrastructure requirements, sustainability standards and requirements, and alternative methods for treating land-use change effects;
- Vehicle policies such as performance standards, feebates, and mandates; and
- Policies and actions that influence consumer purchase and use of vehicles and fuels, including social marketing, vehicle instrumentation, eco-driving, and urban land use.

Notes

1. Our findings are broadly consistent with several recent reports on low-carbon transportation futures. See: D. L. Greene and S. E. Plotkin, *Reducing Greenhouse Gas Emissions from U.S. Transportation*, report prepared for the Pew Center on Global Climate Change, February 2011; International Energy Agency (IEA), *Energy Technology Perspectives*, 2008, p. 650; International Energy Agency (IEA), Transport, Energy, and CO_2, International Energy Agency IEA/OECD, Paris; M. Grahn, J. E. Anderson, and T. J. Wallington, "Cost Effective Vehicle and Fuel Technology Choices in a Carbon-Constrained World: Insights from Global Energy Systems Modeling," Chapter 4 in *Hybrid and Electric Vehicles* (New York: Elsevier, 2010); A. Bandivadekar, K. Bodek, L. Cheah, C. Evans, T. Groode, J. Heywood, E. Kasseris, M. Kromer, and M. Weiss, *On the Road in 2035: Reducing Transportation's Petroleum Consumption and GHG Emissions* (MIT Laboratory for Energy and the Environment, 2008).

2. In the United States, alternative fuels policy over the past 30 years has suffered from the "fuel du jour" syndrome, characterized by short-lived waves of enthusiasm for one fuel after another. The result has been inconsistent "boom and bust" support. "Fuel du jour" is in tune with political desire for a "quick fix" but suffers from a fundamental mismatch between the decadal time frames for changing the transportation energy system, and much shorter political cycles. It sometimes seems that the rate of change of the transportation system (decades) is an order of magnitude longer than the political cycle (a few years), which is in turn an order of magnitude longer than the media cycle (a few weeks).

Acknowledgments: STEPS Program Sponsors

Much of the research presented in this book was drawn from the Sustainable Transportation Energy Pathways (STEPS) program at the Institute of Transportation Studies at the University of California, Davis. STEPS was created in 2007 as a four-year research consortium to explore the prospects for alternative fuels and vehicles. The STEPS program was made possible through the generous support of 23 STEPS consortium sponsors. Our sponsors are global leaders in the areas of alternative fuels and vehicles, bringing diverse perspectives from the automotive and energy industries and the public sector. In addition to providing program funding, STEPS sponsors contributed their ideas, insights and expertise through a series of STEPS workshops and many individual meetings. UC Davis researchers and students benefited greatly from these valuable interactions. We deeply appreciate our sponsors' support and engagement. The work presented in this book would not have been possible without them.

The STEPS program was sponsored by the following 23 organizations:

BMW
BP
California Air Resources Board
Caltrans
Chevron
Conoco Phillips
Daimler
Ford
GM
Honda
Indian Oil
Natural Resources Canada
Nissan
Pacific Gas & Electric Company
Sempra Energy
Shell
South Coast Air Quality Management District
Total
Toyota
U.S. Department of Energy
U.S. Department of Transportation
U.S. Environmental Protection Agency
Volkswagen

Acknowledgments

We would like to thank David Greene (Oak Ridge National Laboratory), Steve Plotkin (Argonne National Laboratory), Anthony Eggert (California Energy Commission), Joshua Cunningham (California Environmental Protection Agency), and Marc Melaina (National Renewable Energy Laboratory) for their insightful expert reviews of sections of this book in draft form. We also thank all the STEPS researchers over the past four years who have contributed to the program's success.

Chapter 1: The Biofuels Pathway

The authors would like to thank Quinn Hart, Peter Tittmann, and Colin Murphy (UC Davis), Richard Nelson (Kansas State University), Ken Skog (USFS), Ed Gray (Antares), and Alex Schroeder (Western Governors' Association) for sharing their expertise through conversations over the years and contributing to the case study presented in the chapter. Additionally, the authors thank Yueyue Fan, Sonia Yeh, and Dan Sperling (UC Davis) for helping develop our understanding in many useful discussions.

Chapter 2: The Plug-In Electric Vehicle Pathway

The authors would like to thank Jamie Davies, Michael Nicholas, and Justin Woodjack for contributing insights from their research to improve this chapter. We would like to acknowledge the STEPS program and the UC Davis PH&EV Research Center for their support of this research. Individual authors would also like to thank the California Energy Commission and the Social Sciences and Humanities Research Council of Canada for funding aspects of this research.

Chapter 3: The Hydrogen Fuel Pathway

The authors would like to thank Daniel Sperling, Mark Delucchi, Sonia Yeh, and Andrew Burke (UC Davis), David Greene, Paul Leiby, and Zhenhong Lin (ORNL), Steve Plotkin and Marianne Mintz (Argonne National Laboratory), Catherine Dunwoody and Bill Elrick (California Fuel Cell Partnership), Anthony Eggert (CEC), Marc Melaina (NREL), Uli Bunger, Christophe Stiller, and Reiner Wurster (LBST), and Alan Lloyd (ICCT) for many useful discussions about hydrogen transitions over the years that have helped inform our thinking. We also acknowledge valuable input from STEPS sponsors, including Phil Baxley, Jim Volk, Duncan Macleod, and Angus Gillespie (Shell), Puneet Verma, Nichole Barber, and Jonathan Weinert (Chevron), Andreas Truckenbrodt (Daimler), Ben Knight, Robert Bienenfeld, and Steve Ellis (Honda), Craig Scott, Tak Yokoo, and Bill Reinert (Toyota), Fred Joseck and Sunita Satypal (USDOE), and Britta Gross and Norm Brinkman (GM).

Chapter 4: Comparing Fuel Economies and Costs of Advanced vs. Conventional Vehicles

The authors would like to thank the UC Davis PH&EV Research Center for partial support, and Lorraine Anderson in particular for her careful edits of this chapter.

Chapter 5: Comparing Infrastructure Requirements

The authors would like to thank Nils Johnson, Peter Tittman, Bryan Jenkins, Mike Nicholas, Steven Chen, and Eric Huang (UC Davis) for their insights on infrastructure designs that helped inform our thinking.

Chapter 6: Comparing Greenhouse Gas Emissions

The authors would like to thank Andy Lentz and Anthony Kwong for compiling information useful for the preparation of this chapter. We would also like to thank Michael Wang, Joan Ogden, Stefan Unnasch, and Robert Williams for many years of useful and productive discussions and information sharing on the topic of the emissions from motor vehicle fuel cycles. We thank the UC Davis PH&EV Research Center for partial support, and Lorraine Anderson in particular for her careful edits of this chapter.

Chapter 7: Comparing Land, Water, and Materials Impacts

The authors would like to thank the contributing authors of the journal articles on which this chapter is largely based, including Sarah Jordaan, Adam M. Brandt, Merritt R. Turetsky, and Sabrina Spatari, as well as David W. Keith for the fossil fuel land-use analysis. The authors would also like to thank the STEPS program for partial funding support, Joan Ogden for useful discussion and inputs, and Lorraine Anderson for helpful edits.

Chapter 8: Scenarios for Deep Reductions in Greenhouse Gas Emissions

The authors would like to acknowledge Ryan McCarthy and Joan Ogden for contributions to the 80in50 modeling and analysis. In addition, we would like to thank the STEPS program for providing support for this research, participants at the Asilomar conference for inspiring this line of research, and Daniel Sperling, Joshua Cunningham, and Nic Lutsey for providing comments and guidance. Finally, we would like to acknowledge Lorraine Anderson for her help in editing this chapter.

Chapter 9: Transitioning the U.S. Light-Duty Sector

The authors thank Daniel Sperling (UC Davis), Anthony Eggert (CEC), Joshua Cunningham (CARB), Wayne Leighty (Shell), and David Greene (ORNL) for useful discussions about light-duty vehicle transitions. Joan Ogden acknowledges the committee members of the NRC transition studies on hydrogen (2008) and plug-in hybrid vehicles (2009), especially Michael Ramage, Ed Rubin, Jim Katzer, Bob Shaw, Gene Nemanich, and Alan Crane, for many useful conversations while she served on these committees. Also, the authors would like to acknowledge Marc Melaina (NREL) for his role in helping develop the hydrogen transition model used in this study while he was a postdoctoral researcher at UC Davis.

Chapter 10:
Optimizing the Transportation Climate Mitigation Wedge

This chapter is largely based on work contributed by Alex Farrell, Richard Plevin, Alan Sanstad, and John Weyant for the U.S. analysis. The discussion of future work—some of which is already ongoing, especially the California model—benefited greatly from discussion and collaboration with Joan Ogden, Chris Yang, Daniel Sperling, and others in the STEPS program. We also thank Lorraine Anderson for her careful edits and useful suggestions that improved the accessibility of the chapter.

Chapter 11: Toward a Universal Low-Carbon Fuel Standard

This chapter is a summary of the work on the California LCFS in the past three years and lays out the foundation for the ongoing work on the national LCFS led by the UC Davis Institute of Transportation Studies. The authors would like to honor their colleague Professor Alex Farrell for his leadership and intellectual contributions in developing the initial design of California's LCFS. Along with Daniel Sperling, he co-directed the initial study design of the LCFS in California and helped conceptualize this chapter before his untimely death in April 2008. We also acknowledge other UC colleagues who have contributed to the California LCFS work, including S. M. Arons, A. R. Brandt, M. A. Delucchi, A. Eggert, B. K. Haya, J. Hughes, B. M. Jenkins, A. D. Jones, D. M. Kammen, S. R. Kaffka, C. R. Knittel, D. M. Lemoine, E. W. Martin, M. W. Melaina, J. M. Ogden, R. J. Plevin, D. Sperling, B. T. Turner, R. B. Williams, and C. Yang; and colleagues who have collaborated with us on the national LCFS implementation and design, including Michael Griffin, Paulina Jaramillo, Haixiao Huang, Madhu Khana, Paul Leiby, Hayri Onal, Joan Ogden, Nathan Parker, Jonathan Rubin, Julie Witcover, and Christopher Yang.

Chapter 12: Key Measurement Uncertainties for Biofuel Policy

The authors would like to thank the contributing authors of the journal articles on which this chapter is largely based, including Brenda Chang and Ben Sharpe for the time accounting analysis; as well as useful feedback by colleagues including Richard Plevin and Michael O'Hare (time accounting). We acknowledge the funding support of the California Air Resources Board, the Energy Foundation, the Packard Foundation, and the STEPS program.

Chapter 13: Beyond Life-Cycle Analysis: Developing a Better Tool for Simulating Policy Impacts

The author would like to thank Richard Plevin of UC Berkeley for inspiring him to write this chapter and for helping develop some of the ideas (although the author takes full responsibility for the material). The author thanks the UC Davis/Chevron Research Program for partial support, and Lorraine Anderson for her excellent editorial suggestions.

Editors

Joan Ogden, *Professor, Environmental Science and Policy; Director, STEPS Program, Institute of Transportation Studies*
Joan Ogden's primary research interest is technical and economic assessment of new energy technologies, especially in the areas of alternative fuels, fuel cells, renewable energy and energy conservation. Her recent work centers on the use of hydrogen as an energy carrier, hydrogen infrastructure strategies, and applications of fuel cell technology in transportation and stationary power production. She has served on California state committees on hydrogen and on California's greenhouse gas regulation AB 32, the U.S. Department of Energy Hydrogen Technical Advisory Committee, the Intergovernmental Panel on Climate Change's panel on Renewable Energy, and on National Academies committees assessing hydrogen fuel cell and plug-in hybrid vehicles. She holds a B.S. in mathematics from the University of Illinois and a Ph.D. in theoretical physics from the University of Maryland.

Lorraine Anderson is a freelance writer and editor with a special interest in nature and sustainability. Her edited works include *Sisters of the Earth: Women's Prose and Poetry about Nature*, *Literature and the Environment: A Reader on Nature and Culture* (with Scott Slovic and John P. O'Grady); and *At Home on This Earth: Two Centuries of U.S. Women's Nature Writing* (with Thomas Edwards). She is the coauthor (with Rick Palkovic) of *Cooking with Sunshine: The Complete Guide to Solar Cuisine*. She lives in Corvallis, Oregon.

Authors

Jonn Axsen, *Postdoctoral Research Associate, PH&EV Research Center*
Jonn Axsen's primary research interest is the nexus between human behavior, energy-using technology, and environmental policy. Research projects have included characterizing the market for plug-in hybrid vehicles, estimating energy impacts, and observing processes of social influence within car buyers' social networks. Jonn earned his Ph.D. in transportation technology and policy at UC Davis, preceded by a master's in environmental management at Simon Fraser University.

Andrew Burke, *Research Engineer, Institute of Transportation Studies*
Andrew Burke has researched and taught graduate courses on advanced electric driveline technologies specializing in batteries, ultracapacitors, fuel cells, and hybrid vehicle design, control and simulation. Since 1974, his career work has involved many aspects of electric and hybrid vehicle design, analysis, and testing. He was the head systems engineer on the U.S. Department of Energy (DOE) funded Hybrid Vehicle project while working at the General Electric Research and Development Center in Schenectady, NY. While a professor of mechanical engineering at Union College in Schenectady, he continued his work on electric vehicle technology through consulting with the Argonne and Idaho National Engineering Laboratories (INEL) on various DOE electric vehicle and battery programs. Andrew was employed from 1988 to 1994 at INEL as a principal program specialist in the electric and hybrid vehicle programs. He has authored more than 140 reports and papers on electric and hybrid vehicles, batteries, and ultracapacitors. Andrew

holds B.S. and M.S. degrees in applied mathematics from Carnegie Institute of Technology, an M.A. in aerospace engineering, and a Ph.D. in aerospace and mechanical sciences from Princeton University.

Joshua Cunningham, *Program Manager, STEPS (2007–2009)*
Joshua Cunningham is currently an engineer at the California Environmental Protection Agency, working in the area of zero emissions vehicle policy analysis. He is also Director of Programs for the California Plug-in Electric Vehicle Collaborative and a key author of their recent report "Taking Charge." From 2007 to 2009 Joshua was program manager of the Sustainable Transportation Energy Pathways (STEPS) research program. Joshua also helped launch the UC Davis Energy Efficiency Center (April 2006) and the CEC PIER-funded Plug-in Hybrid Electric Vehicle Research Center (February 2007). Prior to working at ITS-Davis, Joshua was a systems engineer at UTCFuelCells working on transportation applications. He holds a master's degree in Transportation Technology and Policy (TTP) from UC Davis where he focused on fuel cell systems modeling research, and a bachelor's degree in Mechanical Engineering from Michigan State University.

Mark A. Delucchi, *Research Scientist, Institute of Transportation Studies*
Mark Delucchi specializes in economic, environmental, engineering, and planning analyses of transportation systems and technologies. Mark's research is in seven areas: (1) comprehensive analyses of the full social costs of motor-vehicle use, with special emphasis on the external costs of air pollution, noise, oil use, accidents, and climate change; (2) detailed analyses of emissions of greenhouse gases and criteria pollutants from the life cycle of passenger and freight transport, materials, electricity, and heating and cooking; (3) detailed modeling of the energy use, manufacturing cost, operating costs, and external costs of advanced electric and conventional vehicles; (4) systems analyses of energy, economic, and air-quality impacts of transportation fuels and technologies; (5) design and analysis of a new dual-road transportation infrastructure and new town plan that minimizes virtually all of the negative impacts of transportation; (6) sustainable transportation and energy use; and (7) analyses of supplying 100 percent of the world's energy needs with wind, water, and solar power.

Peter Dempster, *Program Manager, STEPS (2009–2010)*
Peter is an Advanced Technologies Engineer for BMW of North America. He coordinates projects related to sustainable mobility, advanced energy storage devices, market and consumer studies, and innovative business strategies surrounding electro-mobility, including battery second-use and vehicle-to-grid. Prior to joining BMW, Peter managed the Sustainable Transportation Energy Pathways and Toyota Fuel Cell Vehicle Demonstration programs of the UC Davis Institute of Transportation Studies. He also co-managed the Plug-in Hybrid Electric Vehicle Research Project. For three years Peter was a researcher for the California Biomass Collaborative. Peter earned a Master of Science in Biological Systems Engineering and a Bachelor of Science in Aeronautical Engineering, both from UC Davis.

Yueyue Fan, *Associate Professor, Civil and Environmental Engineering*
Yueyue Fan's research interests are in transportation and renewable energy infrastructure system modeling and optimization, critical transportation and energy infrastructure protection, adaptive network routing and resource allocation processes, and stochastic and dynamic system modeling and computational methods.

Brendan Higgins is interested in modeling and experimental research on algal biofuels. Algae are fascinating organisms because they offer significant potential for environmentally friendly fuel production. While completing his M.S. in transportation technology and policy at UC Davis, he studies the life-cycle impacts of using algae to simultaneously treat dairy wastewater and produce electricity. He is also pursuing a Ph.D. in biological engineering. He will experiment with algae cultivation using biomass leachate as the growth medium. Preliminary experiments show that algae remove sugars from the leachate and exhibit high lipid productivity.

Bryan Jenkins, *Professor, Biological and Agricultural Engineering; Director, Energy Institute*
Bryan Jenkins teaches and conducts research in the areas of energy and power, with emphasis on biomass and other renewable resources. He has more than thirty years' experience working in the area of biomass thermochemical conversion, including combustion, gasification, and pyrolysis. His research also includes analysis and optimization of energy systems. He teaches both graduate and undergraduate courses on energy systems, heat and mass transfer, solar energy, and power and energy conversion, including renewable energy and fuels, economic analysis, environmental impacts, fuel cells, engines, electric machines, fluid power, cogeneration, heat pumps, thermal storage, and other technologies. Bryan is a recipient of an Outstanding Achievement Award from the U.S. Department of Energy for exceptional contributions to the development of bioenergy, and the Linneborn Prize from the European Union for outstanding contributions to the development of energy from biomass.

Nils Johnson is a Ph.D. candidate in transportation technology and policy at UC Davis. His research focuses on modeling large-scale infrastructure deployment for CO_2 mitigation technologies, including hydrogen fuel and carbon capture and storage. Nils employs expertise in geographic information systems and techno-economic modeling to evaluate the cost and design of roll-out strategies for these new technologies. He completed a B.A. at Haverford College and master's degrees in both forestry and environmental management at Duke University.

Alissa Kendall, *Assistant Professor, Civil and Environmental Engineering*
Alissa Kendall joined UC Davis in 2007 after completing a multidisciplinary Ph.D. in civil and environmental engineering and natural resource policy at the University of Michigan's Center for Sustainable Systems. Her research evaluates the environmental sustainability of transportation and energy systems from a life-cycle perspective. Her topics of analysis include transportation infrastructure materials and systems, biofuels, agricultural systems, advanced vehicle technologies, and climate change mitigation strategies. In parallel, she conducts research to advance the methods and practice of life-cycle assessment, with a focus on greenhouse gas accounting methods.

Ken Kurani, *Research Engineer, Institute of Transportation Studies*
Ken Kurani is developing approaches and methods to evaluate user responses to new transportation and information technologies. This research includes activity-based approaches applied within quasi-experiments designed around interactive stated preference and reflexive survey methodologies.

Wayne Leighty joined the Institute of Transportation Studies in 2006 as a lifelong technological optimist and tinkerer. He built his first alternative-fuel vehicle at age 14, a battery-electric Honda CRX, and has since converted two diesels to waste-cooking-oil fuel. Wayne is employed as a commercial regulatory analyst with Shell Oil in Anchorage, Alaska. While at UC Davis, Wayne earned two M.S. degrees, an M.B.A., and a Ph.D. His master's thesis investigated the effects of tax structure on optimal oil production and his dissertation modeled dynamic transition paths to deep reductions in greenhouse gas emissions from the California transportation sector. At the Graduate School of Management, Wayne focused on general management, strategy, and emerging technology in the energy and transportation industries. Wayne was a Graduate Automotive Technology Education (GATE) Fellow for two years, the 2008–09 Chevron Graduate Fellow in transportation, and a Dean's Fellow at the Graduate School of Management.

Xuping Li is a Ph.D. candidate in civil and environmental engineering at UC Davis. She holds a master's degree in agricultural and resource economics from UC Davis. Xuping has strong interdisciplinary training in transportation engineering (civil and mechanical disciplines), economics, and policy analysis. Her research interests include modeling and experimenting on sustainable energy systems such as hydrogen energy systems and innovative renewable energy systems, and assessing the technical, economic, and environmental performance of these systems. She is particularly interested in the application of these systems in transportation (alternative fuels), retailers, and residences. She is also dedicated to linking the engineering performance of these systems to policy and economic analyses.

Timothy E. Lipman is an energy and environmental technology, economics, and policy researcher and lecturer with the University of California, Berkeley. He is serving as Co-Director for the campus' Transportation Sustainability Research Center (TSRC), based at the Institute of Transportation Studies, and also as Director of the U.S. Department of Energy Pacific Region Clean Energy Application Center (PCEAC). Tim's research focuses on electric-drive vehicles, fuel cell technology, combined heat and power systems, biofuels, renewable energy, and electricity and hydrogen energy systems infrastructure. Lipman received Ph.D. degree in Environmental Policy Analysis with the Graduate Group in Ecology at UC Davis (1999). He also has received an M.S. degree in Transportation Technology and Policy, also at UC Davis (1998), and a B.A. from Stanford University (1990).

Ryan McCarthy, *Research Analyst, Institute of Transportation Studies*
Ryan McCarthy was chief writer for the strategic plan for plug-in electric vehicles in California, written by researchers at the Institute of Transportation Studies in conjunction with the California Plug-In Electric Vehicle Collaborative. He completed the CCST Science and Technology Policy Fellowship in the office of California Assembly Member Wilmer Amina Carter, where he advised on issues associated with energy, utilities, and the environment, among others. McCarthy holds a master's degree and Ph.D. in civil and environmental engineering from UC Davis, and a bachelor's degree in structural engineering from UC San Diego. His expertise lies in transportation and energy systems analysis, specifically regarding the electricity grid in California and impacts of electric vehicles on energy use and emissions in the state.

David McCollum, *Research Scholar, International Institute for Applied Systems Analysis, Laxenburg, Austria*
David McCollum received his Ph.D. from the Transportation Technology and Policy Program at UC Davis in 2011. His dissertation focused on scenario analysis, developing and using energy-engineering-economy-environment (4E) models to understand the evolution of energy systems over time. Before arriving at UC Davis, David studied chemical engineering at the University of Tennessee and also lived and worked in Japan as an English teacher. His main research interests include the modeling of energy and climate systems, scenario and policy analysis, and assessment of low-carbon technologies.

Marshall Miller, *Research Engineer, Institute of Transportation Studies*
Marshall Miller studies electric and hybrid vehicle propulsion systems and how to integrate these systems in vehicles to optimize performance. He is the technical manager of the Hydrogen Bus Technology Validation Program, which will operate hydrogen-fueled buses in a real transit environment. He also manages the Hybrid Vehicle Propulsion Systems Laboratory where he studies storage battery, ultracapacitor, and fuel cell technology. He develops computer models to simulate the performance of electric and hybrid vehicles using a variety of propulsion systems, using data generated in the lab. Before joining ITS-Davis full-time, he held a joint appointment with ITS-Davis and the Union of Concerned Scientists where he studied technology and policy implications of fuel cell vehicles and hydrogen fuel infrastructure.

Gouri Shankar Mishra, *Researcher, Institute of Transportation Studies*
Gouri Shankar Mishra is involved in research on water impacts of renewable transportation fuels and electricity, and life cycle analysis. He received a master's degree in transportation technology and policy at UC Davis. Before UC Davis, Mishra provided business and technology consulting services to shipping, ports, and railroad operators and was involved in projects related to business process restructuring and automation, project appraisal and valuation, project financing, and corporate communications strategy.

Michael Nicholas, *Postdoctoral Researcher, PH&EV Research Center*
Michael Nicholas completed his undergraduate degree in physics and natural science, and he received his Ph.D. in transportation technology and policy from UC Davis in 2010. He is currently continuing his work as a postdoctoral researcher at UC Davis. His work has centered on understanding refueling behavior from a geographic perspective. His work has assisted in the policy process with respect to hydrogen infrastructure planning, and he is currently researching electric vehicle charging network needs to help future planners and policy makers make informed decisions in this rapidly expanding and poorly understood subject area. He is currently managing a team of researchers in the PH&EV Research Center constructing a GIS toolbox that will allow for a methodical approach to the understanding, planning, and deployment of electric vehicle infrastructure.

Joan Ogden, *Professor, Environmental Science and Policy; Director, STEPS Program, Institute of Transportation Studies*
Joan Ogden's primary research interest is technical and economic assessment of new energy technologies, especially in the areas of alternative fuels, fuel cells, renewable energy and energy conservation. Her recent work centers on the use of hydrogen as an energy carrier, hydrogen infrastructure strategies, and applications of fuel cell technology in transportation and stationary power production. She has served on California state committees on hydrogen and on California's greenhouse gas regulation AB 32, the U.S. Department of Energy Hydrogen Technical Advisory Committee, the Intergovernmental Panel on Climate Change's panel on Renewable Energy, and on National Academies committees assessing hydrogen fuel cell and plug-in hybrid vehicles. She holds a B.S. in mathematics from the University of Illinois and a Ph.D. in theoretical physics from the University of Maryland.

Nathan Parker is a postdoctoral research associate at the Institute of Transportation Studies. His research focuses on modeling the infrastructure needed to enable transitions to alternative fuels. While at UC Davis, Nathan has studied hydrogen pipelines, the production of hydrogen from biomass, and assessments of biofuel supply. Nathan has developed a spatially explicit infrastructure model of national biofuel supply and applied it to analyzing the federal Renewable Fuel Standard. He received his B.S. in physics from Wake Forest University in 2001, his M.S. (2007) and his Ph.D. (2011) in Transportation Technology and Policy at UC Davis.

Daniel Sperling, *Professor and Director, Institute of Transportation Studies; Co-Director, STEPS Program*
Daniel Sperling is a professor of civil engineering and environmental science and policy, and founding director of the Institute of Transportation Studies at UC Davis. He also holds the transportation seat on the California Air Resources Board, where he plays a prominent role in designing and adopting climate policies for vehicles, fuels, and urban travel. He received a 2010 Heinz Award for his "achievements in the research of alternative transportation fuels and his responsibility for the adoption of cleaner transportation policies in California and across the United States."

Since 2009, Dan has been a keynote speaker at thirty universities and conferences; has been featured on *The Daily Show* with Jon Stewart, NPR's *Science Friday*, *Talk of the Nation*, and

Fresh Air, and has published op-eds in the *New York Times*, *Washington Post*, and *Los Angeles Times*. He is author or editor of more than two hundred papers and reports and twelve books (including *Two Billion Cars*, Oxford University Press), has served on thirteen National Academies committees, was selected as a National Associate of the National Academies, recently chaired the Future of Mobility committee of the (Davos) World Economic Forum, testified seven times to the U.S. Congress on alternative fuels and advanced vehicle technology, and was a lead author on the United Nations' Intergovernmental Panel on Climate Change, which shared the 2007 Nobel Peace Prize with former vice president Al Gore.

Yongling Sun, *Ph.D. Recipient, Transportation Technology and Policy (2010)*
Yongling Sun is a research associate at the International Council on Clean Transportation in San Francisco. Her current research interests include life-cycle cost comparison of alternative fuel/propulsion options, benefit-cost analysis of renewable energy options, life-cycle emissions of various hydrogen pathways, and economic and environmental assessments of advanced power systems. She obtained a B.S. in automotive engineering in 1998 at Tsinghua University, an M.S. in vehicle engineering in 2001 at Tongji University, and an M.S. in agricultural and resource economics at UC Davis in 2008, and a Ph.D. in Transportation Technology and Policy at UC Davis in 2010.

Jacob Teter has research interests that include life-cycle analysis and the intersection of environmental policy and resource economics. Currently, Jacob is completing his master's research on Chinese rural vehicles (CRVs) and has begun research on the water-energy nexus with Gouri Mishra, under Sonia Yeh. In summer 2010 he conducted research on the impacts of national industrial and economic policies on CRVs. He visited industry sites and interviewed vehicle owners under an NSF EAPSI fellowship. He recently completed research funded by Shell Hydrogen and hosted by Tsinghua University, for which he used dynamic programming to estimate the societal costs and benefits of building an infrastructure supporting hydrogen fuel cell vehicles in Beijing. Before UC Davis, Jacob was a fellow with the Oberlin Shansi Memorial Association in rural China, where he interviewed farmers about how their agricultural practices had changed since China's reform and opening.

Peter Tittmann is a Ph.D. candidate in Geography at UC Davis. His interests include fire, forest management, bio-energy and bio-fuels, network modeling, remote sensing, Geographic Information Systems (GIS), and regional geography.

Tom Turrentine, *Director, PH&EV Research Center, Institute of Transportation Studies*
Tom Turrentine studies the role of travel and movement in the evolution of culture, society, and lifestyle. He focuses on understanding automobile-based lifestyles, applying anthropological methods and theories to explore potential responses of car users to new technologies and policies aimed at mitigating the negative impacts of automobile infrastructure and use. He has studied consumer responses to electric vehicles, alternative fueled vehicles, micro-vehicles, station car systems, advanced traveler information, and other intelligent transportation systems. Tom also studies travel behavior and road systems in environmentally sensitive areas, focusing on Yosemite National Park and the Sierra Nevada region in California.

Eric Winford conducts research on how resources can be managed to accommodate the needs of humans, wildlife, and ecosystems. Currently, he is working on analyzing the landscape-scale impacts of policies that seek to utilize forest biomass as a fuel. His research utilizes a combination of life-cycle analysis, GIS analysis, and forest growth models to understand the multiple implications of forest management. Eric's research revolves around ecosystem management, energy, and regional planning. Before coming to UC Davis, Eric worked at the Nevada Tahoe Conservation District as an environmental scientist helping homeowners plan, design, and implement conservation practices to prevent erosion, protect against wildfires, and reduce the spread of invasive plant species. Prior to that, Eric was employed by the USDA Forest Service, El Dorado National Forest, as a forestry technician working on recreation and wilderness management issues. Eric is an M.A. candidate in the geography graduate group. He holds a B.S. in journalism from the University of Tennessee, Knoxville.

Julie Witcover, *Postdoctoral Researcher, Institute of Transportation Studies*
Julie Witcover is working on indirect land-use effects of biofuels policy. Her interests include spatial bioeconomic modeling and sustainable development. She received her Ph.D. in agricultural and resource economics from UC Davis in December 2008 (dissertation: *Shaping Land Use Along an Agricultural Frontier: A Dynamic Household Model for Early Small-Scale Settlers in the Brazilian Amazon*), an M.A. in international economics from the Johns Hopkins School of Advanced International Studies, and an A.B. in government from Harvard. Before her Ph.D., she developed her interests in sustainability policies as a research analyst at the International Food Policy Research Institute in Washington, DC; publications included, with Stephen A. Vosti and Chantal Line Carpentier, the research report *Agricultural Intensification by Smallholders in the Western Brazilian Amazon: From Deforestation to Sustainable Land Use*.

Christopher Yang, *Project Scientist, Institute of Transportation Studies*
Christopher Yang is a researcher and the co-leader of the Infrastructure System Analysis research group within the STEPS program. His research interests lie in understanding the role of advanced vehicles and fuels in helping to reduce transportation greenhouse gas emissions through infrastructure and system modeling. He works on hydrogen infrastructure systems, vehicle and electric grid interactions, and scenarios for long-term reductions in greenhouse gases from the transportation sector. He completed his Ph.D. in mechanical engineering from Princeton University and his B.S. and M.S. in environmental science and engineering from Stanford University.

Sonia Yeh, *Research Engineer, Institute of Transportation Studies*
Sonia Yeh's primary research interest is to advance the understanding of future energy systems and their environmental and social impacts, and to seek policy solutions and improve the sustainability of our future energy systems. Sonia's research can be divided into four key tracks: (1) fuels and fuel greenhouse gas emissions regulations, (2) sustainability, (3) energy system modeling, and (4) technological change and learning by doing. She has been involved with the design and implementation of California climate policies, specifically the California Global Warming Solutions Act (AB 32), the Low Carbon Fuel Standard, and the Sustainability Plan for the Alternative and Renewable Fuel and Vehicle Technology Program (AB 118).

Hengbing Zhao, *Research Engineer, Institute of Transportation Studies*
Hengbing Zhao obtained his Ph.D. in 1999 from Zhejiang University. His research has involved many aspects of uninterruptible power sources, distributed power generation systems, fuel cell systems and vehicles, and electric and hybrid vehicles. He was a key contributor to developing battery electric vehicles and distributed power generation systems while working at Myway Labs in Yokohama, Japan. Since 2002, he has worked on fuel cell systems for automotive applications as a research associate at the IFCI-NRC in Vancouver, Canada. He joined the STEPS program as a research engineer in 2007, working on vehicle modeling and evaluation. His recent research includes fuel cell systems and vehicles, hybrid electric vehicles, and applications of batteries and ultracapacitors for electric vehicles.

STEPS Researchers (2007–2010)

STEPS Directors

Joan Ogden, *Professor, Environmental Science and Policy; Director, STEPS Program, Institute of Transportation Studies*

Daniel Sperling, *Professor and Director, Institute of Transportation Studies; Co-Director, STEPS Program*

Research Leaders

Jonn Axsen, *Postdoctoral Research Associate, PH&EV Research Center*

Andrew Burke, *Research Engineer, Institute of Transportation Studies*

Mark A. Delucchi, *Research Scientist, Institute of Transportation Studies*

Yueyue Fan, *Associate Professor, Civil and Environmental Engineering*

David Greene, *Visiting Senior Research Scientist, Institute of Transportation Studies (2008–2009)*

Bryan Jenkins, *Professor, Biological and Agricultural Engineering; Director, Energy Institute*

Alissa Kendall, *Assistant Professor, Civil and Environmental Engineering*

Chris Knittel, *Associate Professor, Economics*

Ken Kurani, *Research Engineer, Institute of Transportation Studies*

Cynthia Lin, *Assistant Professor, Agricultural and Resource Economics and Environmental Science and Policy*

Nic Lutsey, *UC Davis Transportation Technology and Policy Ph.D. 2009; Senior Researcher, International Council on Clean Transportation*

Marshall Miller, *Research Engineer, Institute of Transportation Studies*

Gouri Shankar Mishra, *Researcher, Institute of Transportation Studies*

Michael Nicholas, *Postdoctoral Researcher, PH&EV Research Center*

Nathan Parker, *UC Davis Transportation Technology and Policy Ph.D. 2011; Postdoctoral Research Associate, Institute of Transportation Studies*

Tom Turrentine, *Director, PH&EV Research Center, Institute of Transportation Studies*

Julie Witcover, *Postdoctoral Researcher, Institute of Transportation Studies*

Christopher Yang, *Project Scientist, Institute of Transportation Studies*

Sonia Yeh, *Research Engineer, Institute of Transportation Studies*

Hengbing Zhao, *Research Engineer, Institute of Transportation Studies*

Graduate Researchers

Alexander Allan is an M.S. candidate in Transportation Technology and Policy at UC Davis, and is currently an engineer at PG&E.

Obadiah Bartholomy received an M.S. in Transportation Technology and Policy at UC Davis in 2009, and is currently an engineer at SMUD.

Adina Boyce is a Ph.D. student in Civil and Environmental Engineering at UC Davis.

Joel Bremson is a Ph.D. candidate in Transportation Technology and Policy at UC Davis.

Chien-Wei (Steven) Chen is a Ph.D. candidate in Transportation Technology and Policy at UC Davis.

Yuche Chen is a Ph.D. candidate in Civil Engineering at UC Davis and a graduate student researcher at the Institute of Transportation Studies.

Gustavo Collantes received a Ph.D. in Transportation Technology and Policy at UC Davis in 2007, and is currently with the Department of Transportation, State of Washington.

Jamie Davies is a master's candidate in Transportation Technology and Policy at UC Davis.

Abbas Ghandi is a Ph.D. student in Transportation Technology and Policy at UC Davis.

Reid Heffner received a Ph.D. in Transportation Technology and Policy at UC Davis in 2009. He is currently an analyst at Booz Allen Hamilton.

Brendan Higgins is pursuing a Ph.D. in Biological Engineering at UC Davis.

Ryohei Hinokuma received an M.S. in Transportation Technology and Policy at UC Davis in 2008, and is currently at SF Power.

Eric Huang received a Ph.D. in Civil and Environmental Engineering at UC Davis in 2010. He is currently a senior associate the Energy and Environment Division at International Resources Group in Washington DC.

Jonathan Hughes received a Ph.D. in Transportation Technology and Policy at UC Davis in 2008. He is currently an assistant professor in economics at the University of Colorado, Boulder.

Nils Johnson is a Ph.D. candidate in Transportation Technology and Policy at UC Davis.

Matt Jones is a Ph.D. student in Transportation Technology and Policy at UC Davis.

Wayne Leighty earned four graduate degrees at UC Davis: two M.S. degrees (in Agricultural and Resource Economics in 2008 and in Transportation Technology and Policy in 2007), an M.B.A. in 2010, and a Ph.D. in Transportation Technology and Policy in 2010.

Xuping Li is a Ph.D. candidate in Civil and Environmental Engineering at UC Davis.

Zhenhong Lin received his Ph.D. in Civil and Environmental Engineering at UC Davis is 2008. He is currently a research engineer at Oak Ridge National Laboratory.

Ryan McCarthy received his Ph.D. in Civil and Environmental Engineering at UC Davis is 2009. He is currently a special assistant to the Chairman of the California Air Resources Board.

David McCollum received his Ph.D. in Transportation Technology and Policy at UC Davis in 2011. He is currently a Research Scholar, International Institute for Applied Systems Analysis, Laxenburg, Austria.

Geoff Morrison is a Ph.D. candidate in Civil and Environmental Engineering at UC Davis.

Colin Murphy is a Ph.D. student in Transportation Technology and Policy at UC Davis.

Kalai Ramea is a Ph.D. student in Transportation Technology and Policy at UC Davis.

Brent Riffel received his M.S. in Transportation Technology and Policy at UC Davis in 2007. He is currently an analyst at Lifecycle Associates.

Yongling Sun received a Ph.D. in Transportation Technology and Policy at UC Davis in 2010. She is currently a research associate at the International Council on Clean Transportation in San Francisco.

Jacob Teter is completing his master's research in Transportation Technology and Policy at UC Davis.

Peter Tittmann is a Ph.D. candidate in Geography at UC Davis.

Guihua Wang received his Ph.D. in Transportation Technology and Policy at UC Davis in 2008. He is currently an air quality engineer at the California Environmental Protection Agency.

Jonathan Weinert received his Ph.D. in Transportation Technology and Policy at UC Davis in 2007. He is currently an engineer at Chevron Technology Ventures.

Brett Williams received his Ph.D. in Transportation Technology and Policy at UC Davis in 2007. He is currently a researcher at the Transportation Sustainability Research Center (TSRC), UC Berkeley.

Eric Winford is an M.A. candidate in the Geography Graduate Group at UC Davis.

Justin Woodjack is master's student in Transportation Technology and Policy at UC Davis.

Sahoko Yui is a master's candidate in Transportation Technology and Policy at UC Davis.

Jane Zeng is a master's student in Transportation Technology and Policy as well as in Statistics at UC Davis.

Jessie Zheng is a Ph.D. student in Civil and Environmental Engineering at UC Davis.

Program Management

Paul Gruber, *Program Manager, STEPS and NextSTEPS (2010–present)*

Peter Dempster, *Program Manager, STEPS (2009–2010)*

Joshua Cunningham, *Program Manager, STEPS (2007–2009)*

Made in the USA
Charleston, SC
12 October 2011